Energy Crisis

Volume 1

1969-73

Energy Crisis

Volume 1

1969-73

Edited by Lester A. Sobel

Contributing editors: Joseph Fickes, Hal Kosut, Chris Hunt, Barry Youngerman, Mary Elizabeth Clifford, Gerry Satterwhite

FACTS ON FILE, INC.   NEW YORK, N.Y.

Energy
Crisis
Volume 1
1969-73

Library of Congress Catalog Card No. 74-75154
ISBN 0-87196-278-0

9 8 7 6 5 4 3 2 1

**PRINTED IN
THE UNITED STATES OF AMERICA**

Contents

Introduction

ENERGY IS WIDELY REGARDED as the basis of modern industrial progress.

This progress had its origin in the far past when man discovered how to use fire, learned to domesticate animals and found out that wind could propel ships and that wind or running water could turn mill-wheels. He thus acquired means of transportation that far exceeded the capability of his own two legs and animate or inanimate servants whose strength multiplied by many times his capacity for work.

Over the millenia man increased his ability to use mechanical servants. This was a slow process until recently. But today's housewife, consuming 1,000 kilowatt hours of electricity a month, has at her command the equivalent in energy of 15 men working for her household day and night.

A major factor in the rapid industrial development of the past century has been the enormous increase in the use of fossil fuels, or, as Admiral Hyman G. Rickover expressed it June 8, 1972, in an address to the Pensacola Area Chamber of Commerce, "the shift...from an economy...based largely on renewable resources to one depending almost entirely on nonrenewable resources."

Rickover said: "In 1850—to give an example of the magnitude of this shift—fossil fuels supplied 5% of the world's energy; men and animals 94%. By 1950 the percentages had reversed themselves, 93% coming from coal, oil and natural gas, 1% from water power and only 6% from the labor of men and animals.... Through nearly all of his history on earth, man has lived almost

1

entirely on renewable resources. These left him energy poor but fairly secure in the little he had. We are energy rich today but find ourselves in the unenviable position of the prodigal son who lives luxuriously on his inheritance but after a brief moment of glory is left with nothing to his name.... Half the coal ever mined has been taken out of the ground in the last three decades or so; ...more than half the world's [petroleum] production has occurred in the last 15 years...."

This book is a record of America's—and the world's—recent concern with the depletion of nonrenewable energy resources.

Concern over such depletion, however, is not a recent phenomenon.

"Our energy resources are not inexhaustible, yet we are permitting waste in their use and production," President Franklin D. Roosevelt wrote to Congress Feb. 15, 1939. "In some instances, to achieve apparent economies today future generations will be forced to carry the burden of unnecessarily high costs and to substitute inferior fuels for particular purposes." This warning was part of a Presidential message accompanying an energy resources study that Roosevelt had ordered a year earlier. This 1939 report, prepared by the National Resources Committee, made presumably disregarded suggestions for, in Roosevelt's words, "policies, investigations, and legislation necessary to carry forward a broad national program for the prudent utilization and conservation of the Nation's energy resources."

According to a report of the Senate Interior Committee, "Senate concern for energy and development dates to 1943 with the introduction by Senator [Joseph C.] O'Mahoney [D, Wyo.], Chairman of the Senate Interior Committee, of the Synthetic Liquid Fuels Act. A similar measure was introduced by then Representative Jennings Randolph of West Virginia in the House of Representatives. Subsequent enactment of this measure in 1944 initiated an eight-year program for the construction and operation of demonstration plants to produce synthetic fuels from coal, oil, shale, agricultural and forestry products, and other substances in order to conserve and increase the oil resources of the United States."

Yet in 1951, O'Mahoney, quoting from the prewar National Resources Committee report, told the Senate that "we must still give consideration to the formulation of a broad 'national fuels policy to meet the needs of the United States in times of peace and war.' "

National energy policies, however, were in the process of formation in Canada and elsewhere. Senator Vance Hartke (D,

Ind.) told the Senate April 21, 1960 that "other countries... are not so naive as we. Many of them have already realized the importance of energy supplies to their economic progress and have consequently formulated fuels policies to guide them." Citing specifically the Soviet Union, Hartke quoted from a 1959 *Izvestia* article: "In the seven-year plan for the development of our national [Soviet] economy, it is extremely important to find a practical solution to the problems connected with the integrated use of natural fuels resources."

Conservationists and some critics of American life styles attribute at least part of the energy problem to excessive use, or "waste," of energy. The U.S. is described as an "energy glutton." Six percent of the earth's population live in the U.S., but in 1972 this 6% used nearly 40% of all the energy consumed by all human beings on the planet. Former Interior Secretary Stewart L. Udall noted in the May 8, 1973 edition of *World* magazine that the "209 million Americans use about as much energy for air conditioning alone as the 800 million mainland Chinese use for all purposes—and Americans waste each year almost as much energy as the Japanese (105 million people) consume annually."

Senator Henry M. Jackson (D, Wash.), chairman of the Senate Interior Committee and a Congressional leader on energy matters, discussed at the National Press Club in Washington March 22, 1973 the explosive growth in American energy use and in U.S. energy problems. Jackson said:

"Between 1940 and 1965 the consumption of energy in the United States doubled. If present trends continue, consumption should double again by 1980. The rate of energy consumption increased twice as fast in 1972 as it did in 1971.

"In 1965 we consumed nearly 12 million barrels of oil a day; in 1972 we consumed more than 16 million barrels a day. In 1965 we had three million barrels a day spare producing capacity; in 1972 we had none....

"At a time when energy demand is greater than ever, there is not one single new refinery under construction in the United States....

"Since 1968 we have been using twice as much natural gas as we can find.... Potential gas users were denied service in 21 states in 1972...."

A major contributor to the energy predicament is petroleum's displacement of coal as the U.S.' major fuel. In 1900 only 8% of the fuel used in the U.S. was obtained from petroleum while coal constituted 89%. By 1973 petroleum provided about 45%

and coal 20%. And U.S. petroleum consumption rose at the same time to a level at which it exceeded domestic production. America thus found itself increasingly in danger of being at the mercy of foreign oil producers who could raise prices at whim, reduce production and exports when they saw fit or embargo shipments for political reasons. Domestic production capacity of the displaced coal, meanwhile, remained sufficient for hundreds of years of foreseeable U.S. energy needs.

This book on the world's energy predicament begins with the year 1969, when the public was becoming aware of the matter, and continues through 1973, when the Arab oil embargo made the issue one of overriding concern to truckdrivers, automobile plant workers, housewives and the rest of a previously unheeding population who suddenly began to perceive that their way of life was threatened by an "energy crisis." The material in this book is taken largely from the printed record compiled by FACTS ON FILE in its weekly coverage of world events. Where changes, deletions or additions were made, the purpose was usually to clarify, to eliminate duplication or to provide necessary amplification. A conscientious effort was made to record all developments without bias.

1969

The growing shortage of oil, natural gas and other energy sources became increasingly apparent during 1969. Planning was continued in a program to bring oil from Alaska's tremendous North Slope deposits to the "lower 48" states. In the biggest merger in oil industry history, Sinclair Oil became part of the Atlantic Richfield Co. Attacks on oil-industry tax "loopholes" and other "privileges" continued. A Peruvian military government seized the U.S.-owned International Petroleum Co., and a Bolivian military government did the same with the U.S.-owned Bolivian Gulf Oil Co. Arab guerrillas blew up part of the Trans-Arabian Pipeline in Syria's Israeli-occupied Golan Heights. The U.S.S.R. arranged to finance Iraqi oil operations and made plans for selling its own oil and natural gas to the West. Soviet progress was reported in research on using thermonuclear reactions—the power of the H-bomb—for civilian energy needs.

ENERGY CONSUMPTION AND LIVING STANDARDS

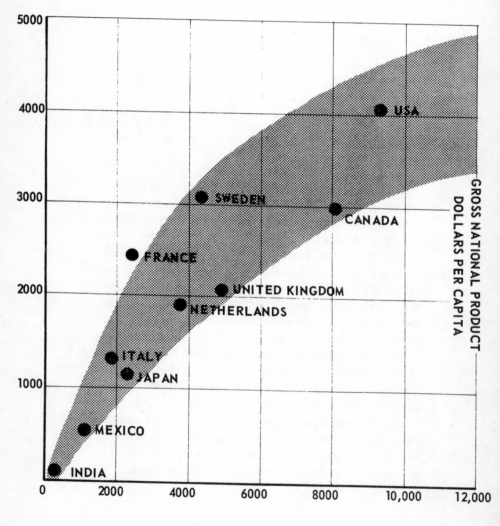

Dwindling Energy Supply

U.S. Energy Consumption Exceeds Domestic Output

By 1969 it was obvious to the U.S. public as well as to industry and government circles that a serious energy shortage was imminent. During the previous year (1968) the U.S. had produced sources of 57,495 trillion BTUs (British thermal units) of energy, principally in the form of petroleum, natural gas, coal and other fuels, but had consumed 62,143 trillion BTUs. The difference had to be imported, and it was generally agreed that as the anticipated "energy crisis" grew, increasing amounts of fuels, almost exclusively in the form of petroleum, petroleum products and natural gas, would have to be bought abroad.

The U.S. output of petroleum and crude oil products in 1968 amounted to the equivalent of 19,308 trillion BTUs, but the U.S. consumption of oil and petroleum products that year exceeded by 50% the amount added to its proved reserves. This was the ninth successive year in which consumption outpaced additions to proved reserves.

Although coal is the U.S.' most abundant usable energy source, in 1968 bituminous coal and lignite demand outran production for the first time in U.S. history. During the same year U.S. natural gas consumption exceeded natural gas discoveries in the U.S. for the first time. U.S. consumption of natural gas in 1968 totaled a record 19,351 trillion BTUs and accounted for 31.1% of all U.S. energy consumption.

U.S. energy consumption increased by 5.1% in 1968 and 5.3% in 1969, when the level was 51.2% higher than it had been 10 years previously. In the decade ending in 1969 U.S. consumption of dry natural gas increased by 75.4%, and these other increases in consumption also were recorded: petroleum and natural gas liquids 44.4%, bituminous coal and lignite 37.7%, waterpower 55.8%. Anthracite consumption, however, declined by 49.8% in the decade.

Petroleum supplied 43.2% of the U.S.' energy needs in 1969. Other sources were: natural gas 32.1%, bituminous coal and lignite 20.1%, waterpower 4%, anthracite .4% and nuclear energy .02%. Coal was still the major fuel for generating electric power, and the demand for coal increased by 1.1% in 1969 to 505 million tons.

The demand for residual-type fuel oil, used primarily in electric utilities and industry, in space heating and in ships' bunkers, had been growing at an average of 1.7% annually during 1960–68, but the demand spurted by more than 8% in 1969. The U.S. imported some 63% of its residual oil needs during 1969. The daily average imported was about 1,300,000 barrels, all but a small fraction of it going to the East Coast.

These were among 1969's developments involving energy in general, the growing energy shortage and the efforts of the U.S. and other nations to appease the increasing hunger for energy.

Oil from Alaska & Canada. In mid-1968 the Atlantic Richfield Co. of the U.S. had announced the discovery of petroleum deposits at Prudhoe Bay on Alaska's North Slope. Investigations indicated that this was the biggest oil field ever located on the continent, and plans were formulated for the construction of a pipeline to get the oil to the "lower 48" (the 48 states south of Canada).

Canadian Prime Minister Pierre Elliott Trudeau said at a news conference at the National Press Club in Washington March 25 that the joint Canadian-U.S. oil policy was threatened by the recent discoveries in Alaska. Under the current program, Canada exported to the Western U.S. about as much oil as its Eastern provinces imported from Venezuela. Formal quotas had existed since 1959 for oil imported from countries other than Canada. In 1967, the U.S. and Canada had agreed to voluntary limits on Canadian oil exports to the U.S. east of the Rocky Mountains.

In a note to a U.S. cabinet task force reviewing U.S. oil policy, the Canadian government asked for an easing of restrictions on U.S. imports of Canadian oil. The note, released Aug. 6, said Canada attached "the highest importance" to continuing the exemption from U.S. import controls and asked for unrestricted access to the U.S. for its oil on a "normal commercial basis."

In its note, the Canadian government warned that the continued availability of Canadian natural gas to the U.S. would depend on availability of U.S. markets for Canadian oil.

Stressing the mutual interdependency in trade and defense of the two countries, Canada said that policies should be formulated to give both suppliers and users of Canadian oil reasonable expectations on which to base expansion plans.

(It was reported April 30 that Canada's National Energy Board had asked U.S. buyers of Canadian oil to cut back on purchases in order to maintain the levels set under voluntary restraints. The New York Times reported May 16 that Western Canadian producers, feeling the effects of the cutback, had urged that Ottawa enforce its rule that cheaper Venezuelan crude oil not be marketed west of Quebec Province.)

Plans for a 2,080-mile natural gas pipeline to deliver gas reserves from Alaska's North Slope to U.S. and Canadian markets were announced June 22. The Westcoast Transmission Co. of Vancouver, British Columbia, and the Bechtel Corp. of San Francisco, joint sponsors of the venture, estimated the

---

WORLD CRUDE OIL-IN-PLACE DISCOVERABLE AND DISCOVERED
[In billions of barrels]

	Existing [1]	Discoverable [2]	Discovered	
			January 1, 1962 [3]	January 1, 1966 [4]
United States	1,600	1,000	346	386
Canada, Mexico, Central America, and Caribbean	500	300	50	77
South America (including Venezuela)	800	500	214	238
Total Western Hemisphere	2,900	1,800	610	701
Europe	500	300	21	26
Africa	1,800	1,100	56	139
Middle East (excluding Turkey)	1,400	900	793	928
Southern Asia	200	100	7	8
U.S.S.R., China, Mongolia	2,900	1,800	90	122
Indonesia, Australia, et cetera	300	200	21	25
Total Eastern Hemisphere	7,100	4,400	988	1,248
Total world	10,000	6,200	1,598	1,949

[1] Total oil-in-place (Hendricks, 1965).
[2] Total discoverable oil-in-place (Hendricks, 1965).
[3] Original oil-in-place in known reservoirs as at January 1, 1962 (Torrey, Moore and Weber, 1963).
[4] Hypothetical calculation of original oil-in-place in reservoirs as known at January 1, 1966 (D.C. Ion, 1967).

Source: U.S. Energy Outlook 1971–85, National Petroleum Council, vol. II.

cost of the line at $1.2 billion. A first phase of the pipeline would reach from Canada's Northwest Territories to the U.S. border. A second extension, to be completed by 1978, would carry the line to Prudhoe Bay.

In another plan to get at the oil, the 115,000-ton icebreaking tanker S.S. Manhattan set sail from Chester, Pa. Aug. 24 on an Atlantic-to-Pacific voyage through the Arctic north of Canada to open a commercial shipping route through the Northwest Passage. The three-month, 10,000-mile journey would test the feasibility of a year-round tanker route through the passage to tap the oil deposits and transport the oil profitably to the eastern U.S. and Canada. The Manhattan conquered the Northwest Passage Sept. 5–14 and sailed Sept. 14 into the Amundsen Gulf of the Beaufort Sea.

An announcement of the projected voyage had been made June 3 by officials of the Humble Oil & Refining Co., sponsors of the project. Charles F. Jones, Humble president, said the tanker route could "bring crude oil to the Atlantic seaboard at a substantial cost advantage." Jones estimated that the total oil flow could reach 2 million barrels a day by 1980. He said that the demand would require doubling the U.S.-flag tanker fleet and would bring on the biggest shipbuilding boom since World War II. The voyage would also reshape world trade patterns by establishing the long-sought northern route from Europe to the East. The development of Canadian and Alaskan Arctic regions was also a major possibility.

Humble had chartered the Manhattan from Seatrain Lines, Inc. in January at a charge of $10 million for two years.

The $40 million expedition was headed by Stanley B. Haas, 45, a Humble chemical engineer. The 126-man roster included a crew of 54 scientists, Canadian and U.S. government representatives, oil firm officials and press representatives. (The Atlantic Richfield Oil Co. announced Sept. 8 it had placed an order with Bethlehem Steel Corp. for three 120,000 deadweight-ton U.S. flag tankers to carry North-Slope oil to West-Coast U.S. markets.)

Sinclair-Atlantic Richfield Merger. The Sinclair Oil Corp. formally merged into the Atlantic Richfield Co. March 4, one day after the Justice Department and Atlantic Richfield had reached an unusual agreement to proceed with the combination subject to certain restrictions. The agreement provided for the department to continue its suit against the merger. (Federal District Judge Frederick vanPelt Bryan in New York Feb. 17 had issued a preliminary injunction against the merger, "pending final determination" of the charge of antitrust violation.)

Under the agreement, Atlantic would maintain the trademark "Sinclair" in areas where Sinclair was currently marketed and would promote the company as a competitive corporation pending the completion of all legal actions involving the merger. Atlantic Richfield also agreed to sell all of Sinclair's 10,000 retail gasoline outlets in the northeastern and southeastern states to the British Petroleum Co., Ltd. (Originally, Atlantic Richfield had planned to keep the outlets in the Southeast.)

Immediately after the signing of the agreement for the $1.6 billion merger, considered the largest in the history of the oil industry, British Petroleum bought the Sinclair outlets for $400 million. At a meeting following the merger and sale, O. P. Thomas, ex-president and chief executive officer of Sinclair, was named chairman of the 13-man executive committee of the merged company. Robert O. Anderson remained chairman and chief executive of Atlantic Richfield, and Thornton F. Bradshaw remained president.

Natural Gas in Arctic Islands. A consortium composed of 20 private companies and the Canadian government announced Nov. 6 that preliminary testing in the Canadian Arctic Islands had discovered a large natural gas reservoir. Panarctic Oils Ltd. said a test of a well on Melville Island had released a flow rated at 10 million cubic feet a day.

An exploratory oil well on Melville had "spudded in" April 19, Panarctic said.

Oil Price Dispute. Texaco, Inc., the U.S.' third largest oil company and largest domestic seller of gasoline, announced Feb. 24 a .6¢-a-gallon raise in the wholesale price of gasoline for the continental U.S. (The N.Y. City price thus rose to 18.7¢ a gallon.) Texaco also announced a 20¢-a-barrel increase in the price it would pay for crude oil nationwide. This was the first nationwide crude oil price increase since the 1957 Suez Canal crisis. (The 1968 price for a barrel of crude oil averaged $2.94.) A Texaco statement said that the increase had been made "to partially offset cost increases and to encourage exploration for and development of new oil reserves."

Phillips Petroleum Co. and the American Oil Co., a subsidiary of Standard Oil Co. (Indiana), Feb. 26 raised wholesale gasoline prices .7¢ a gallon. They suggested that dealers raise their retail prices 1¢ a gallon. Both companies attributed the increase to rising labor costs. Standard Oil Co. of California raised the price it paid for crude oil by 15¢–20¢ a barrel March 1 and its wholesale price of gasoline by .6¢–.7¢ a gallon. The general rise in gasoline prices became nearly unanimous by March 3, when the Humble Oil & Refining Co., the chief domestic subsidiary of Standard Oil Co. (N.J.), raised wholesale prices .5¢–.7¢ a gallon (effective March 1). Most companies also had raised crude oil prices 14¢–21¢ a gallon.

Sen. Edward M. Kennedy (D, Mass.) urged in a letter to President Nixon March 5 that the Administration take action to force a reversal of the oil price increase. Kennedy April 1 made public a letter in which Bryce N. Harlow, assistant to the President for Congressional relations, indicated that the Nixon Administration would examine the possibility of raising foreign oil quotas to reduce domestic prices. (Since 1959, oil imports had been limited to about 12% of domestic production.) Harlow said in his letter, dated March 19, that the Office of Emergency Preparedness would consider the 1¢-a-gallon increase in its review of the oil import program. Kennedy asserted that domestic oil prices "are propped up to extraordinarily high artificial levels only because of governmental intervention and support, and thus the government has not only the right, but the obligation to prevent those prices from going even higher." The new increases were costing consumers $2 million a day, he declared.

Fourteen Senators—13 Democrats and one Republican—petitioned Attorney General John N. Mitchell to investigate the possibility of collusion among the 14 oil companies that had raised their gasoline prices. The petition, reported April 2, said: "Because of the stringent laws against price fixing, we think the fact that all these oil companies raised their prices by almost the same amount at the same time merits close consideration."

Sen. William Proxmire (D, Wisc.) had charged March 24 that a "secret" 1967 U.S.-Canadian agreement to limit Canadian oil imports had been designed to maintain high oil prices, not to protect U.S. national security. Proxmire said that the pipeline supply of Canadian crude oil "is more secure than any production from Alaska or from our offshore wells. If national security is really the underlying theme of the oil import program, the development of such oil should be encouraged." Proxmire claimed that the oil import restrictions were costing U.S. consumers $4 billion annually. (The Interior Department had said the cost was about half this figure.)

Testifying before the Senate Antitrust & Monopoly Subcommittee April 2, Walter Adams, an economics professor and acting president of Michigan State University, described the U.S. oil industry as "a government cartel" and "a honeycomb of artificial restraints, privilege and monopoly." Adams said that the expansion of oil import quotas would have the double effect of lowering consumer prices and, by conserving U.S. reserves, of promoting national security.

The Nixon Administration Feb. 27 had delayed a decision on Maine's application for a free trade zone. Commerce Secretary Maurice H. Stans said no decision would be made until the Administration completed its review of the oil import program (ordered by Mr. Nixon Feb. 20). Maine, supported by the other New England states, had sought the foreign trade zone at Portland and a subzone at Machiasport, where a refinery would be built to refine duty-free foreign oil. Sen. George D. Aiken (R, Vt.) charged that the Administration's delay represented a veto. He said : "It's the first rift in the honeymoon. The Republicans, like the Democrats, are working with the oil companies."

In testimony before the Senate Antitrust Subcommittee April 1, Professor Joel B. Dirlam of Rhode Island University said that federal oil policy, including restrictions, was costing New England residents $325 million annually. Dirlam estimated the "excess costs" at $167 million for gasoline and $158 million for heating oil. Dirlam claimed that the proposed refinery at Machiasport could save New England consumers $158 million a year.

Attack on oil's tax 'loopholes.' Sen. Proxmire, leading a fight against oil-industry tax "loopholes," noted in the Senate June 26 that oil "is a very healthy and powerful industry by any criteria. In 1968 the combined net profits of the 12 largest U.S. oil companies was just a fraction under $5 billion. Each of those 12 companies, moreover, has set new profit records in each of the last four years."

In an article entitled "Oil and Politics" in the September issue of the Atlantic Monthly, Ronnie Dugger reported on What Proxmire described Sept. 9 as federal "subsidies to the oil industry":

The oil industry has three distinct kinds of privileges, or, if you prefer, incentives: proration, import controls, and tax advantages. . . .

An oil industry program to get a production-limitation movement going in the 1920s was aborted when the U.S. Attorney General ruled in 1929 that the plan, advanced in the name of conservation, would be an attempt to win immunity from the antitrust laws. The companies then turned to the states. In 1930 the great East Texas field came in, leading to overproduction, waste of oil, and very low oil prices. As W. J. (Jack) Crawford, the tax administrator of Humble Oil, says, "We had to let a president of Humble quit to become governor to establish proration," that is, production control. Governor Ross Sterling, the former Humble president, ordered state troops into the oilfield to stop all that wasteful free enterprise. . . .

The system, which is most significant now in Texas and Louisiana, tells producers exactly how much oil, and no more, they may legally produce each month. . . .

Apart from straight protectionism, the only argument for import controls has been the contention that lower U.S. prices would cause high-cost U.S. producers to quit and exploration in the United States to decline. Oil and gas now provide three fourths of the nation's energy supply, and visions are conjured of enemies torpedoing our tankers and America "running out of oil" during a national emergency. The traditional protection-

ist solution would have been a sharp increase in the ten-cents-a-barrel tariff. . . .

Instead, there somewhat mysteriously appeared, in the 1958 Trade Agreements Act, an authorization for the President to establish mandatory controls over oil imports if he found that they threatened to impair the national security. . . .

In 1962 the approved import percentage became 12.2 percent of the demand east of the Rockies. Today about one fifth of the nation's oil consumption is provided by imports. . . .

The theory of the oil depletion allowance depends on the assertion that an oil venture's capital is not the same thing as its capital investment. Oil, it is explained, is an irreplaceable "wasting asset," a limited quantity of an elusive substance that is being depleted. Therefore, we are told, when oil is discovered in the ground, it becomes capital that should be valued on the basis of its selling price. Once capital investment has thus been eased over into the more adaptable idea of capital, the conclusion is drawn that the income tax, being a tax on income, should not apply to that portion of the income from the oil which is a return of capital. . . .

Instead of deducting all their costs, oilmen can ring up, free of tax, 27.5 percent of their gross income on oil and gas production, up to a limit of 50 percent of their net income from it. In practice this means that in the oil- and gas-producing business, between 40 and 50 percent of net profit is tax-free, year after year as long as the production continues. . . .

In 1960 the oil industry received 44 percent of its pre-tax income tax-free. The Treasury's 1958–1960 depletion survey indicated an even higher rate of about 55 percent tax-free, with the realized rate as a percentage of the gross running just a point or two below the 27.5 percent maximum. . . .

Of probably greater practical significance is the foreign tax credit, which is available to all U.S. firms abroad but is uniquely important in oil because of the high foreign levies on oil companies. . . .

It is a fair question to ask how much U.S. income tax, on both foreign and domestic profits, is paid by the five international majors that dominate the U.S. industry, Jersey, Texaco, Gulf, Mobil, and California Standard. The answer is that during the five years 1963–67 these companies paid, on total net profits before income tax of about $21 billion, federal income tax of about $1 billion. Their five-year U.S. income tax rate was 4.9 percent. They paid foreign governments more than five times as much. . . .

Jersey's five-year rate was 5.2 percent, Texaco's a percent, Gulf's 8.4 percent, Mobil's 5.2 percent, and California Standard's 2.5 percent. On the basis of figures in George Spencer's *U.S. Oil Week,* the twenty-three largest refiners in the United States, including the big five, paid in the same period $2 billion U.S. income tax on total earnings before income taxes of $30 billion. The aggregate rate for the twenty-three firms figures out to about 7.2 percent. . . .

The effective income tax rate for all manufacturing is 43.3 percent; for petroleum, it is 21.1 percent. . . .

The industry counterattacks with figures showing that excluding excise taxes, it pays about the same percentage of its gross income in taxes (in the range of 5 percent) as other industry does. . . .

In recent years the minerals industry has been making tax use of contrivances called "carve-outs" and "ABC transactions." Both are based in "production payments," which are rights that are "sold" to profits from future production from a well or mine. Johnson's and Nixon's Treasury people agree that these devices are now costing the United States $200 million a year, and they are spreading. . . .

The main objections to depletion and other U.S. oil policies now under criticism are that they are unfair to taxpayers and consumers and enhance monopoly trends in the oil industry. The central line of defense is that the national security requires a strong domestic industry, which in turn requires protections and tax advantages. Representative Lloyd Meeds of Washington says, "The salient fact is that the individual must bear the burden of this loss." Representative Charles A. Vanik of New York estimates that since it was first allowed, depletion has cost about $140 billion, "paid at the expense of almost all of the other taxpayers of the country." . . .

Their theme runs through the debate of the last two decades. Back in 1950 President Truman told of an oil millionaire who raked in $14 million over a five-year period, but paid only $80,000 income taxes for the period, about half of one percent, compared to the lowest rate of 20 percent for a single man who earned less than $2000 a year. Senator George Aiken, the Vermont Republican, said in the 1957 debate that others must "dig into their pockets" to make up the money lost by depletion. Now Proxmire says the ordinary taxpayer must wonder why he pays 14, or 16, or 20 percent of his hard-earned money to the federal government when a company like Atlantic Richfield pays nothing whatever. . . .

As chairman of the powerful Finance Committee, Russell Long is the most strategically placed congressional ally of the oil industry. He is probably the shrewdest and most active defender of depletion in the Senate. . . .

Senator Long is also an oilman. He readily acknowledges that he inherited valuable oil and gas properties and that he has participated in drilling thirty or forty wells. He says his income from the inherited properties exceeds what he makes in the Senate. . . .

Depletion allowance cut. Under the Tax Reform Act of 1969, passed by Congress Dec. 22 and signed by President Nixon

Dec. 30, the oil depletion allowance was reduced from $27\frac{1}{2}\%$ to 22%.

California Fleet to End Gasoline Use. California Gov. Ronald Reagan announced Dec. 22 that the state would begin converting its motor vehicle fleet from the gasoline engine to a dual fuel system using smog-free compressed natural gas. Reagan said that 175 of the state's 28,500 vehicles would be converted immediately and the rest would switch "on a programmed, continuing basis." According to Reagan's office, natural gas engines had a life-expectancy of 400,000 miles but could travel only 125 miles on a tank of gas.

Latin America

Peru seizes IPC assets. The Peruvian military government announced Feb. 6 that it was seizing all remaining assets of the International Petroleum Co. (IPC), a subsidiary of Standard Oil of New Jersey. Most of IPC's assets already had been taken over.

In a televised speech to the nation, President Juan Velasco Alvarado stated: ". . . This morning the government notified the International Petroleum Company that it had begun the collection of debts amounting to $690,524,283 and that it had requested adoption of precautionary measures to assure the payment of this great sum to the state. Thus we have arrived at that end which all proud nations long for—their sovereignty." The dispute stemmed from Peruvian claims that IPC's title to the oil fields at La Brea y Parenas was invalid and that the company owed the government for oil extracted over a period of 45 years.

U.S. State Department spokesman Robert J. McCloskey said Feb. 7 that the U.S. regretted the Peruvian junta's action. "This development does not appear to be leading to a resolution of the problem in accordance with international law," he declared. U.S. officials, however, also emphasized that they recognized Peru's right to expropriate the company, and contested only its failure to pay adequate compensation. (J. Wesley Jones, U.S. ambassador to Peru, had been recalled to Washington Feb. 4 for consultations.)

In his announcement Gen. Velasco (who had officially retired from the army Jan. 31) emphasized the uniqueness of the IPC case and indicated that the junta was open to other foreign investment.

The junta's action Feb. 6 climaxed four months of conflict between it and the company. Following its October 1968 coup, the junta had siezed the Talara refinery and the La Brea y Parinas oil fields. The government oil company, Empresa Petrolera Fiscal (EPF), Jan. 8 had reportedly billed IPC $14.4 million for products delivered to the company after the refinery and oil fields were taken over. In an advertisement published in the New York Times Jan. 31, IPC claimed that the Peruvian government had attached its assets and bank accounts in Lima Jan. 16 to collect the claim. Then, on the ground that the attachment had not produced the amount sought, EPF took over IPC's headquarters in Lima Jan. 28 and barred about 20 American employes from their jobs; EPF threatened to auction $65 million worth of company holdings if the bill was not paid by Feb. 5. Velasco's announcement confirming the seizure of the rest of IPC's holdings followed the next morning.

The Peruvian government denied rumors that other U.S. properties were to be expropriated. Fernando Berckemeyer, Peruvian ambassador to the U.S., declared Jan. 30: "In relation to rumors concerning the expropriation of Cerro Corp. and other U.S. properties in Peru, I would like to state that at no time has the government of Peru indicated that it would expropriate these properties." The rumors had focused on Cerro Corp., a mining company, Xerox Corp., and Peruvian Telephone Co., a subsidiary of International Telephone & Telegraph Corp. Spokesmen for the three companies confirmed that they were in no danger of expropriation.

(The junta Jan. 27 was reported to have granted oil exploration rights to seven U.S. companies. The rights, covering about 6.3 million acres, were granted to Texas Petroleum, Peruvian Gulf Oil, Occidental of Peru, Sulana, Ltd., Continental Oil of Peru, Resources Betteous and New Resources, Inc.

(Gen. Jorge Fernandez Maldonado, minister of mines and energy, revealed April 17 that the government had decided to end the system of concessions used so far in granting oil exploitation rights. "From now on," he declared, "the exploitation of the country's oil and mining wealth will be made on the basis of direct contracts between the state and national or foreign companies."

(The government announced May 5 that it had granted a five-year exploration and exploitation concession to the Peruvian Gulf Oil Co. The concession covered an area of 11,520 acres in eastern Loreto Province.)

Bolivian oil & gas controversy. A World Bank loan to aid the construction of a pipeline system to transport natural gas from Bolivia to Argentina raised strong nationalist feelings in Bolivia and stirred a conflict that threatened to develop into an important political controversy.

The $23.25 million loan, announced July 2 by the International Bank for Reconstruction & Development (World Bank), was granted to the Compania Yacibol Bogoc Transportadores (YABOG) and guaranteed by the Bolivian government and YABOG's two owners. (YABOG was owned equally by the Bolivian government oil firm Yacimientos Petroliferos Fiscales Bolivianos [YPFB] and a wholly-owned subsidiary of the U.S.-owned Gulf Oil Corp., the Bolivian Gulf Oil Co.; after a 20-year period, full ownership of YABOG was to pass to the Bolivian government.)

The World Bank loan was to finance about half the cost of a 329-mile, high-pressure trunk line from Santa Cruz, Bolivia to the Argentine border town of Yacuiba and about 100 miles of lateral pipelines connecting several gas fields with the trunk line. The New York State Common Retirement Fund was to finance the other half of the cost.

A sales agreement between YABOG and an Argentine government firm, Gas del Estado, provided for deliveries through the pipeline of 9.5 million cubic meters of natural gas to Argentina over a 20-year period at a fixed price in U.S. dollars; the agreement was expected to net Bolivia an estimated $340 million in gross revenues.

According to World Bank regulations, contract assignments for major construction jobs required international bidding and prequalification of bidders along guidelines established by the bank. Williams Brothers, a U.S. pipeline construction company, which had often obtained contracts with Gulf and its affiliates, was among six firms found qualified to build the pipeline, and the World Bank subsequently awarded the contract to Williams. The Bolivian National Council of Petroleum & Petrochemicals approved the contract assignment.

With the award of the contract, an already-simmering conflict over the status of Gulf's assets in Bolivia was re-ignited. Bolivian nationalists claimed that Williams, in previous work contracts with YPFB, had increased the costs to Bolivian authorities and had produced work of questionable quality. In addition, nationalists claimed that Gulf's concessions authorized it to develop oil, but not gas, and that the company had no warrant to sell gas abroad. The anti-Gulf campaign spread, with nationalists demanding the revision of Bolivia's petroleum code; some called for expropriation of Gulf's assets.

The magazine Economist for Latin America reported July 7 that President Luis Adolfo Siles Salinas, faced with sharp criticism in Congress, had asserted that his government supported the contract assignment only because World Bank regulations ruled out any modifications of the agreement. He announced July 15 that the government "will prepare shortly a new petroleum code" and that negotiations with Gulf would be opened with the intention of increasing the state's participation in the firm's profits. Siles added that YPFB would be granted the rest of the oil area not conceded.

(Gulf, the only private oil firm in Bolivia, had investments in Bolivia totaling about $141 million and produced about 12 million barrels of oil there annually. In recent years, petroleum had become Bolivia's second most important export, next to tin. The current petroleum code, which established a general policy of cooperation between the state and private firms, was decreed in 1955 by former President Victor Paz Estenssoro and had continued with little modification under the late President Rene Barrientos Ortuno. Barrientos had authorized the Argentine pipeline agreement in 1968.

(Of the proved oil and gas reserves in Bolivia, YPFB held letters of credit on 6.5 million cubic meters of petroleum and 8.3 billion cubic meters of natural gas, while Gulf had claim to 33 million cubic meters of petroleum and 82.2 billion cubic meters of gas.)

The Siles government ordered Bolivian Gulf Oil to provide the entire country with free natural gas for domestic consumption for 10 years starting Oct. 1. The order, approved by the Bolivian National Council of Petroleum and Petrochemicals, said that Bolivian Gulf was to deliver 20 million cubic feet of gas daily without charge. Announcing the move Sept. 3, President Siles asserted that it "constitutes a reparation for the damage caused by the offer the company made only on a regional scale." (A Gulf spokesman confirmed Sept. 3 that Bolivian Gulf had voluntarily offered to provide 10 million cubic feet per day for 10 years in the Santa Cruz area of Bolivia.)

(Bolivian Gulf was also involved in a heated dispute with the state oil firm, Yacimientos Petroliferos Fiscales Bolivianos, the New York Times reported Sept. 2. YPFB was demanding payment from Gulf of an alleged debt of $307,000 resulting from commitments to joint projects. Gulf, however, reportedly had recognized only $85,000 of the debt.)

The Siles government was overthrown by the Bolivian armed forces Sept. 26 in a bloodless coup d'etat, and Gen. Alfredo Ovando Candia, the armed forces commander-in-chief, was named president.

Ovando had resigned as commander-in-chief of the armed forces Sept. 22, in the wake of well-publicized accusations by a Socialist deputy that he was receiving $600,000 in campaign contributions from Gulf Oil Corp. in exchange for pledges to grant concessions and contracts after his election. Ovando denied the accusation and asserted his determination "to unmask the nation's enemies who have put themselves in key positions to serve the foreign occupation."

In his first specific act as president, Gen. Ovando Sept. 26 announced the invalidation of the nation's petroleum code, a move that would directly affect the Bolivian Gulf Oil Co., the only major foreign oil firm operating in Bolivia. Ovando said that the original petroleum code, promulgated in 1956, constituted "a restriction on the income the state should receive from profits and royalties." Under the old code, a foreign firm paid 33.5% of its profits and delivered 11% of its petroleum production to the government. (Gulf spokesmen said the company had invested about $140 million in oil production in Bolivia in 1957–68.)

(Siles also had announced his intention of revising the petroleum code and changing the state's relationship with Bolivian Gulf.)

In a press interview Sept. 29, President Ovando asserted that the new government was not the enemy of private industry. However, he added, "it will be necessary for private companies to accommodate their thinking to the needs of Bolivia." (He also pointed to the similarity between the situations in Bolivia and Peru. "Their problems differ and their approaches to these problems differ," he said, "but fundamentally our revolution is the same as Peru's." And he added: "Remember always that we like what we have seen in Peru.")

During recent months, political pressures had increased for a greater Bolivian share in the profits of Bolivian Gulf. The New York Times reported Sept. 2 that the Bolivian Senate had sent to committee a bill that would nationalize all of the country's natural gas resources; Bolivian Gulf was the only potential commercial producer of natural gas. Passage of the bill would have jeopardized the construction of the controversial $46 million natural gas pipeline being financed in part by the World Bank.

Bolivian Gulf nationalized—The Ovando military government seized and nationalized the Bolivian Gulf Oil Co. Oct. 17. Ovando issued a formal nationalization decree several hours after troops and police had occupied, without incident,

Bolivian Gulf's offices and installations in Santa Cruz and La Paz. The decree stated that the Bolivian state oil enterprise Yacimientos Petroliferos Fiscales Bolivianos (YPFB) would immediately take over exploitation of the firm's oil fields.

(The State Department confirmed Oct. 18 and 20 that Bolivian troops had also seized the offices of Parker Drilling Co. of Oklahoma, a subcontractor to Gulf.)

The government decree also nationalized the partly World Bank-financed natural gas pipeline under construction between Santa Cruz, Bolivia and Argentina.

In a nationwide broadcast Oct. 17, President Ovando claimed that Bolivia had been "cheated and fleeced" by "international blackmail" and expressed dissatisfaction with "the modest fiscal share received by the government compared with the large profits Gulf made." He pledged "not [to] permit the private accumulation of economic power at a level that endangers the economic independence of the state." Ovando announced the appointment of a government commission, headed by Mines and Petroleum Minister Marcelo Quiroga Santa Cruz, to determine the compensation to be paid the firm.

Information Minister Alberto Bailey Gutierrez said Oct. 18 that "Bolivia fully intends to pay for the Gulf properties." He explained that nationalization had been necessary because Gulf had been unwilling to make changes in its tax position that would be acceptable to the government. He added that the decision to nationalize had been made one week before the action was taken.

Bolivian Gulf officials said Oct. 17 that they had no warning of the takeover, but leftist pressures for nationalization of foreign firms had been increasingly strong in recent months.

The government decreed a national holiday Oct. 20 in honor of the Bolivian Gulf takeover. Called a "Day of National Dignity," the holiday was marked by parades, demonstrations and speeches. President Ovando told an enthusiastic crowd: "We are at war and must join together in common cause against imperialism."

(However, according to the Washington Post Oct. 21, the government was seeking Gulf's aid in handling oil shipments at the same time it was denouncing the firm. Bolivia reportedly was negotiating through diplomatic channels in La Paz to get Gulf to continue, at least temporarily, its tanker shipments of oil. The Post reported that the Santa Cruz oil fields produced about 33,000 barrels of oil per day or 111 million barrels a year, about 75% of which was sold to the U.S. following shipment by pipeline from Santa Cruz to the Chilean port of Arica and transport by tanker to the U.S. Storage space in both Santa Cruz and Arica was not adequate to handle continued production if the oil was not kept moving to overseas markets.)

(The Associated Press reported that Bolivia and Rumania had established diplomatic relations Oct. 18. The report indicated that Bolivia was seeking Rumanian help in operating the Gulf properties.)

Board chairman E. D. Brockett of Gulf Oil Corp. Oct. 30 denounced Bolivia's nationalization of the Bolivian Gulf Oil Co. and said that U.S. sanctions were in order. Brockett called it "a straightforward case to which the Hickenlooper Amendment can be applied." (The Hickenlooper Amendment to the U.S. Foreign Assistance Act required suspension of aid to countries nationalizing U.S. assets unless the country took "appropriate steps" to provide compensation within six months. The Nixon Administration had refrained from applying the amendment in the Peruvian expropriation of International Petroleum Corp.)

Brockett rejected as unacceptable what he described as Bolivian proposals to "pay compensation from the profits of our own crude oil." He added: "Until Gulf reaches a reasonable solution with the Bolivian government, we feel we must protect the investment we have made there. It may be that oil and gas produced from Bolivian Gulf's concessions will be offered for sale by others than Gulf Oil Corp. In this event, Gulf will find it necessary to take all measures available to it to preserve its title or other rights."

(Mines Minister Marcelo Quiroga Santa Cruz had said Oct. 21 that the government would compensate Bolivian Gulf for its properties and physical assets, but that it would not pay for the gas and oil reserves since they "belong to the people of Bolivia.")

Brockett said Gulf tankers continued to boycott the oil stored in the Chilean port Arica and that storage facilities in Bolivia were also full, forcing the curtailment of production.

(A Bolivian government statement Oct. 29 said oil production would be cut to serve only the needs of Bolivia and Argentina, but that no wells would be shut down completely. The statement explained that production would be regulated until Bolivia could find foreign markets for the oil.

(Bolivian Information Minister Alberto Bailey said Nov. 8 that Peru had offered to rent three petroleum tankers to Bolivia. Bailey said the offer had been received with "satisfaction at a time Gulf is trying to strangle the national economy with an economic blockade.")

Bolivian Ambassador to the U.S. Julio Sanjines Goytia Oct. 30 called Brockett's remarks "improper" and accused Gulf Oil of "intervention in the affairs of the Bolivian state." Asserting that the Hickenlooper Amendment should not be considered since Bolivia was planning compensation, the ambassador said: "According to our laws Bolivia has the right to nationalize any wealth that will benefit the economy of the country. This right of nationalization has been recognized in international law and by the United Nations."

Williams Brothers reported that it had suspended construction of the Bolivia-Argentina natural gas pipeline Nov. 5. The Bolivian government had charged that Gulf had halted in Argentina delivery of necessary construction equipment.

Parker Drilling Co. of Tulsa, Okla., which had been reported taken over along with Bolivian Gulf, Nov. 13 affirmed an earlier assertion that it had not been nationalized by the government. A spokesman announced, however, that the firm was cutting its "personnel down to the minimum," but added that the "work suspension is temporary. We expect to reach an agreement with the government in the coming weeks."

Ecuador. The government announced March 13 that the Texaco and Gulf oil companies had accepted Ecuador's demands for higher royalties and the return of part of the land they had been exploiting since 1964. Operating in Ecuador as a consortium, the two American companies agreed to return 2.5 million acres of land and to increase petroleum royalties to the government from 6% to 11%, a government spokesman reported. The companies also promised to seek public bids for construction of a 180-mile pipeline from the Pacific Ocean to oil fields in eastern Ecuador; to pay land taxes of 25 cents per hectare (about 2.5 acres), increasing to 40 cents in 40 years, and to invest $55.5 million in development projects in Ecuador. President Jose Maria Velasco Ibarra had demanded the return of the land on grounds that the existing concessions were unfair and must be changed. The companies had rejected the demands March 6, but had apparently reconsidered their position.

Colombian oil pipeline opened. Colombian President Lleras Restrepo May 11 officially opened a 194-mile oil pipeline from the Amazon jungle across the Andes Mountains to the Pacific Ocean. The pipeline's most inaccessible sections had been laid by helicopter. The line reached 11,500 feet at its highest point and had an initial capacity of 50,000 barrels a day. Sponsored by the Texas Petroleum Co. and the Colombian Gulf Oil Co., the line pumped crude oil from the Orito oil field, the Ecuador border, to the Pacific. (The two companies shared equal interest in the Orito concession, although President Lleras confirmed in his opening speech that the contracts with the Colombian government were being renegotiated.) Fifty workers, among them three helicopter pilots, were said to have been killed during construction of the pipeline.

■Venezuela signed an agreement with Colombia Aug. 22 for joint utilization of electric power plants in their adjoining border areas.

Prospecting 'Mexicanized.' In what was described as the most important nationalization act of President Gustavo Diaz-Ordaz's administration, the government-owned oil firm Pemex took over prospecting rights granted to four U.S. oil companies in 1949–1951, the Journal of Commerce reported June 9. The cancellation of the firms' contracts involved payment of $18 million to the four companies— Continental Oil, Pauley Petroleum, Pauley Pan American Petroleum and American Independent Oil—and the return to Mexico of exclusive rights to explore, drill and exploit oil resources along its Gulf coast.

Brazilian power loan. The Inter-American Development Bank announced approval of a $21.3 million loan Aug. 28 to help finance a 220,000 kilowatt hydroelectric power plant in Rio Grande do Sul, Brazil's southernmost state. The loan, extended to Brazil's power corporation Eletrobras, was to help achieve the Brazilian goal of expanding the nation's electric power generating capacity from 8 million kilowatts in 1967 to 11.2 million in 1970.

Loans to Argentina. The Inter-American Development Bank (IDB) announced approval of a $20 million loan Sept. 11 for construction of a natural gas pipeline from fields in the western Neuquen Province to the eastern city of Bahia Blanca. At Bahia Blanca the 354-mile pipeline was to feed into the major north-south trunk system which supplied Buenos Aires and other major consumer areas. The project, part of a nation-wide 10-year plan, was designed to meet the growing demands for natural gas in the Buenos Aires area and to help the nation save foreign exchange by reducing imports of liquid gas.

The International Bank for Reconstruction and Development (World Bank) Oct. 8 approved a $60 million loan for electric power development in the Buenos Aires area. The loan was extended to Servicios Electricos del Gran Buenos Aires, S.A. (SEGBA) to finance the bulk of the foreign exchange required for its $247 million 1970–72 expansion program. SEGBA would change its role from a utility supplying self-generated electricity to a distributor of large amounts of energy generated by other utilities outside the Buenos Aires metropolitan area.

Arab Oil

Guerrillas hit pipeline. Arab guerrillas of the Popular Front for the Liberation of Palestine (PFLP) May 30 blew up and heavily damaged a section of the Trans-Arabian Pipeline, owned by the Arabian-American Oil Co. (Aramco), in the Israeli-occupied Golan Heights of Syria.

The explosion and resultant fire blocked the flow of oil through the 1,000-mile pipeline connecting Dharan, Saudi Arabia on the Persian Gulf to Sidon, a Lebanese port on the Mediterranean. The pipeline provided Saudi Arabia, Jordan, Syria and Lebanon with millions of dollars in royalties and transit fees each year.

A PFLP spokesman in Amman, Jordan said May 31 that an explosive charge had been placed in the Baniyas River where the pipeline runs along the river bed. The purpose of the blast, the spokesman said, was to spill oil into the river and pollute the water it supplied to Israeli settlements and fisheries in the Huleh Valley. As a result of the blast, oil was reported seeping into the northern part of the Sea of Galilee, an important water source for Israel, but without causing dangerous pollution. Oil slicks also were observed on the Jordan River. Israelis fought the pipeline blaze 14 hours before extinguishing it May 31. (Aramco's oil had flowed uninterrupted through the 25-mile Golan Heights section of the line despite Israel's capture of the area in 1967.)

The PFLP was sharply criticized in the Arab world June 1 for having blown up the pipeline. The Egyptian newspaper Al Ahram said the PFLP's "incomprehensible" action had inflicted no harm on Israel but had violated a "logical framework of Arab principle and interests." Arab policy, the newspaper reminded PFLP, was to permit the free flow of oil to bolster Arab economies.

The Saudi Arabian government, which stood to lose most from the pipeline attack, charged in a Mecca radio broadcast that the PFLP commandos had served Zionism by their sabotage.

In Jidda, Saudi Arabia, the newspaper El Bilad asserted that the pipeline assault "undermines the economies of the transit countries."

The Beirut newspaper Al Hayat said the PFLP was serving Israeli interests by its attack.

Israeli Premier Golda Meir said June 3 that Israel would keep the pipeline closed until it was assured that there would be no repetition of the guerrilla sabotage and until Aramco gave "us guarantees against any damage we may suffer from their oil." "This time," Mrs. Meir said, "we were saved by a miracle." She was referring to the minor damage suffered by Israeli water sources as a result of the pollution.

Israel and Aramco July 9 signed an agreement for the pipeline to be repaired.

An Israeli Foreign Ministry spokesman said July 10 that the agreement compensated Israel for the damage (estimated at $500,000) caused by the resultant oil leakage and "provided maximum safeguards" against future terrorist raids on the pipeline. The precautionary measures were said to include fencing, guards, burying of certain sections of the pipeline and greater access to valves for the Israelis.

Soviet Aid Pacts. Lt. Gen. Saleh Mahdi Amash, Iraqi deputy premier and interior minister, announced July 6 that the U.S.S.R. had agreed to lend Iraq $70 million for development of the North Rumailah and Pakkawi oil fields. The loan reportedly was part of a new pact under which the U.S.S.R. would help Iraq build a dam on its part of the Euphrates River. (The Soviet Union was already aiding Syria to build a dam on its part of the Euphrates.) Amash also said that the new agreement provided for Soviet assistance to Iraq's state-controlled National Minerals Company for the exploration of natural gas and iron ore deposits.

The two countries had signed another agreement June 21 under which the U.S.S.R. would provide the National Minerals Company with all machinery needed to begin oil production. Both Soviet loans were to be repaid with crude

ARAB & WORLD PETROLEUM RESERVES

(Estimate, in billions of barrels, as of end of 1969. Source: Oil & Gas Journal.)

Abu Dhabi	0. 016
Bahrain	. 343
Dubai	1. 000
Iraq	27. 500
Kuwait	68. 000
Muscat-Oman	5. 000
Neutral Zone	13. 000
Qatar	5. 500
Saudi Arabia	140. 000
Syria	1. 500
Turkey	. 650
Algeria	8. 000
Egypt	5. 000
Libya	35. 000
Total	310. 509
United States	38. 700
Free world	463. 780
Communist world	60. 000
Total, world	523. 780

oil. The agreements were expected to undermine the internationally-owned Iraq Petroleum Company, which was the country's only producer of oil. IPC had paid the government $400 million annually in royalties, which accounted for 70% of the nation's income.

(Iraq and East Germany signed a long-term economic and technical cooperation agreement June 25. Under the agreement, East Germany would lend Iraq $80 million for 13 major projects. The loan was to be repaid partly with goods produced by factories to be built in Iraq with the East German loan and partly in Iraqi oil.)

UAR Plans Pipeline. The Egyptian government signed an agreement with a French-led consortium in Cairo July 13 to construct a $175 million oil pipeline from Port Suez to Alexandria, bypassing the blocked Suez Canal.

The consortium was composed of five French companies, three Italian and one each from Britain, Holland and Spain.

Russia & Eastern Europe

Pipeline to West Germany Proposed. The West German government disclosed May 5 that the U.S.S.R. and East Germany had proposed building oil and natural gas pipelines to West Germany. The separate proposals were made to West German Economy Minister Karl Schiller at the Hanover Trade Fair in late April by Nikolai S. Patolichev, Soviet foreign trade minister, and Heinz Behrendt, East German deputy trade minister.

The oil line presumably would be an extension of the "Friendship Pipeline" running from Soviet oil fields to the petrochemical center at Schwedt, East Germany. (A related natural gas pipeline was planned.) The center currently received about four million tons of oil annually. Sources in Bonn said that the U.S.S.R. was ready to send over six million tons to West Germany annually.

The pipeline offers were made after Patolichev had signed an agreement April 28 with the Thyssen Roehrenwerke for the construction of two pipe factories near Moscow. It was indicated at that time that Thyssen pipe would be

ARAB OIL EXPORTS & CONSUMER OIL IMPORTS
(Thousands of barrels daily, 1969)

To—	Arab oil exports	Consuming area oil imports	Percent imports from Arab countries
United States and Canada	607	3, 875	15. 7
Other Western Hemisphere	173	785	22. 0
Western Europe	9, 018	11, 740	76. 8
Africa	64	570	11. 2
Southeast Asia	247	830	29. 8
Japan	1, 858	3, 455	53. 8
Australasia	359	530	67. 7

Source: BP Statistical Review of the World Oil Industry, 1969.

used to build the new extension if Bonn agreed to the proposals.

Soviet exports of crude oil to Western markets had totaled about 530,000 barrels daily during 1968, down from 550,000 barrels daily in 1967, the Wall Street Journal reported May 8. The decline was attributed to inadequacies in crude oil reserves. The U.S.S.R.'s 1968 crude oil production was reported to have totaled 309 million tons.

West Germany signed a 20-year agreement with the Soviet Union Nov. 30 to purchase natural gas from Siberia in exchange, in part, for large-diameter steel pipes for use in the Siberian gas fields. It was reported that East German Communist Party First Secretary Walter Ulbricht was displeased by the agreement because it demonstrated Soviet willingness to deal with the Bonn government and because a new pipeline for delivery of the gas would be laid through Czechoslovakia, depriving East Germany of any revenues. The first shipment of half a billion cubic meters (17.5 billion cubic feet) would be in 1973.

Bonn announced Dec. 8 that a consortium of West German banks was preparing a $300 million credit for the U.S.S.R. to buy 1.2 million tons of 100-inch gas pipes from two West German firms for transmission of natural gas from Siberia to West Germany in the 1970s.

Vlastimil Plechac, a deputy chairman of the Czechoslovak chemical industry, said Nov. 23 that the U.S.S.R. had told its East European allies to begin buying crude oil in North Africa and the Middle East because Soviet production would be shifting from the Volga-Urals to the Siberian fields during the 1970s, adding greatly to transmission costs. At present, Rumania was the only Soviet-bloc state to purchase oil from the Middle East; the other countries relied on shipments from the Volga-Urals fields via the Friendship Pipeline.

Italy buys Soviet gas. Italy's state-owned oil and gas agency Ente Nazionale Idrocarburi (ENI) signed a 20-year agreement with the U.S.S.R. Dec. 10 to purchase 3.5 trillion cubic feet of natural gas in exchange for pipe, valves, compressors, and telemetric and telecom-

munications instruments. Soviet deliveries were expected to begin in three years at an annual rate of about six billion cubic meters (211.8 million cubic feet). Deliveries would be made from the Ukraine to the Austrian-Czechoslovak border through an existing pipeline; it would then be transferred to Tarvisio in northeastern Italy via a new 230-mile pipeline to be constructed by an Italian-Austrian-Soviet consortium. The agreement was valued at $3 billion.

Rumania. Venezuela and Rumania signed an oil agreement which provided for the purchase of four million tons of crude oil from Venezuela between 1971-75, it was reported Sept. 16. Rumania reportedly also proposed to buy up to 11 million tons by 1980.

Among Rumanian achievements in 1968: Coal production was 17 million tons, slightly below the goal of 17.1 million tons. Oil production showed a slight increase over 1967 to 13.28 million tons in 1968.

Czechoslovakia & the Soviet Union. Following a nine-day visit by Prague leaders to the Soviet Union Oct. 20-28, the two countries issued a joint communique Oct. 28 strengthening their economic, political and social relations.

The communique said the U.S.S.R. would increase its trade with Czechoslovakia for 1970-75. "The Soviet Union," it said, "will increase above the planned quotas its deliveries of petroleum, cast iron, cotton and other important kinds of raw material . . . as well as equipment Czechoslovakia greatly needs."

The communique said an agreement had been reached on expanding cooperation in nuclear power engineering. The Soviet Union agreed to supply equipment for the construction of atomic power stations in Czechoslovakia, and both countries were to cooperate in the manufacture of other equipment for the power stations and for computer facilities.

Soviet power output up. Among improvements reported in the Soviet economy during 1969 (as compared with 1968): the output of electricity increased

8%; oil went up 6%; gas rose 7%; coal production was up 2%. But the Natural Gas Ministry failed to fulfill its quota.

Other Areas

Oil Under East China Sea. Japanese geologists reported that huge oil deposits had been discovered in the area of the Senkaku Islands in the East China Sea during a survey conducted June 4 to July 13. The islands, currently administered as part of the Ryukyu Island chain by the U.S., were located about 100 miles northeast of Taiwan and 50 miles from the Chinese mainland. Dr. Hıroshi Niino of Tokai University said the underwater oilfield could be one of the 10 largest in the world. At present, Japan had almost no petroleum deposits and imported nearly $1 billion in oil annually, almost all from the Middle East. The government announced Aug. 29 that $280,000 would be allocated in next year's budget for exploration and development of the oil fields.

If a major oil strike was made, it was expected that both Communist and Nationalist China might insist on rights. The Taiwan government had claimed July 17 exclusive rights to all natural resources in the continental shelf on which the Senkakus lay.

Angola. Petroleum production had begun during December 1968 in the Cabinda district of Angola, the N.Y. Times reported Jan. 12. Some 30 offshore oil wells were reported producing 20,000 to 25,000 barrels of high-quality, low-sulfur crude oil a day. All the concessions in Cabinda, separated from the main territory of Angola by a thin strip of the Congo (Kinshasa), were held by Cabinda Gulf Oil, a subsidiary of the Gulf Oil Corp. Cabinda Gulf had spent about $125 million on its facilities. It recently completed the Malongo Terminal Complex, capable of storing 1,370,-000 barrels of oil and loading it on 200,000-ton tankers.

Ceylon. The World Bank announced a $21 million loan to the Ceylon Electricity Board July 23. The loan was to be used to finance a $31.5 million program for the installation of 115 megawatts of electric generating capacity and associated transmission facilities. The loan was granted for 25 years at 6½% interest, with a five-year grace period.

Mozambique. The Portuguese government said Sept. 2 that a consortium of French, German, Swedish and South African firms had been awarded the contract for construction of the Cabora Bassa hydroelectric power project. The contract was signed Sept. 19.

Atomic Power

U.S. To Sell Three Uranium Plants. President Nixon announced Nov. 10 that he had instructed the Atomic Energy Commission (AEC) to place its three uranium enrichment facilities under separate management in order to prepare them for sale to private enterprise. The gaseous diffusion plants, located at Oak Ridge, Tenn., Paducah, Ky., and Portsmouth, Ohio, were valued at $2.3 billion, although they had cost the government an estimated $4 billion to build. They produced enriched uranium for U.S. nuclear weapons and for privately operated electric power plants.

Stating that the government's responsibility for operating the plants should eventually terminate, the President said, "These facilities should be transferred to the private sector, by sale, at such time as various national interests will best be served, including a reasonable return to the Treasury." He added that the government's needs for enriched uranium would be relatively small and "national needs . . . are now largely commercial."

Rep. Chet Holifield (D., Calif.), chairman of the Joint Committee on Atomic Energy, said Nov. 10 the sale would require Congressional authorization, and he doubted whether the sale would result in enough money to "provide the additional enrichment capacity" to fuel nuclear plants of private industry. He further asserted "there is a very substantial question whether there is any likelihood for effective price competition in what is now, and in all probability for some years to come will be a highly concentrated industry."

(The AEC had announced March 11 that as of April 1, it would permit utilities to buy any enriched uranium they currently leased from the commission on condition they bought uranium concentrates on the open market to reimburse AEC enrichment stocks. The AEC intended to save money by reducing its own purchases of uranium concentrates from mining companies.)

Uranium Demand Rises. An increase of 60%–70% in uranium production was needed in the next five years to meet rising demands for nuclear power fuels, according to a joint report of the International Atomic Energy Agency and the European Nuclear Energy Agency. The report, made available in Vienna March 20, said that no significant additional capacity was likely to become available unless new reserves capable of being processed at low cost were discovered. Current annual output was estimated as 23,500 tons of uranium oxide.

British-Dutch-German plants. Britain the Netherlands and West Germany agreed in London March 11 to build two plants to produce enriched uranium by a new gas centrifuge method. The agreement called for establishment of two organizations, one to operate the two plants to be built in Britain and the Netherlands, and the other to supervise manufacture of centrifuges and construction of enrichment plants in all three countries.

Japan Develops Enriched Uranium. Scientists at the Institute of Physical and Chemical Reseach announced in Tokyo March 31 that they had produced enriched uranium in Japan for the first time. The uranium had a concentration of .742%. A spokesman for the Atomic Energy Agency said that a 2–3% concentration was needed for nuclear fuel used in power generation, while 90% concentration was needed for producing nuclear bombs. An article in the newspaper Asahi Shimbun said March 31 that "the fact that it has become technically possible here to produce enriched uranium . . . is seen by some observers as

indirectly strengthening Japan's diplomatic position." (Japan's 1946 constitution forbade the country from acquiring or maintaining any "military potential.") Japan had one nuclear power station in commercial use: five others were under construction.

(Japan's first atomic power industrial group, FAPIG, signed an agreement with the British Nuclear Power Group May 22 to introduce advanced gas-cooled British reactors to Japan.

New U.S.-Euratom uranium contract. The U.S. AEC and the European Atomic Energy Community (Euratom) signed a multi-lease contract Nov. 7 for enriched uranium. The new contract, which became effective Nov. 1 and was to run to Dec. 31, 1970, allowed for the leasing of special fissile materials without the AEC's prior written consent. Euratom presently had $19 million of leased enriched uranium from prior contracts.

France to Adopt U.S. A-Plant Design. The French government announced Nov. 14 that, beginning in 1970, France would build atomic plants based on the U.S. design and dependent on U.S. enriched uranium for fuel. The decision reversed a 15-year program of attempting to attain nuclear independence by developing graphite, gas-cooled reactors that used natural uranium for fuel. A statement said research on natural uranium and on fast-breeder reactors would continue.

The decision was taken amid strikes by members of the Atomic Energy Commissariat across France to protest an announced reduction of 2,600 employes by 1971 as a result of a 9% budget cut. The strikes of the 31,000 nuclear workers had begun Nov. 13.

Use of the U.S. reactor design had been urged by Marcel Boiteux, director of France's national power agency (Electricite de France), who said Oct. 16 that his country's current reactors were unprofitable, although their design had been proven technically successful. "Taking into consideration the international market," he said, "Electricite de France considers it desirable, both for itself and French industry, to concentrate its efforts on the construction of centers using the 'American system' of enriched uranium."

USSR Builds Breeder Reactor. The Soviet Union announced Jan. 3 that an experimental breeder reactor had been constructed at the atomic reactor development center at Melekess, in the Volga River Valley. The breeder, designated BOR-60, was a 60,000 kilowatt installation.

Soviet H-Power Progress. Prof. Lev A. Artsimovich, the Soviet scientist in charge of research on control of thermonuclear reactions, reported April 11 that his laboratory had taken an important step toward producing electric power from the energy of the H-bomb reaction.

Artsimovich, currently at the Massachusetts Institute of Technology as a visiting professor, said that equipment at Moscow's Kourchatov Atomic Energy Institute had sustained a significant thermonuclear reaction for one thirty-third of a second at a temperature of 9 million degrees.

Prof. David Rose, an MIT nuclear engineering expert, said the Soviet report indicated "the best combination of experimental ingredients yet achieved by any nation" in the 20-year effort to harness the power of the H-bomb. It was hoped that controlled thermonuclear reactors could be used to produce unlimited amounts of cheap electric power.

The major British science journal Nature reported Nov. 1 that British scientists had verified Soviet claims of harnessing hydrogen power for civilian use. Using five tons of equipment flown to Moscow, the scientists demonstrated that a Soviet machine, known as the Tokamak 3, had generated hot electrified gas at temperatures and pressures near those at which atoms like hydrogen fuse together to release large quantities of energy. (In reactors currently in use, energy was made by fission—the splitting of large atoms.)

Other Soviet developments. Among other Soviet atomic developments:

The Soviet Union announced Jan. 8 that it was developing a major mining complex for production of tantalum, a rare metal with new uses in electronics and atomic energy. The new complex was being developed in the Chita Province of southern Siberia, Tass reported. Because of the strategic character of the metal, location of other Soviet deposits had been kept secret in recent years. (The leading producers of tantalum were Nigeria, Canada and Brazil. U.S. production was negligible.)

Sources in Helsinki reported April 20 that Finland was obtaining its first atomic power plant from the Soviet Union. The purchase of the plant and a supply of uranium to fuel it was to take place under an agreement on peaceful uses of nuclear energy between the two countries.

Tass reported April 14 that a small atomic plant had been designed by Soviet scientists to provide heat and power for remote settlements in Arctic regions. The plant's atomic elements, designed to operate a 1,500-kilowatt turbine, would provide power 60% to 90% more cheaply than standard fuels.

The Soviet Union announced April 22 it had opened a new atomic research center near Tashkent, Uzbekistan "to analyze the chemical components of ores, metals and other materials."

South Korean Reactor. The Westinghouse Electric Corp. announced Jan. 31 that it had contracted to supply South Korea with its first nuclear power plant. The contract, worth about $96 million, provided for the sale of a 565-megawatt water reactor to the Korea Electric Co.

New West German A-Plants. Scientific Research Minister Gerhard Stoltenberg announced March 18 that West Germany would build five new nuclear power stations valued at $550 million in the next three years. Other plans, he said, called for construction of 14 additional nuclear power plants, valued at $1 billion, by 1975.

A-plant in India. India's first nuclear power station, at Tarapur, 70 miles north of Bombay, began test operations April 1. The station was built by the General Electric Corporation with an $80 million U.S. loan. (President Zakir Hussain had reported Feb. 17 that India's first uranium mine and mill, in Bihar State, had begun production of uranium concentrates.)

Taiwan to get A-plants. The U.S. and Japan approved loans to help finance a nuclear power plant in Taiwan, it was reported Sept. 3. The U.S. Export-Import Bank approved a $79.7 million loan at 6% annual interest. Westinghouse Electric Corp. and General Electric Co. would supply the equipment, valued at $62.6 million; $17.1 million would be used to finance the cost of uranium and to produce the initial fuel core in the U.S.

The Japanese loan of $20 million for the Chinshan plant was made by a group of banks led by Mitsubishi and Sumitomo. The total cost of the 550-megawatt plant was estimated at $157.5 million, and it was expected to be in operation by December 1975.

Canada agreed to sell a 40-megawatt nuclear reactor to Nationalist China Sept. 16 for $28 million. The reactor was to be installed at Huaitzupu.

A-station for Australia. Australian Prime Minister Gorton announced Oct. 8 that Australia would begin building its first nuclear power station within three years. The plant would be located at Jervis Bay on the New South Wales southern coast.

Safety pact. Canada and Pakistan signed a safeguard transfer agreement with the International Atomic Energy Agency in Vienna Oct. 17. Under the agreement, the Karachi Nuclear Power Station would become the first atomic plant subject to IAEA safeguards in Southeast Asia.

A-plant for Chile. The Chilean embassy bulletin reported Nov. 3 that Chile had agreed to purchase an atomic reactor from Great Britain.

U.S.-Soviet Talks on Peaceful Uses. Representatives of the U.S. and the Soviet Union met April 14 in Vienna to exchange information on peaceful uses of nuclear explosions. The talks, which were the outgrowth of the 1968 agreement on a nuclear non-proliferation treaty, were held in Vienna because that city was headquarters site for the International Atomic Energy Agency (IAEA).

The U.S. delegation was led by Gerald F. Tape, member of the Atomic Energy Commission. Yevgeny K. Federov, a member of the Soviet Academy of Sciences, led the Soviet delegation.

After three days of closed door sessions, the delegates issued a joint statement April 16 that held out hope for the peaceful uses of atomic blasts: "The parties were of the view that underground nuclear explosions may be successfully used in the not so far off future to stimu-

late oil and gas production and to create underground cavities. It may also be technically feasible to use them in earth-moving works for the construction of water reservoirs in arid areas, to dig canals and in removing the upper earth layer in surface mining etc."

At a joint press conference April 16, the chief delegates agreed that use of such explosions for the benefit of non-nuclear countries was still five years away. Tape said that the use of atomic explosions to stimulate the release of natural gas would probably come before 1974. Federov indicated his agreement.

The director general of the IAEA warned April 15 against premature optimism on the peaceful uses of nuclear explosions. Speaking in Vienna, Dr. Sigvard Eklund said the immediate value of such explosions had been exaggerated for what he called "political reasons." "For the time being," Eklund said, "our knowledge of how to conduct nuclear explosions for civil engineering is just in its infancy. The technology is still not completely known, and how much radioactivity would be produced by a specific project still has to be learned."

Project Gasbuggy Yield. Three 30-day production tests carried out under Project Gasbuggy, a project to stimulate natural gas production by means of nuclear explosions, had yielded 109 million cubic feet of natural gas, the Atomic Energy Commission, El Paso Natural Gas Co. and the Bureau of Mines announced March 19.

The tests brought total natural gas production from the explosive emplacement hole to 167 million cubic feet since January 1968. By comparison, the project participants noted, an existing conventional well about 400 feet away had produced 85 million cubic feet of natural gas from the same formation during a nine-year period.

The nuclear explosion for the Gasbuggy experiment had been detonated 4,200 feet underground Dec. 10, 1967.

Natural Gas Research. A 40-kiloton underground nuclear explosion, Project Rulison, was conducted Sept. 10 under Battlement Mesa in western Colorado, 40 miles northeast of Grand Junction.

The explosion, 8,442 feet below the surface, created a 180-foot hole 1½ miles underground. Officials of the Atomic Energy Commission (AEC) said no radioactivity had been released above ground. The site of the explosion was sealed for six months to allow the heat produced to be dissipated.

Project Rulison, the first nuclear test to take place in Colorado, was part of the AEC's Plowshare Program for developing peaceful uses of atomic energy. The purpose of the test was to pulverize heavy Mesa Verde formations of shale and sand and trapping the underground natural gas. The area contained an estimated 6–10 trillion cubic feet of gas.

The project was sponsored jointly by the AEC and the Austral Oil Co. of Houston, Tex., and was managed by the CER Geonuclear Corp. of Las Vegas, Nev. It was the second joint industry-government nuclear test designed to free natural gas.

Opponents of Project Rulison had attempted to block the explosion by court orders. Federal Judge Alfred A. Arraj of Denver, however, had refused Aug. 27 to issue a temporary injunction to halt the blast. He said: "I am impressed with the fact that the government has exercised extreme caution and care to protect persons, animal life, plant life, water supply and other things that could be adversely affected." Arraj's ruling was upheld Sept. 2 by 10th U.S. Circuit Court of Appeals Judge Delmas C. Hill in Wichita, Kan. Hill said the opponents had presented "a very weak case" that "irreparable damages" might be caused by the explosion. Supreme Court Justice Thurgood Marshall refused Sept. 3 to overturn the lower court ruling. Judge Arraj denied another injunction request by the Colorado Open Space Coordinating Council, Inc. Sept. 3.

Nuclear Ships. The Italian Communist Party newspaper L'Unita reported Jan. 17 that Communist China had launched two nuclear-powered merchant ships. The vessels, said to have been built at the Tientsin shipyards, were the 20,000-ton Zan Tahn, and the Bac Puhn, with a 210 megawatt reactor. L'Unita gave no source for its report.

The U.S. Navy launched the world's first nuclear-powered deep-diving research submarine Jan. 25 at Groton, Conn. The NRI (Nuclear Research One), built by the Electric Boat Division of the General Dynamics Corp., had a hull 140 feet long and was manned by five officers and men and two scientific observers. The NRI was built at a cost of $99.2 million (the original 1965 cost estimate had been $30 million). The Navy said that the NRI was "designed as a multipurpose deep submergence vessel—capable of conducting military missions as well as Navy and civilian oceanographic missions."

U.S. report for 1969. In its annual report for 1969 released Jan. 30, the U.S. Atomic Energy Commission said that a fire at its Rocky Flats plutonium plant near Golden, Colo. May 11 had caused $45 million in damages (excluding the cost of the plutonium). It said the plant was "the primary facility in the AEC production complex for fabrication of plutonium weapons parts."

The AEC also reported: it had completed a facility at its Savannah River plant in South Carolina to recover components from nuclear weapons stockpiles; it had completed ground tests of the Nerva experimental nuclear engine in Nevada; and it was continuing work on the development of two new nuclear attack submarines—the high-speed submarine and the electric-drive submarine.

The report said construction of new nuclear energy power plants in 1969 had slowed down, mainly due to public fear. It said 1969 was a year "in which the growth in nuclear energy so urgently needed to meet the power requirements of the future was incongruously slowed down by public apprehension, construction delays and difficulties in plant equipment manufacturing capacity."

1970

As the U.S. energy deficit grew, the Nixon Administration decided during 1970 to continue the quota
program for restricting oil imports. After at first
ordering further curbs on imports of oil from Canada,
the U.S. administration ultimately decided to permit
the entry of all the oil Canada was prepared to sell.
Northeastern areas of the U.S. suffered "dimouts"
during the summer of 1970 as a heat wave spurred a
record use of air-conditioning and severely taxed increasingly inadequate power supplies. The U.S. ended
a 19-month moratorium on the sale of leases for offshore oil production despite protests of environmentalists. The construction of new electric-power
plants was hampered by environmental considerations. Oil companies were accused of monopolizing
energy sources. The West was warned not to depend
too heavily on the politically unreliable Middle
East for vital energy. Algeria and Libya nationalized foreign oil facilities.

Consumption & Scarcity Increase

U.S. Energy 'Gap' Widens

The U.S. energy deficit, the difference between the quantity of fuel the U.S. produced and the amount it used, continued to grow throughout 1970.

Although Americans consumed 5.4 billion barrels of liquid petroleum products during 1970, U.S. oil wells supplied only 3.5 billion barrels of crude oil. The U.S. produced and used 22 trillion cubic feet of natural gas in 1970, a year in which petroleum supplied 43.4% and natural gas 32.5% \ of the energy used by the U.S.

Only coal was in surplus, the U.S. production of 590 million tons in 1970 providing the 520 million tons needed for domestic use plus 70 million tons for export. Coal supplied 20% of the energy consumed in the U.S. in 1970, waterpower provided 3.8% and atomic energy .3%.

The deepening of what was being described as an incipient "energy crisis" supplied added fuel for the long-standing dispute over the U.S. policy of restricting petroleum imports in order to provide incentives for domestic oil producers.

Oil import change deferred. Any major change in the oil import control program was deferred by President Nixon Feb. 20 pending further consultations with other oil producing countries. The announcement was made with release of a report on the subject by a Cabinet task force headed by Labor Secretary George P. Shultz.

A majority of the task force recommended a major change to an ordinary tariff from the company-by-company import quotas in effect for 10 years. The tariff recommended would result in a 30¢ decrease in the $3.30 a barrel domestic crude oil price and a resultant reduction of about one-half cent in the price of gasoline or home heating oil. This would apply in full to Middle Eastern oil, which would be limited to 10% of U.S. imports. The new tariff would be lower for Venezuela and eliminated for Canada and Mexico.

The report said that "in 1969 consumers paid about $5 billion more for oil products than they would have paid in the absence of import restrictions."

However, two of the majority members—Secretary of State William P. Rogers and Defense Secretary Melvin R. Laird—while favoring a change in the policy, advocated a delay pending international talks. The two members of the panel in the minority—Commerce Secretary Maurice H. Stans and Interior Secretary Walter J. Hickel—opposed elimination of import controls while favoring improvements and an increase in the permitted imports.

The three remaining members—Shultz, Treasury Secretary David M. Kennedy and George A. Lincoln, director of the Office of Emergency Preparedness—urged an early change to a new system.

Lincoln was named by President Nixon Feb. 20 to head a new Oil Policy Committee to consider "interim and long-term adjustments that will increase the effectiveness and enhance the equity of the oil import program." The new group included all the task force members except Shultz, as well as Attorney General John N. Mitchell and Paul W. McCracken, chairman of the Council of Economic Advisers.

Nixon noted in his statement that all of the original task force members agreed "that a unique degree of security can be afforded by moving toward an integrated North American energy market."

The President's decision to delay a change in the oil import system generally was supported by the oil industry and opposed by New England senators, nine of whom made public Feb. 20 a letter to Nixon urging a new program to bring equitable treatment for the Northeast, where hardship allocations of home heating oil had been required.

U.S. cuts oil imports from Canada. President Nixon March 10 issued a proclamation restricting U.S. imports of Canadian oil to 395,000 barrels a day (b-d). The quotas affected oil imported to the area east of the Rockies.

The new curbs replaced voluntary controls agreed to in 1967. These had set informal limits at 332,000 b-d, but violations of the controls had brought the import rate to 550,000 b-d by February 1970. Philip H. Trezise, assistant U.S. secretary of state for economic affairs, called this rate "excessive" and "disorderly." He said the formal quotas had been introduced because the voluntary controls were "not workable."

The 395,000 b-d ceiling would be retroactive to March 1 and would remain in effect through 1970. Trezise, in explaining the presidential order, disputed the claims of some Senate opponents that the limitation would raise prices to U.S. consumers. Trezise added that the limitation would benefit oil interests in Venezuela and the Middle East, as well as southwestern U.S. oil producers.

The new quotas were based on a determination that the U.S. national security might be impaired by the unrestricted flow of oil from Canada. Trezise expressed the U.S. government's concern that "in an emergency," oil fields in western Canada would be "almost without reserves" to supply U.S. needs. (Hollis Dole, assistant secretary of the interior for mineral resources, told a House subcommittee hearing on oil imports March 10, that Canada's "dependence upon foreign sources for approximately half its own oil consumption" was the basis for the decision on security.) The President's order changed the membership of the Oil Imports Board to substitute a Justice Department member in place of the Defense Department member previously on the board.

The Administration decision had been hinted in press reports March 3. The reports said the imposition of mandatory controls would represent a transitional move toward fulfillment of the recommendations of the President's task force on the oil import quota system. Observers speculated that the move was aimed at strengthening the U.S. bargaining position for an overall energy agreement with Canada.

In Ottawa, Energy, Mines & Resources Minister J. J. Greene contended March 10 that the U.S. decision was "a mistake," but Greene predicted that Canadian oil producers would have better access to U.S. markets by the start of 1971. "Apparently the President believes that by the end of the year a more permanent arrangement can be achieved," the minister said. He reaffirmed Canada's position that Canadian oil should have an open market in the U.S. and emphasized that negotiations would continue over the next few months. (The Washington announcement had come as a bipartisan parliamentary delegation was on its way to Washington to discuss energy policies with U.S. senators and congressmen.)

The Toronto Globe & Mail reported March 14 that "some prominent Northern [U.S] senators" were challenging the legality of the President's decision. The Globe & Mail cited Sens. Walter Mondale (D, Minn.), Edward Kennedy (D, Mass), William Proxmire (D, Wis.) and Jacob Javits (R, N.Y.) as arguing that the level of imports before the quotas were set did not pose a threat to national security, as the President had contended. In a letter to George Lincoln, director of the Office of Emergency Preparedness and president of Nixon's new oil policy committee, Kennedy said it was "not clear what evidence the President considered which demonstrated that the cutback was necessary to serve our national security."

■Twenty-four Canadian MPs and senators from all parties met March 11 with 13 U.S. senators and 15 congressmen in Washington for a discussion on energy policies. A communique issued at the conclusion of the meeting said the two groups had disagreed on the status of the waters between the Arctic islands and on whether water should be included in a common energy policy. The Canadian delegation saw the Arctic waters as Canadian and rejected a U.S. proposal that fresh waters be part of an overall energy pact. The meeting was for an exchange of views; neither delegation was empowered to negotiate.

Canadian Prime Minister Pierre Elliott Trudeau and Energy Minister J. J. Greene had pledged Jan. 15 that Canada would maintain full control over water resources even if the U.S. were to seek shares jurisdiction in the context of an overall energy agreement. A U.S. Labor Department task force had recommended to Nixon that the level of Canadian oil exports to the U.S. be increased in exchange for Canadian agreement to a continental approach to energy development.

Asked in Commons to assure that no control of Canadian sovereignty over its water resources would be traded for greater oil markets in the U.S., Trudeau said "I think I can give that assurance." Canadian External Affairs Minister Mitchell Sharp said Feb. 19 that Canada claimed jurisdiction over all waters of the Northwest Passage and between the islands of the Arctic archipelago. Sharp's statement was the strongest claim to Arctic sovereignty yet made by the Canadian government. He declared: "These are

our waters. . . ." Sharp added that the fact that the voyage of the U.S. icebreaking tanker Manhattan was conducted "with our acquiescence and our cooperation is evidence of our position." In questioning in the House of Commons Feb. 20, Prime Minister Trudeau indicated that Canada stood ready to bar a second voyage of the Manhattan if it were determined that the journey might constitute a pollution hazard.

■Transport Minister Donald Jamieson said March 7 that Humble Oil and Refining Co. would be required to post a bond to pay the clean-up costs of any oil spillage during the voyage of the tanker Manhattan. Humble would have to concede before the voyage that it would be responsible for all pollution damage. In addition, Transport Department specialists would inspect the ship in drydock before the voyage to make certain all anti-pollution specifications were observed.

■A May 13 report said Ottawa was asking the U.S. to rescind its March 10 cutback on Canadian oil imports as a condition for increased commitment of Canadian natural gas to the U.S.

Canadian gas sale to U.S. The Canadian government Sept. 29 approved the largest single increase in exports of natural gas to the U.S. since the pipeline exports began. Energy Minister Green announced the approval of licenses for the export of an additional 6.3 trillion cubic feet of natural gas over the next 15 to 20 years—an increase of more than 50% over current levels.

The licenses granted to four companies would increase the amount of Canadian gas committed to the U.S. to 18.3 trillion cubic feet from the current 12 trillion cubic feet. However, the government decision reduced the quantity sought by almost one-third, shortened the duration of export commitments and introduced regulations for periodic price reviews to protect Canadian interests.

In another move Sept. 29, the U.S. government announced that the level of daily oil imports from Canada to the area east of the Rockies could be raised by an estimated 40,000 barrels a day without altering President Nixon's quota.

U.S.-Canadian accord on oil. Canadian and U.S. cabinet officers agreed "in principle" Nov. 24 that the U.S. would buy all the oil Canada was prepared to sell. The agreement was reached during talks in Ottawa Nov. 23-24 on trade and commerce. The talks were held in an atmosphere of stiff Canadian opposition to protectionist sentiment in the U.S. Congress.

That opposition was voiced in the Commons by Prime Minister Pierre Trudeau the week of Nov. 16. Trudeau said he had informed President Nixon of his concern over the protectionist sentiment in the U.S. and would use the two-day ministerial conference "to reiterate our firm belief in benefits of multilateral trade."

The oil accord came on the second day of talks, with the U.S. agreeing in principle to give Canada full and unimpeded access to U.S. markets. The U.S. said that in 1971 its oil requirements would take up the full export capacity of the Canadian pipeline system. This was expected to lead to a 100,000 barrel-a-day rise in Canadian oil exports to the U.S. east of the Rockies, a 20% increase that would bring extra revenues of over $125 million to Canadian producers.

Oil quotas system retained. George A. Lincoln, head of the President's Oil Policy Committee, disclosed at a White House briefing Aug. 17 that President Nixon had decided not to shift oil import controls from quotas to tariffs, as suggested by one Cabinet-level study group. Lincoln, also director of the Office of Emergency Preparedness, said the President accepted the committee's recommendation, offered in a letter from Lincoln Aug. 13, that the government improve the present quota system.

Noting that he had originally supported a tariff system that might have reduced the domestic oil price, Lincoln indicated that his change of mind had been influenced by considerations of an interrupted flow of oil to Europe, uncertainty over the availability of Alaskan oil, the effects of various environmental programs, increased demand for petroleum products and decreased domestic supplies. He reported "new estimates" that "indicate we have a more

severe problem than we estimated six months ago in preventing an unwise dependence on relatively insecure sources of supply by even as early as 1975."

(Nixon Dec. 22 authorized an increase of 100,000 barrels a day in oil imports for 1971. The President's proclamation also included exemptions from the control program for several products, including natural gas liquids from Canada, ethane, propane, butane and asphalt.)

Labor Secretary George P. Shultz, testifying March 3 in his capacity as chairman of the Cabinet Task Force on Oil Import Control, had told the Senate Antitrust & Monopoly Subcommittee that "on the basis of most industry submissions, we calculate that oil imports at present domestic prices will have to supply about 27% of our requirements by 1980 unless there is an unforeseen breakthrough in the technology of producing oil from synthetic sources.... The present percentage is 19%, and just about everybody agrees that this will have to be increased."

Shultz, whose group had called for a shift from quotas to tariffs, said that under his task force proposals, the U.S. "would draw the bulk of our imports from Canada and other acceptably secure Western Hemisphere sources and would strictly limit Eastern Hemisphere [largely Middle East] imports to a maximum of 10% of U.S. demand." He noted that his task force's report "strongly recommends that we should examine with our allies the measures that might be taken—such as increased emergency storage in Europe and Japan—to assure adequate supplies to them in the event of sustained curtailment of Eastern Hemisphere supplies." Shultz reported his task force's conclusion that "the most serious contingency" that could cause an interruption of foreign oil supplies "is not military but political and arises out of Arab-Israeli tensions."

Sen. William Proxmire (D, Wis.), also urging a change from quotas to tariffs, told the Senate April 10: "The present oil import quota program is costing the American consumers about $5 billion a year now, and the cost will increase to about $8.4 billion a year by 1980.... President Nixon's task force concluded that if we removed all import controls, letting the price of domestic oil fall to the price of world oil, we could meet 104% of our needs in 1980 without any rationing even if the Middle East cut off all their oil for one whole year.... [A tariff system] has several immediate advantages ... [over quotas]: it can be geared to be more responsive to our national security needs by favoring secure sources of oil; it would stop the handout of over one-half billion dollars a year of import tickets to the major oil companies; ... and it would place a ceiling over oil prices rather than a floor under them like the present program, thereby helping to combat inflation...."

A statement opposing a shift from quotas to tariffs was made before the House Ways & Means Committee June 3 by Sun Oil Co. Chairman Robert G. Dunlop, speaking in behalf of the American Petroleum Institute: "... [The tariff system] is not directed principally toward volumetric control of foreign oil flowing into the United States, which is our basic need. While the quota system achieves this goal very precisely, the tariff system does not and ... cannot. The cabinet task force recognized this ... when it recommended that imports from the Eastern Hemisphere be limited to 10% of domestic demand. In effect, the task force has superimposed a tariff plan on the quota system...."

Sen. John Tower (R., Tex.), supporting the quota system, told the Senate July 6: "... As increased imports drive the price of crude down, the oil and gas exploration segment of the domestic industry would rapidly become nonexistent. There would be an accompanying loss of 268,000 jobs directly connected with the exploration of oil and gas.... We would be forced to import nearly all our needs of ... gas [at an additional cost of more than $5 billion a year].... Further, the foreign supplies of crude and liquified natural gas could be shut off at will by the producer governments or even be halted through acts beyond the control of those governments...."

Alaska oil pipeline halted. U.S. District Court Judge George L. Hart Jr. in Washington, D.C., ruling on a suit brought by 66 Alaskan Indians, granted a preliminary injunction April 1 against the construction of a $900 million oil pipeline through land claimed by the Indians. The injunction prevented the Interior Department from granting a right-of-way through Stevens Village, Alaska to Trans-Alaska Pipeline Systems (TAPS), an oil combine planning to build the 800-mile pipeline.

An attorney for the Indians indicated that his clients were willing to let the pipeline go through if they were compensated for their land with money or jobs. The Indians had argued that the project threatened the ecological balance of the area, but Hart confined his first ruling to whether the pipeline "is being done legally."

The conservation suit was brought by Wilderness Society, Friends of the Earth and Environmental Defense Fund Inc. Scientists had argued that since much of the pipeline as planned would be built underground, the 180° oil in the pipe could cause dangerous melting of the permafrost—a permanently frozen layer below the earth's surface in frigid regions.

Interior Secretary Walter J. Hickel, in an Earth Day speech at the University of Alaska (College, Alaska), said April 22 that he would issue a right-of-way permit for the pipeline, but he added that construction would be authorized "only after a thorough engineering and decision analysis," and he pledged not to approve "any design based on the old and faulty concept of 'build now, repair later.'"

Hickel said perhaps a year or more would elapse before TAPS would present a construction plan that could be approved. He said surveys indicated that 40–50% of the pipeline would have to be suspended above ground to prevent dangerous melting of the Alaskan permafrost by hot oil traveling through the pipeline.

Hickel's decision to proceed with the pipeline was denounced April 23 by Sen. Edmund S. Muskie (D, Me.) and conservation groups who charged that Hickel appeared to have already made up his mind about the project without waiting for results of scientific studies on the pipeline's environmental effects.

Alaska Gov. Keith Miller (R), accompanied by more than 100 prominent businessmen from the state, arrived in Washington April 26 to lobby for prompt approval of the pipeline and road permits. The group argued that Alaska faced "unprecedented economic disaster" if the pipeline was delayed because of investments already tied up in the project.

William T. Pecora, director of the Geological Survey, told the Alaskan group April 27 that his agency was in "unanimous agreement" that it was unsafe to place most of the pipe below ground. Pecora criticized TAPS for choosing a route without consulting the Interior Department or the State of Alaska and for failing to present structural design plans. Pecora said the Geological Survey had seen only route maps.

'Brownouts' dim Northeast. Large areas of the Eastern seaboard were brought to the brink of a massive electrical blackout Sept. 23 as a prolonged heat wave taxed already low power reserves. Power in many areas was drastically rationed and voltage was reduced by as much as 8%. In the hardest hit areas of Philadelphia and New York, utilities resorted to selective blackouts "to avoid a complete closedown."

New York faced a continuing summer power crisis. Voltage reductions of up to 8% during a sweltering July 23-24 weekend by the Consolidated Edison Co. were credited with averting a blackout. The Tennessee Valley Authority channeled several hundred thousand kilowatts of power to the city to ease the critical shortage that had been caused by the breakdown of Con Edison's largest generator July 21. Nevertheless, the continuing drain caused by the constant use of air conditioners forced other voltage reductions during periods in July and August.

(Mayor John V. Lindsay approved a partial expansion of Con Edison's generating plant in a residential section of the city Aug. 22. Critics opposed expansion on the grounds that it would contribute to air pollution.)

The utility commissioners of eight Eastern states had urged Secretary of the Interior Walter J. Hickel Aug. 17 to replenish the nation's dwindling natural gas reserves by fostering more rapid development of new supplies. Hickel agreed to study long-range pro-

posals for offshore drilling and to carry out more programs to induce onshore development, according to one official.

The American Public Power Association urged President Nixon Aug. 31 to impose rationing and price controls on fuel oil and coal. The association, an organization of 1,400 publicly owned local electric utilities, also warned that unless "drastic prompt action" was taken, power blackouts and "brownouts" could be expected during the winter.

Ban on oil lease sale ended. Interior Secretary Walter J. Hickel announced the end of a 19-month moratorium on the sale of U.S. off-shore oil leases Oct. 16. At a GOP fund-raising dinner in Houston, Hickel said that leases on 593,000 acres off the coast of Louisiana would be put up for sale Dec. 15 in New Orleans.

Hickel had canceled the sale of any new off-shore oil leases in March 1969 after a disastrous oil spill near Santa Barbara, Calif. Hickel said Oct. 16: "This sale and those to be proposed later are the results of long study and careful planning. We urgently need to increase our energy reserves, particularly natural gas."

The Interior Department Dec. 15 opened bids for off-shore drilling rights on 127 tracts off the Louisiana coast. An Interior official warned, however, that the department "is not completely satisfied with the safety record of off-shore drilling operations." (While the sale proceeded, Shell Oil Co. continued to fight a fire that had broken out on a Gulf of Mexico drilling platform Dec. 1. The well was capped Dec. 30, curbing the blaze by 30% to 40%. Stringent rules to curb oil pollution, proposed by Hickel July 24, went into effect Sept. 10 when they were published in the Federal Register.)

Oil companies fined. Three major oil companies were fined more than $500,000 Dec. 2 in U.S. District Court in New Orleans after pleading "no contest" to charges that they failed to install safety valves on off-shore oil wells in the Gulf of Mexico.

The charges were contained in November indictments brought by the federal government against Humble Oil Co., Union Oil Co. and Continental Oil Co. Shell Oil Co., also charged in the indictments, pleaded innocent, and Judge Fred J. Cassiery said he would set a trial date later.

The indictments grew out of a grand jury investigation of a massive oil spill and fire in February in wells owned by Chevron Oil Co. Chevron had been fined $1 million Aug. 26 on a plea of "no contest", to 500 violations. In the cases settled Dec. 2, Humble was fined $300,000 on 150 offenses involving 33 wells; Union, $24,000 on 12 counts involving eight wells; and Continental, $242,000 on 121 offenses involving 40 wells.

The Chevron fire had started Feb. 10 and was put out March 10. The damaged wells then began pouring about 42,000 gallons of oil a day into the Gulf before the biggest leak was capped March 31.

■ The government filed charges Nov. 20 against three other companies for safety failures on 74 oil wells in the Gulf. The charges, filed in New Orleans, were against Shell Oil Co., Continental Oil Co. and Union Oil Co. As in the Humble and Chevron cases, the government contended that the wells lacked devices to shut off oil flow when there was a sudden change in pressure.

Pollution problem. Bitter controversy was already raging between environmentalists and industry leaders and the partisans of both over the damage done to air, water and land as a by-product of supplying fuel and energy. There was constant risk of environment-destroying oil spills in drilling for off-shore oil and in transporting oil by tanker. Coal- and oil-burning electric plants caused health-destroying smog, as did gasoline-burning autos. Nuclear energy plants posed the threat of radioactive contamination of the air, and they were accused of threatening water life by heating the water of streams they use for cooling. Strip-mining of coal gouged ugly gashes in the earth and denuded it of plant-nourishing top soil.

In his State-of-the-Union message Jan. 22, President Nixon stressed the need to make progress without abusing the environment.

In response to the President's proposals, Sen. Edmund S. Muskie (D, Me.) Jan. 23 presented a legislative program to control pollution. Muskie was chairman of the Senate Subcommittee on Air & Water Pollution and author of most of the antipollution legislation currently enacted. To deal with oil leaks along the California coast, Muskie said he would propose legislation to authorize "the compensated acquisition by the federal government ... of all oil leases in the Santa Barbara Channel" and the removal of all drilling platforms in the channel. Muskie said he had scheduled hearings on a bill to assure public participation in the selection of electric power plant sites and to set up environmental standards for power facilities.

■Rep. Chet Holifield (D, Calif.), chairman of the Joint Committee on Atomic Energy, Jan. 12 released 1,108 pages of testimony taken during hearings that began Oct. 28, 1969 on the environmental effects of producing electrical power. Commenting on testimony pointing to increasing dangers of water and air pollution from conventional means of producing electrical power, Holifield contended that nuclear reactors could make "a meaningful contribution towards the reduction of pollution now attributed to the electric power industry."

■Former Interior Secretary Stewart L. Udall, during a Jan. 14 panel discussion at the New School for Social Research in New York City, linked pollution problems with population and family planning issues. Citing the increasing need for power plants to serve an expanding population, Udall said "in the interest of national survival we cannot permit this haphazard growth that has been destroying our environment." Udall, who served in both the Kennedy and Johnson Administrations, currently headed the Overview Group, an environmental consultant concern.

■A number of state governors stressed environmental problems in messages to legislatures. Gov. Ronald Reagan (R, Calif.) told the California Legislature in Sacramento Jan. 6 that "the good life will be no good at all if our air is too dirty to breathe, our water too polluted to use, our surroundings too noisy and our land too cluttered and littered to allow us to live decently." He proposed a program Jan. 22 that would "provide the teeth needed" to enforce the state's "already tough controls on smog." Gov. Nelson A. Rockefeller (R, N.Y.), in his annual message to the state legislature in Albany Jan. 7, urged the creation of a new department of environmental conservation that would combine the conservation and pollution control responsibilities of a number of New York agencies. Gov. Marvin Mandel (D, Md.) told Maryland legislators Jan. 21 that he planned to submit "the most comprehensive environmental control program ever to come before you."

■A General Services Administration (GSA) official told a Senate subcommittee on environment Jan. 27 that more than 1,000 federal vehicles would be converted in 1970 to a dual-fuel system as part of an experimental program. The vehicles would use fume-free natural gas for operation in metropolitan areas and regular gasoline on longer trips.

Florida coast damaged by slicks—Up to 10,000 gallons of crude oil spread a 20 square mile slick along Florida's Old Tampa Bay after a Greek tanker, the Delian Apollon, ran aground Feb. 13. The same ship had dumped 4,000 gallons of oil earlier in the week at Port Everglades on the east coast of the state.

As the slick spread along a five-mile stretch of beach near St. Petersburg, city workmen spread straw and Polyurethane foam to soak up oil, and students worked to save oil-soaked birds. The Audubon Society estimated that more than 1,000 birds had died. The Florida attorney general filed a $2 million negligence suit Feb. 16 against Shipping Development Corp. of Panama, owner of the tanker which was under Greek registry.

A spill of 7,000 gallons of oil in Florida's St. Johns River threatened a wildlife preserve near Jacksonville after a Danish freighter collided with a barge. Most of the oil was cleaned up by Feb. 28.

(Massachusetts officials reported Feb. 8 that an unexplained oil slick on the ocean side of Martha's Vinyard Island off Cape Cod had killed hundreds of birds. Another unexplained oil slick washed up on the beaches of the New Jersey shore early in March.)

Oil slicks kill Alaskan wildlife—Federal officials traveled to King Salmon, Alaska

April 27 to trace the source of a nearby oil slick that apparently had killed tens of thousands of birds, at least two whales and a number of seals and other marine animals. Ray Morris, chief of the Anchorage oil pollution branch of the Federal Water Quality Administration, said the wildlife deaths were caused by the "highly toxic properties" of the oil rather than the size of the slick. The spill was first reported April 22 in an isolated area northeast of the Aleutian chain and about 300 miles southwest of Anchorage.

Federal officials said March 6 that tankers dumping dirty ballast apparently were responsible for an earlier slick off Alaska's Kodiak Island along the shoreline of Southwestern Alaska that killed at least 10,000 birds. The area was so remote that officials said little could be done to save the birds.

A boarding party from a U.S. Coast Guard cutter boarded a Soviet tanker in the Gulf of Alaska April 13 and arrested the ship's master on charges of spilling a mile-long oil slick near Kodiak Island. In Washington, D.C., a State Department spokesman said April 14 that the ship's master had admitted that an oil spillage had occurred in U.S territorial waters while the tanker was refueling Soviet shrimp boats.

Slick in Gulf of St. Lawrence—A big oil slick moved through the Gulf of St. Lawrence towards the New Brunswick coast after the sinking Sept. 7 of the 2,300-ton unmanned barge Irving Whale, which was carrying 875,000 gallons of bunker oil when it sank 60 miles north of Prince Edward Island.

U.S. moves against thermal pollution—The Department of Justice filed suit in U.S. District Court in Miami March 13 to halt present and future thermal pollution of Florida's Biscayne Bay by the Florida Power and Light Co. The suit, filed upon recommendation of Interior Secretary Walter J. Hickel, alleged that heated water being discharged by two power plants was ruining marine life in the bay, including an area designated by Congress as the Biscayne National Monument. The suit claimed that damage would be even greater when two planned nuclear power plants were installed at the existing sites.

The government, in its first suit against thermal pollution, said the plants drew water from the bay at a rate of almost 550,000 gallons a minute for cooling and condensing purposes; that the water, which left the plant at about 10 to 20 degrees above the normal temperature of the bay, raised the bay water to "temperatures substantially higher than their natural condition"; that 300 acres were damaged in 1968 and 300 more acres in 1969; that when the nuclear reactors were completed, heated water equal in volume to all of the bay would pass through the plant in less than month; and that the nuclear plant intake would disturb the bay bottom and disrupt marine life.

The government also asked that the company be permanently enjoined against operating "its presently existing fossil-fueled power plants and its nuclear reactors now under construction as to increase the temperature or otherwise adversely affect the quality of the waters of Biscayne Bay or to adversely affect marine life of the Biscayne Bay National Monument and the lands therein."

Hudson River plant attacked—New York filed suit in State Supreme Court May 12 against Consolidated Edison Co., charging the power company with "creating serious conditions of thermal and chemical pollution in the Hudson River" and "endangering the ecology." The suit, which concerned the company's nuclear generating plant at Indian Point, asked that the plant be closed until "suitable methods" to protect the Hudson were found and sought $5 million in damages for fish kills resulting from operation of the plant.

Fish kills from the Indian Point plant were reported as long ago as 1966, and two separate fish kills were reported in 1970. In January, an estimated 150,000 fish were killed, requiring a change in a filtering device designed to screen fish out of the plant's water intake system. In March, 120,000 fish were killed by what officials said were "unknown reasons."

Consolidated Edison officials expressed concern over the suit because the Indian Point plant was counted on to pro-

duce "about 3% of our total power generating capacity."

In New York the Federal Power Commission Aug. 19 authorized a license for the controversial hydroelectric project at Cornwall on the Hudson River. The two-million kilowatt project was originally granted a license in 1965 but this was later overturned by a federal court in a suit brought by conservationists.

Lake Michigan policy—The Federal Water Quality Administration proposed a new thermal pollution policy in Chicago May 7 that would forbid the dumping of water into Lake Michigan that would raise the temperature of the lake by more than one degree Fahrenheit at the point of discharge. The states of Illinois, Indiana, Michigan and Wisconsin were to be given the opportunity to incorporate the new level into their state water quality standards.

Murray Stein, the agency's chief enforcement officer, said if the states refused the standard, the federal government could adopt the one-degree level as a federal standard.

Power plants under attack. The Joint Congressional Committee on Atomic Energy noted June 17 that the electric-power industry was facing growing public attacks on most proposals for building new plants:

The Committee has noted with concern in the past year the increased public opposition, not a little of it wholly unreasoning, to the construction of all electrical generating sources and their transmission systems. While a preponderance of this opposition has of late been concentrated on nuclear powerplants, coal and other fossil-fueled plants, hydro-electric projects and pumped storage facilities also have experienced significant difficulties in this regard. *It seems that what should be a genuine, legitimate, and quite understandable concern about the effects on the environment of large powerplants has been transformed in many instances into an insistence on pristine purity that will brook no balancing of two worthwhile but somewhat competing values; namely, the goal of clean air and water and natural beauty, on the one hand, and the objective of abundant, economical, and reliable electric power on the other.*

Neither of these goals can be achieved without some impact on the other. The task confronting the responsible and the informed is to harmonize these contending goals and, to the greatest extent possible, minimize the effects on the environment that inevitably flow from industrial growth. As the chairman of this Committee observed recently (Representative Chet Holifield, D.-California), unless the demands for clean air and clean water are kept in perspective—that is, unless there is a reasonable and fruitful union between industry and the environment—the anti-technologists and single-minded environmentatlists may find themselves conducting their work by the light of a flickering candle.

Unfortunately, it may require several serious electrical "brownouts," or even worse one or more blackouts on the order of the Great Northeast Blackout of 1965 before the full realization dawns on the average homeowner and businessman, and on the community wishing to attract new industry through assurances of plentiful electric power, that the price to be paid in environmental effects from a new powerplant in the area may be quite small in relation to the price to be paid in other ways for not making adequate provisions for accepting that plant. The latter price may include the well-being of the community concerned—a loss at least as direct and adverse as any that might have accompanied continued progress in the community in terms of its standard of living as well as in its industrial growth. Moreover, one fundamental factor often overlooked is that once a shortage of electrical energy is permitted to exist, corrective action to eliminate the shortage and accommodate the normal growth in demand of power would in all likelihood consume the major portion of a decade. *The serious implications for our national welfare of such a sustained period of shortage of energy are manifest.* Accordingly, sufficient advance planning to take into account the long lead times which exist in the electrical utility business is required if adequate capacity is to be available when and as needed. Irresponsible actions of a few persons cannot be permitted to interfere with installation of such capacity after being properly planned. . . .

In the eyes of the Joint Committee the problem of electric generating plant siting has already passed the "worrisome" stage—it is becoming critical. . . .

A half century ago, procedures were initiated for rational, regional development of the power potential of our Nation's rivers. Now more meaningful efforts should be directed toward applying similar comprehensive planning principles to systematic solution of siting difficulties associated with the burgeoning number of thermal stations and their power transmission systems. *Reliance on ad hoc, plant-by-plant arrangements has already proved inadequate; a bold new approach is long overdue. . . .*

U.S. inquiry on oil price hikes. The government announced Nov. 12 a major investigation into the price increase of crude oil posted Nov. 11 by the Gulf Oil Corp. and Nov. 12 by the Atlantic Richfield Co. The 25¢ a barrel increase for crude oil raised prices from $3.10 to $3.35 and was the first increase since early 1969. The action included a companion boost of .7¢ a gallon for wholesale gasoline.

The Nixon Administration expressed its concern in the investigation by assigning Paul W. McCracken, chairman of the Council of Economic Advisers (CEA), a "major role" in the inquiry.

George A. Lincoln, director of the Office of Emergency Preparedness (OEP) and chairman of the Oil Import Policy Committee, said the inquiry into "the reasons for and consequences of the increase" was mandated under the 1959 oil import proclamation requiring him as chairman to keep a constant surveillance of oil imports "in respect to national security." The announcement by Lincoln did not specify what recommendation would be made to the President if the price boost was found to be unwarranted.

Gulf Oil said Nov. 15 that it would "cooperate fully" in the investigation, although it had not yet heard directly from OEP which was conducting the inquiry.

The company said it believed that the facts would prove the increase "fully justified and long overdue" in view of "the industry's ability to supply the energy necessary" to meet the public's need without compromising national security.

Nixon acts on oil. President Nixon announced in a speech before the National Association of Manufacturers in New York Dec. 4 that he had taken two steps that, he anticipated, would have the effect of increasing the supply of oil and thereby restraining the rise in oil and gasoline prices.

Nixon said he had ordered an end to "state restrictions on [oil] production on federal off-shore leases [that] have held down the supply of crude oil." He "also directed that companies importing Canadian oil be permitted to use their overseas allocations for the purchase of more crude oil from Canada."

Cost alert scores oil move. The Administration's second inflation alert, issued by the President's Council of Economic Advisers Dec. 1, specifically criticized wage or price actions in the auto, oil, rail and construction industries.

The report admonished the Texas state oil authorities for permitting a rollback in the oil production rate for December, thereby reducing the output of crude oil supply, merely on the request of one major oil company that complained of excessive oil inventories.

Among findings noted in the first inflation alert, which had been issued Aug. 8:

■Bituminous coal registered the most striking single price increase (up at a 56% annual rate in the first half of 1970), which the report attributed to a swift increase in demand rather than higher wage settlements.

■Electric power rates for consumers increased slowly during 1970, despite the coal price hikes and higher interest rates, but would probably quicken their pace in the near future.

Monopolization of energy? S. David Freeman, then director of the energy policy staff of the White House Office of Science & Technology, testified before the House Antitrust & Monopoly Subcommittee May 15 that there was a "trend toward acquisition of coal companies by petroleum firms" and "greatly expanded participation in uranium production by the petroleum industry." "Many of the major oil and gas producers have become energy companies," he declared.

Asserting that "interfuel competition can be an effective force for consumer protection in the energy field," Freeman said that "the speed and scale of dominance of the independent coal industry by their petroleum company competitors raises a question of whether the public interest is being best served." He noted that "only two of the 10 largest coal companies are independently owned any longer.... The remaining eight, which account for 42.9% of industry production, are owned by oil companies ... or other large industrial concerns with other stakes in energy. These same eight companies ... opened over 67% of the new mines in the United States in 1968...."

As for uranium, Freeman reported, "during 1968 oil companies drilled 44% of the exploratory and development footage" and in 1969 drilled 37% of such footage. "Of the 15 significant uranium discoveries in 1969," Freeman said, "eight were oil companies."

Sen. William Proxmire (D, Wis.) inserted in the Congressional Record Sept. 14 the text of a speech in which Lawrence G. Meyer, director of the FTC's Office of Policy Planning & Evaluation, noted that "in recent years a number of major [oil] companies have acquired leading independent gasoline marketers and refiner-marketers." "Basically," Meyer said, "the structural characteristic of the petroleum industry setting the competitive tone is the dominance of this industry by a relatively small number of large, vertically integrated major companies, marketing a well established brand of gasoline which is extensively advertised. Twenty such firms control over 80% of the value of shipments in the refining industry—with the top eight of these firms accounting for about 55%."

Oil lobbies. Erwin Knoll reported in the New York Times Magazine March 8 that the oil and gas "industry's interests are served [in Washington and in state capitals across the country] by a costly and complex but closely coordinated lobbying apparatus" whose principal components include:

The American Petroleum Institute, whose membership roster of 400 companies and 8,000 individuals represents about 85 per cent of the total production, refining and marketing volume in the oil and gas industry....

The Independent Petroleum Association of America, with some 5,000 members representing about 60 per cent of the independent oil producers....

The National Petroleum Refiners Association, composed of domestic refining companies and representing about 90 per cent of the refinery production in the United States.

The Independent Natural Gas Association of America, representing major pipeline companies. Its executive director is a former Texas Representative, Walter E. Rogers. He served in Congress as Chairman of the House Subcommittee on Communications and Power, which handles gas-pipeline legislation. He gave up his Congressional seat in 1966 and registered as a lobbyist in 1967 to represent 12 pipeline companies in a vigorous—and successful—effort to water down a pending bill that would have established strict Federal safety standards for the nation's 800,000 miles of gas pipelines.

A formidable array of regional and state groups—among them the Mid-Continent Oil and Gas Association, the Western Oil and Gas Association, the Texas Independent Producers and Royalty Owners Association and the Kansas Independent Oil and Gas Association—augments the national contingent.

Knoll reported that Sen. Russell B. Long (D, La.), chairman of the Senate Finance Committee, had asserted that by "represent[ing] the oil and gas industry, I . . . represent the state of Louisiana," which "produces more oil and gas per acre than any state in the Union." Knoll wrote:

According to records of the Louisiana Mineral Board, Senator Long has received income of $1,196,915 since 1964 from his interests in four state oil and gas leases, and almost $330,000 of that income has been exempt from Federal income taxes because of the oil-depletion allowance. The Senator is also a trustee of family trusts that have collected $961,443 from holdings in state leases since 1964; and he has an interest in at least seven private leases whose royalty reports are not available for public scrutiny.

Few of his colleagues can match Senator Long's oil holdings, but many share his solicitous concern for the industry's welfare. Among those on whom the oil moguls can generally count for unstinting support are Senators John G. Tower of Texas, Gordon Allott of Colorado, Clifford P. Hansen of Wyoming, Henry L. Bellmon of Oklahoma, Roman L Hruska of Nebraska, Robert J. Dole of Kansas, Peter H. Dominick of Colorado, Allen J. Ellender of Louisiana, Theodore F. Stevens of Alaska, George Murphy of California and Karl E. Mundt of South Dakota.

Most—but not all—of oil's fast friends in the Senate are stanch conservatives. Nonetheless, such liberal heroes as J. William Fulbright of Arkansas and Eugene J. McCarthy of Minnesota can usually be counted on to see oil's side. . . .

McCarthy, who voted consistently against oil privileges during most of his first Senate term, cast his first vote in favor of depletion in 1964 and has generally favored the industry's positions since. There were published reports in 1968 that he had raised about $40,000 for his Presidential campaign in one day at the Petroleum Club in Houston. . . .

Oil ship ruling suspended. The Treasury Department suspended a ruling March 10 favorable to a tanker company formerly headed by Presidential aide Peter M. Flanigan. The suspension came

after Sen. Joseph D. Tydings (D, Md.) said in the Senate March 9 that the ruling would provide a $6 million "windfall" for the company, and Rep. Edward Garmatz (D, Md.), chairman of the House Merchant Marine Committee, announced an inquiry into the situation. Garmatz canceled the inquiry after suspension of the ruling.

The March 2 ruling granted a waiver for the Barracuda Tanker Corp.'s tanker Sansinena, registered in Liberia, to carry oil for the Union Oil Co. from Alaska to California. Foreign flag ships were barred by law from sailing between American ports; exceptions were permitted for national security reasons. Flanigan was president of Barracuda until his White House appointment, when he resigned and put his 308 shares of stock (about 4%) in trust with his father. (These shares were sold five days before the Treasury ruling to the new president of Barracuda for $20,020, according to Flanigan March 10, a price that "did not in any way reflect the possibility that the ship might at some time be used in coastwise trade." Tydings had figured the "windfall" at $6 million by calculating that the value of the tanker would rise from $4.5 million to $11 million because of the opening into the lucrative coastwise trade. The corresponding increase in the value of the Flanigan shares, based on such a calculation, would be approximately $250,000.)

White House Press Secretary Ronald L. Ziegler said March 9 Flanigan "was unaware" of the Treasury decision.

Treasury Secretary David Kennedy defended the waiver March 10 on the ground it involved "national defense implications of the development of Alaskan oil."

Elk Hills plan opposed. Sen. Thomas J. McIntyre (D, N.H.) announced to the Senate Aug. 20 his opposition to the June 11 request of the Interior Department that, in McIntyre's words, "oil be pumped from the Elk Hills [Calif.] Naval Petroleum Reserve in order to compensate some, though not all, of the oil companies holding drilling leases in the Santa Barbara Channal."

McIntyre pointed out that the Elk Hills reserve was "the second largest proved oil field in the United States, one of the 20 largest in the world," and that it was "a total national defense reserve ... vitally essential to our overall strategic needs for petroleum sufficiency."

The Middle East

Situation described. Dr. Wilson Laird, director of the Office of Oil & Gas, warned U.S. Congress members that political considerations made the Middle East an undependable source of the U.S.' oil supply. In a briefing inserted in the Congressional Record by Sen. Clifford P. Hansen (R, Wyo.) July 31, Laird provided this data on the situation:

"Of the known oil reserves of the non-Communist world today, more than 80% are contained in the North Africa/Middle East area. Nearly every country of the free world is dependent to some degree on oil from [this area].... Western Europe relies on oil from these areas to the extent of 85% of its total oil supply and more than half of its total energy supply; Japan relies on the Middle East for nearly 90% of its oil supply and 60% of its total energy. The United States relies on Middle East/North Africa sources to meet only about 3% of its oil needs. However, about 1/5 of the oil supply of our East Coast refineries originates in the Middle East or North Africa. These sources also supply a substantial portion of our overseas military oil needs. Oil produced in the Middle East and North Africa is currently supplying more than 45% of the oil requirements of the non-Communist world."

Algeria. The Wall Street Journal reported Jan. 27 that the Algerian government had completed its takeover of the oil properties of Sinclair Mediterranean Petroleum Co., a subsidiary of Atlantic Richfield Co. The report said Algeria's government-owned oil company, Sonatrach (Societe Nationale Algerienne des Hydrocarbures), would own the properties.

The Algerian government announced June 15 that five more foreign oil-producing companies would be nationalized "to recuperate natural riches for the benefit of the national community." The takeover affected interests of Royal Dutch-Shell, Phillips Petroleum Co., Elwerath-Sofrapel of West Germany and Ausonia Mineraria of Milan, Italy. The seized properties were placed under Sonatrach.

In a related action June 18, the government warned foreign oil companies still operating in the country that their operations might be seized unless they invested a substantial part of their profits in exploration.

An announcement May 15 in Algiers disclosed that Soviet experts would help Algeria conduct research aimed at boosting its annual oil production to 100 million tons from the present 43 million.

Algeria July 21 raised the tax reference price on crude oil, forcing French oil companies to increase their tax payments by about 50% annually. The unilateral decision brought a new strain to Algerian relations with France, since negotiations on the 1965 oil accord between the two nations were then in process.

The reference price on which the export tax was based was raised 77¢ to $2.85 a barrel. Paris sources estimated the move would cost French oil companies about $100 million a year. The price rise was made retroactive to January 1969.

The major companies affected were the French government-owned Entreprise de Recherches d'Activite Petrolieres and the Compagnie Francaise des Petroles, in which the French government held 35% of the shares. France was the major consumer of Algeria's 900,000 barrel-a-day output.

It was disclosed in a Nov. 6 report that Consolidated Natural Gas Co., the third largest gas utility in the U.S., had signed a letter of intent for deliveries of liquefied gas to be imported from Algeria. Consolidated would purchase the gas from the importer, El Paso Natural Gas. The deal would involve a level of 200 million cubic feet a day by 1976. Purchases would begin in 1974 and run for 25 years.

At the 20th General Conference of the Organization of Petroleum Exporting Countries June 24, Algerian President Houari Boumedienne told the oil-exporting nations to bypass international oil companies and deal directly with the governments of the consumer countries. Libya backed Algeria June 25 in a drive aimed at curbing the refining and marketing róles of U.S. and other international companies in the 10 nations at the conference.

Libya. The Libyan revolutionary government July 4 ordered the nationalization of all four oil distributing companies in the country. The companies were taken over by the Libyan National Oil Corporation. Oil production was not affected.

The government July 5 nationalized domestic marketing operations of three foreign oil companies—Standard Oil Co. (N.J.), Royal-Dutch Shell group and Italy's Agip S.p.A. The 100 service stations affected were renamed Braga Co.

In a June 15 action, the government ordered production cutbacks for Texaco Inc. and Standard Oil Co. of California by 100,000 barrels a day. Oasis Oil Co., the nation's largest oil-producing group, was ordered to cut its output by 150,000 barrels daily, according to a July 13 report. (Oasis was owned a third each by Continental Oil Co. and Marathon Oil Co., and a sixth each by Amerada Hess Corp. and Royal Dutch-Shell group.)

(In another action, the government June 16 blocked shipments of liquefied natural gas from the Standard Oil Co. [N.J.]'s plant in a dispute over the taxes to apply to the gas exports.)

The government concluded agreements with major oil companies for increased tax and royalty payments from oil production.

Libya had demanded the increases in February, but the oil companies had balked. Libya, with more than $2 billion in oil-earned reserves, applied pressure by cutting its 3.3-million-barrel-a-day output to $2\frac{1}{2}$ million barrels.

The first surrender was an accord announced Sept. 4 with Occidental Petroleum Co. Occidental refused to disclose terms of the agreement, but sources in Tripoli said it called for an immediate 30¢-a-barrel increase in the posted price (the figure on which taxes were based) and subsequent annual increases of 2¢ a barrel to a ceiling of 40¢ a barrel by 1975. This indicated that Occidental had agreed to a new tax rate of 58%, or a boost of 17¢ a barrel in tax and royalty payments, rising to 22¢ in 1975. Occidental announced that production from its Libyan concession had been restored Sept. 2. An Oct. 9 announcement disclosed that the government's deal with Occidental was retroactive to 1965. The arrears due the government as a result of the backdating would be met over five years by means of a tax increment.)

The government then moved to deliver an ultimatum to the Oasis group to comply with its demands for increased tax and royalty payments. In letters delivered to the four companies of the Oasis group Sept. 14, the government "invited" the companies to agree to a price increase of 30¢ a barrel, rising to 40¢ over a five-year period. Sources in Tripoli said the letters hinted at militant action if the new demands were not agreed to within six days.

Marathon Oil Co. announced Sept. 22 that it had agreed to lift its posted price from $2.23 to $2.53 as demanded. The new accord provided for an increase in tax rate of four points to 54%. Marathon also agreed to the 2¢-a-barrel annual rise through 1975. The government, in its own announcement of the Marathon agreement, said the accord would "apply to the other oil companies." The government announcement also said Marathon, Amerada Hess and Continental had agreed to payment of back taxes and royalties since 1965 totaling $243 million.

In another development, Texaco Inc. and British Petroleum Co., Ltd. announced Sept. 28 that they had reached agreements with the Libyan government for increases on posted prices for oil. B.P. said it had agreed to a 30¢-a-barrel increase; Texaco would not disclose details of the accord.

The government Oct. 4 announced price accords with California Asiatic Oil Co., Texaco Overseas Petroleum Co., Bunker Hunt Co., Gelsenberg A. G., Grace Petroleum Corp., Libyan American Oil Co. and Amoco. The companies agreed to payment of $157 million in arrears on profits earned from 1963. This represented the amount the companies had discounted for oil containing wax (five cents a barrel). Deputy Premier Abdel

Salam Jalloud said the government had refused to recognize this discount operation.

Shell, the last of the international oil companies to sign with the government, announced its acceptance Oct. 16. The company had opposed Libya's demand for retroactive payments. Shell agreed to raise the posted price by 30 cents a barrel to $2.53 a barrel retroactive to Sept. 1, plus another 2 cents a barrel annually till 1975. The settlement also raised the tax rate by 4% to 54%.

Latin America

Argentina. The Inter-American Development Bank May 7 approved two loans totaling $30 million for the expansion of electric power systems in the Argentine interior. The loans, $9 million from the IDB's ordinary capital resources and $21 million from the Fund for Special Operations, were to help finance the first stage in a 10-year (1970–1980), $1.2 billion power expansion program.

Bolivia. President Alfredo Ovando Candia announced Sept. 10 that Bolivia would pay Gulf Oil Corp. $78.6 million in compensation for the nationalization of its Bolivian Gulf Oil Co. Oct. 17, 1969.

Gulf officials called the compensation agreement "fair and equitable under the circumstances," but added that "greater benefits would have accrued to both parties had the Bolivian government elected not to nationalize Bolivian Gulf Oil Co."

Explaining the terms of the agreement, Ovando said the compensation, to be paid with no interest over a 20-year period beginning in 1973, would come from 25% of the income Bolivia received from the oil fields. In addition, the payments were to end after the 20-year period even if the total had not been paid and would end completely if the oil reserves ran out. The payments were also conditional on the transfer by Gulf of all technical plans and documents concerning its Bolivian operations and a guarantee that the $46.5 million gas pipeline to Argentina would be completed.

Gulf officials added that the agreement included payment by Bolivia to Gulf for all funds advanced or expended by Gulf on material and equipment before the nationalization and for loans made to the Bolivian government and to the Bolivian state oil firm YPFB. Gulf also said that it would purchase oil from Bolivia for export through the Chilean port Arica to Gulf's U.S. refineries.

(Gulf had originally valued its Bolivian investments at about $150 million, but later confirmed that this figure included "sizable expenditures" on concessions relinquished before the nationalization took place. French experts contracted by the Bolivian government had estimated Gulf's investments at slightly more than $101 million. According to a Miami Herald report Sept. 12, the $101 million figure was reduced to $78 million after the government had subtracted for taxes allegedly owed by Bolivian Gulf to Bolivia.)

In a related development, President Ovando had said Sept. 8 that the firm Camba, jointly backed by the Spanish and Bolivian governments, would receive $400,000 annually for three years to handle the marketing of Bolivian crude oil and natural gas. Ovando added that the Bolivian oil firm retained its autonomy in negotiating sales contracts and in establishing prices.

An agreement between the Spanish government-owned conglomerate National Industrial Institute [INI] and Bolivia to operate the Bolivian Gulf properties and market the oil had been signed March 9. The agreement reportedly provided for the formation of Camba, S.A. to operate the field and ship the oil to the Chilean port of Arica, probably to be sold and shipped to Gulf Oil's refineries in Los Angeles, Calif. as before. The pact would also provide for the completion of the natural gas pipeline from Santa Cruz to Argentina.

Brazil. The International Development Bank May 20 announced its approval of an $80 million loan to Furnas, one of the major utilities supplying power in the south-central region of Brazil, which accounted for 40% of the nation's agricultural production, 70% of its industrial production and 80% of its electricity consumption. The current loan included construction of a large hydroelectric plant on the Rio Grande and transmission lines to link the plant with Rio de Janeiro.

Chile. President Eduardo Frei acted Aug. 11 to authorize the government's $81.3 million purchase of Boise Cascade Corp.'s total interest in Chilectra, the nation's largest privately owned electric power company, which supplied 27% of Chile's power needs. Boise Cascade held about 75% of Chilectra's stock through its subsidiary South American Power.

Dominican Republic. The government's cancellation of petroleum and mining concessions extended to three U.S. firms was reported Aug. 27. The action, described as preliminary to new policies for the nation's mineral resources, affected Tenneco Oil Co., Consolidated Petroleum Co., and Bellomar, Inc.

Guatemala. President Julio Cesar Mendez Montenegro officially opened a section of the nation's largest hydroelectric plant in January. The plant, which would have a capacity of 60,000 kilowatts on its completion, was to increase the nation's output of electricity by 80%.

Honduras. The World Bank and International Development Association June 17 announced approval of two loans totaling $5.5 million each to extend the electrical power transmission system in Honduras to cover northern and southern towns of growing industrial and agricultural importance.

Mexico. The World Bank Feb. 25 announced a $125 million loan to Mexico. The loan, to be combined with about $40 million from export financing institutions in major supplier countries, was to be applied to the foreign exchange costs of Mexico's nationwide electric power expansion program during 1970–71.

Panama. A British-German-Italian consortium signed an agreement with the Panamanian government Jan. 19 to finance and build a coast-to-coast oil pipeline. The $80 million Panamanian-owned pipeline would be able to pump up to 700,000 barrels of oil daily across the isthmus from Alaskan and Pacific coast fields to Atlantic coast markets. The consortium was composed of the International Management and Engineering Group Ltd. of London, Thyssen Stahlunion-Export of West Germany, and Cia. Italiana Trans-Oceanica of Italy.

The World Bank March 11 announced its approval of a $42 million loan to assist Panama in its first large hydroelectric project. The project, with a total estimated cost of $58.3 million, would more than double the generating capacity of the nation's public power supply.

Uruguay. The World Bank Nov. 25 announced its approval of an $18 million loan to help finance additional generating and power distribution capacity in the electrical networks serving the larger cities of Uruguay.

Venezuela. President Rafael Caldera paid a state visit to the U.S. June 2–5. In his talks with U.S. officials, Caldera urged the adoption of a more liberal policy toward American imports of Venezuelan oil.

Addressing Congress June 3, Caldera noted the deterioration in the position of Venezuelan oil in the U.S. market. He urged "just and nondiscriminatory treatment" of Venezuelan oil imports, not only as a commercial matter, but also as "a condition for the fulfillment of the development programs of a neighboring and friendly country and a key to the direction that future relations between the United States and Latin America will take."

Caldera expressed his conviction that the future of Latin America depended on the extent to which the U.S. reached "a decision to become a pioneer in social international justice." He asserted that the "formula for achieving cordial relations . . . cannot be the merciless attempts at forever lowering the prices of our goods while increasing the price of the commodities we have to import. The thesis that more trade diminishes the need for aid is correct as long as the trade is a just one and is converted into a greater possibility for attaining the urgently needed changes in developing nations."

Following a conference between Presidents Nixon and Caldera June 4, the White House announced that imports of Venezuelan oil would be increased. White House spokesman Ronald L. Ziegler said that "Caldera was told the United States will be announcing oil import quota measures which will insure an increase in imports during the second half of 1970." Ziegler added: "The United States recognizes the importance to Venezuela of sharing the growth of the United States petroleum market and that Venezuela consistently has been a secure and stable oil source. These factors will be taken into account in formulating our long-range oil program." (The announcement of increased import quotas for foreign crude oil was made June 17.

In a speech to the National Press Club in Washington June 2, President Caldera again detailed Venezuela's views on the oil question. Asserting that Venezuela wanted "a reasonable participation in the U.S. oil market and its expansion," he criticized any policy that would favor Canada over Venezuela. "If Canada has been a secure supplier of oil to the United States, Venezuela has also proved . . . that it has been a secure supplier of oil to the United States."

The Venezuelan Senate Dec. 15 completed legislative action on a bill to increase taxes on oil profits from 52% to 60%. The bill, retroactive to Jan. 1, also authorized the president to unilaterally increase tax reference prices, the minimum per-barrel price to be used for computing taxes.

Finance Minister Pedro Tinoco estimated that the increase in 1970 revenues would be about $155 million. Oil companies' payments to Venezuela in 1969 had totaled $1.2 billion. The Wall Street Journal reported Dec. 17 that Creole Petroleum Corp., Venezuela's largest oil producing firm (95% owned by Standard Oil Co. of New Jersey), had estimated that the new law would cost it about $49 million in additional 1970 tax obligations.

A 230-kilometer natural gas pipeline linking gas fields in Anaco, about 200 miles east of Caracas, with Puerto Ordaz in a new industrial complex under development in the Guayana region, was inaugurated in early September. The pipeline, to carry about 150 million cubic feet of natural gas per day, was built by the state petroleum enterprise (CVP) at a cost of about $13 million.

The largest fuel oil desulfurization plant in the world was inaugurated recently in the northern Venezuela town of Amuay, according to Oct. 4 reports. The $120 million plant, located at the Creole Petroleum Corp. refinery in Amuay, had a capacity of 160,000 barrels of fuel oil per day, with 1% or less sulfur, and could produce smaller quantities of oil with sulfur contents ranging down to 0.3%.

Other Areas

Ceylon. The government Dec. 31 nationalized property and bunkering services of the Shell, Esso and Caltex oil companies. The properties were to be operated by the state-owned Ceylon Petroleum Corp.

The government July 24 had canceled the oil prospecting contract of the French Institute of Petroleum.

Ghana. It was announced June 17 that an American consortium had struck oil on the continental shelf eight miles off the coast of Ghana. The consortium was composed of the Signal Exploitation and Development Company, Occidental of Ghana and Amoco Ghana Exploitation Company. It was Ghana's first oil strike.

Great Britain. The discovery of a major oilfield in the North Sea 110 miles northeast of Aberdeen, Scotland was announced Oct. 19 by the British Petroleum Co. The firm said one drilling rig already was in operation and was producing 4,700 barrels a day. Oil experts predicted further strikes in the area, which they said might ultimately match the Ekofisk field discovered off Norway earlier in the year. Britain's oil needs currently totaled two million barrels a day.

Greece. The government was reported May 31 to have signed an agreement with Texaco for undersea oil prospecting in the Aegean sea. It was the sixth agreement with U.S. firms in 17 months.

Nigeria. The Wall Street Journal reported Nov. 6 that petroleum production had begun at two Nigerian oil fields at the rate of 20,000 barrels daily, shared equally by Phillips Petroleum Co. and Nigerian Agip Oil Co., an affiliate of Italy's state-owned oil agency. The two firms jointly held a 1.3 million-acre concession in Nigeria. The announcement said production would rise to 25,-000 barrels daily by the end of 1970 and to 40,000 barrels daily by mid-1971.

The government Nov. 11 began a four-year development plan under which a National Oil Corporation was to be established to participate in the exploration for and mining, refining, distribution and marketing of oil.

Soviet Union. Soviet petroleum output rose from 750,000 barrels daily in 1950 to 2,960,000 in 1960 and 7,060,000 in 1970, while domestic consumption grew from about 670,000 barrels daily in 1950 to nearly 5 million in 1970. During the last half of the 1960s the U.S.S.R. had exported about a third of its oil production, most of it to East and West Europe but some to Africa, Asia, Cuba and other Latin American countries and even small amounts to the U.S.

Coal had provided 80% of the U.S.S.R.'s energy fuel needs and oil and gas 18% in 1950. By 1960 coal provided 64% and gas and oil 35%. But in 1970 oil and gas had become the major fuel, supplying 57% of Soviet fuel needs, while coal provided 40%.

Zambia. The state industrial organization, Indeco, and two companies of Ente Nazionale Idrocarburi, the Italian national hydrocarbons agency, were reported Feb. 15 to have signed an agreement calling for the organization of a joint company to build an oil refinery in Zambia. The plant, to be capable of processing 1.1 million tons of crude oil per year, would meet Zambia's entire fuel requirements, including expected consumption increases.

Atomic Power

Indian developments. Prime Minister Gandhi dedicated the Tarapur atomic power station, India's first (and Asia's largest) nuclear power plant, Jan. 19. The plant, located on the coast of the Arabian Sea north of Bombay, was built to provide nearly 400 megawatts of power to western India. The plant, which had been operating for several months, was built by the International General Electric Co. The U.S. had supplied $90 million of the $125 million cost by means of a loan and supplies of enriched uranium.

The prime minister April 20 rejected a plea in the Lok Sabha (lower house of parliament) that India begin using nuclear engineering technology for peaceful purposes. Mrs. Gandhi pointed out that blast technology was still in an experimental stage. She denied that the government had hampered nuclear scientists in conducting their experiments.

In a subsequent development May 25, Dr. Vikram Sarabhai announced plans to expand India's atomic research program for peace. Although Sarabhai, chairman of India's Atomic Energy Commission, stressed that the expansion was for peaceful purposes, the plans included a project to develop rocket and space technology, a guidance system, and missile and tracking systems. The 10-year plan included construction of four nuclear power stations with a total generating capacity of 1,700 megawatts.

India and the Soviet Union were reported June 14 to have signed an agreement to exchange nuclear physicists under a bilateral cultural program for 1970–71.

TVA planning new units. The Tennessee Valley Authority Aug. 27 disclosed plans for a fourth new nuclear power plant to meet the requirements of the seven-state region supplied by the federal power agency. Plans for a third plant were announced Aug. 24. Both plants were scheduled to begin operations in 1976.

Swedish-Soviet accords. Sweden and the Soviet Union signed two agreements covering scientific collaboration and the peaceful uses of nuclear energy during the week of Jan. 11–17. Under the nuclear accord, signed in Moscow, the Soviet Union would undertake to refine uranium for Sweden.

U.S.S.R. to build Finnish plant. The Soviet Union announced conclusion of an agreement with a Finnish firm for construction and equipment of Finland's first atomic power station, the Wall Street Journal reported June 11. Under the agreement, the Soviets were to provide the power plant, two turbo-generators, auxiliary systems and nuclear fuel; in addition the U.S.S.R. would aid in the construction of the station, in Loviis Town. The station was to have a capacity of 440 megawatts and to begin operations in 1976.

Canadian-Soviet deal. Canada and the Soviet Union concluded an agreement for Atomic Energy of Canada, Ltd. to purchase 55 tons of heavy water from the U.S.S.R., it was reported Nov. 24. Canada would pay $3.3 million in cash— about $30 a pound—for the water, which would be delivered in 1971. The agreement marked the first time that the Soviet Union had sold a product connected with nuclear power to a major western country.

U.S.-Soviet talks on blast uses. In talks held in Moscow Feb. 12–17, U.S. and Soviet delegations agreed that greater possibilities existed for peaceful use of nuclear explosions. A joint statement issued after the talks said there had been "an identity of views" on the use of nuclear explosions for such projects as oil and gas extraction and the excavation of reservoirs and canals. (The New York Times reported March 5 that Soviet delegates to the talks had disclosed that the U.S.S.R.'s program for peaceful use of nuclear explosions was more extensive than the corresponding U.S. Atomic Energy Commission Plowshare Program. The Soviet program reportedly had included 13 experiments with nuclear explosives for a wide range of projects. The U.S. had conducted 10 such experiments and eight other tests.

Gas centrifuge pact signed. Great Britain, the Netherlands, and West Germany signed a treaty March 4 in Almelo, Holland for cooperative development of the gas centrifuge process of producing uranium. The process could lead to a relatively inexpensive means of separating, according to density by centrifugal force, fissionable uranium isotope 235 from the nonfissionable uranium isotope 238.

The treaty, signed by Dutch Foreign Minister Joseph M. A. H. Luns, West German Foreign Minister Walter Scheel, and Britain Minister of Technology Anthony Wedgwood Benn, provided for the construction, commissioning and operation of plants at Almelo and at Capenhurst in Cheshire, England.

Spokesmen for the three powers emphasized that the treaty provided adequate safeguards against military application of the process and that it in no way contravened the rules of Euratom, by which the Dutch and German governments were bound. (Scientists had warned that centrifuge separation could become an easily-hidden process for production of weapons grade fission materials.)

South African uranium plant. South Africa entered the nuclear fuel market July 20 with an announcement by Prime Minister John Vorster that its scientists had invented a new process for enriching uranium. Vorster said the process would be put to work in a new plant that would compete with western markets.

The process, it was announced, would produce uranium more cheaply than anywhere else in the West. A spokesman for the South African Atomic Energy Board said July 27 that the new method was derived from entirely new principles and was not based on the diffusion process developed in the U.S. or the centrifuge process developed by a British-Dutch-German combine.

The government July 23 presented a Uranium Enrichment Act which would provide for the establishment of a government corporation to utilize the process. Mines Minister Carel De Wet said July 27 that the uranium industry was planning for an expected increase in world demand to 80,000 tons a year by 1980.

In his July 20 announcement, the prime minister emphasized that "our sole objective in the further development and application of the process is to promote the peaceful application of nuclear energy." He said the enriched uranium would be used for the generation of power and for de-salination of sea water. Vorster said South Africa was prepared to subject its nuclear activities to safeguards, including inspection.

Niger uranium program. It was disclosed June 14 that Niger, the French Atomic Energy Commission and a Japanese consortium had concluded a tripartite deal to cover the exploration and development of a uranium deposit in Niger, estimated at 30,000 tons.

AEC waste procedure scored. Sen. Frank Church (D, Ida.) charged March 6 that the Atomic Energy Commission had failed to release a 1966 report by a panel of the National Academy of Sciences critical of AEC procedures for storing radioactive waste.

The report, made available to the press March 6, said that the four major AEC plants where waste was stored were located in poor geological areas for such uses, that the practice of storing waste in the ground could pose a hazard through build-up, and that there was no uniform standard among the plants for determining the degree of radioactivity of waste materials. The panel had said that it saw no immediate danger in the way the AEC was storing or disposing of nuclear wastes.

John Erlewine, AEC assistant general manager for operations, was reported March 6 to have said that the report was only one of many not published by the AEC. He also said that while the AEC was putting waste into the ground it did not necessarily intend to leave it there and its operations were within the standards laid down by the Federal Radiation Council.

The Washington Post reported March 13 that four government agencies had told the AEC that it was careless in the way it disposed of radioactive wastes.

The paper cited a report that had been prepared for Sen. Church by the Bureau of Radiological Health, the Bureau of Sport Fisheries and Wildlife, the U.S. Geological Survey and the Federal Water Pollution Control Administration. The agencies recommended that the AEC keep at least two feet of clay or gravel between atomic burial pits and the exposed basalt formations below (the AEC had required no minimum protective barrier) and that the commission take steps to keep atomic burial sites free of flood waters.

New radiation rules proposed. Atomic Energy Commissioner James T. Ramey said March 27 that the AEC was proposing regulations to require industry to take advantage of improvements in technology to minimize the amount of radiation released by new nuclear plants. (Current regulations required only that emissions of radiation be within AEC safety limits.)

Ramey said the proposed rules would not reduce the present maximum permissable limits of radioactivity emitted from nuclear power plants. He denied there was any connection between the proposed changes and a recent case in Minnesota in which the state sought limits on radioactive discharges that would be more rigorous than the AEC's. (In the Minnesota case, still pending, the Northern States Power Company, located about 30 miles north of Minneapolis, was granted a permit by the Minnesota pollution control agency for a "stack release" from its water reactor that would limit the firm to about 2% of the radioactivity permitted by AEC standards. The power company had brought suit to challenge Minnesota's authority to set standards more rigid than the AEC's. Four states had petitioned the Federal Government for permission to file amicus curiae briefs supporting Minnesota's position; seven had asked to be added to Minnesota's brief.)

Controversy over the effect of radiation on the environment marked a three-day symposium on engineering with nuclear explosives Jan. 14–17. Dr. Theos J. Thompson, a member of the AEC, said at the opening of the sessions, sponsored by the American Nuclear Society in cooperation with the AEC, that concern by environmentalists about radiation hazards was reaching almost ridiculous levels. He added that such persons did not evaluate the benefits of programs to develop peaceful uses of atomic energy, but had concentrated on very low level effects "whose extreme extrapolation might be detrimental."

Rep. Chet Holifield (D, Calif.), chairman of the Joint Committee on Atomic Energy, criticized the Administration's decision to cut funds for Project Plowshare, a 13-year program by the AEC to find peaceful uses for atomic explosions, from $29 million to $14 million in 1969. Holifield predicted more cuts in the program in the 1971 fiscal budget.

Dr. Edward L. Teller, a leading developer of the hydrogen bomb, said Jan. 15 that radiation from peaceful uses of atomic energy "can be easily guarded against." Teller said of the AEC's programs that "no big-scale enterprise has ever been carried out with more assurance [of public safety] than the atomic energy enterprise."

Robert B. Miller, an attorney for the American Civil Liberties Union, charged in a federal district court in Denver Jan. 12 that safety standards of the AEC "grossly underestimate" potential damage from radiation. Miller made the charge at the reopening of a suit to prevent natural gas from being freed from an explosion conducted Sept. 10, 1969— Project Rulison—in Colorado. He also asserted that the AEC had violated its own standards by conducting the test when wind directions had not been correct.

Soviet A-accident. Travelers to Moscow reported Feb. 20 that an explosion had occurred in the Soviet Union's main nuclear submarine plant near Gorky, 250 miles east of Moscow. Several workers were said to have been killed and the Volga river polluted with radioactive waste. According to the Washington Post Feb. 25, Soviet sources said that radioactive material had leaked from a container in a shed at the plant, killing two employees and injuring three others. The accident, according to the Soviet sources, had occured in January, had been confined to the shed, and had not polluted the Volga.

1971

The world search for energy sources accelerated during 1971 as U.S. oil production declined from its 1970 peak and U.S. gas reserves reached a 15-year low. President Nixon, in a special message on energy, stressed the need to find non-polluting fuels. The U.S. continued with plans to bring Alaskan North Slope oil to the "lower 48" states, and a trans-Canadian pipeline was proposed as a rival to the projected trans-Alaskan line. Heated controversy continued over the harm done to the environment by the different branches of the energy community. Oil-producing countries exacted higher payments from Western oil companies. Expropriations took place in Algeria, Libya and Venezuela. The Soviet Union announced its development of a method of converting nuclear energy directly into electric current.

Intensified Search for Energy Sources

U.S. Program

By 1971 the world's developed nations were mounting an intensified hunt for additional sources of energy, and the U.S. was leading the search. Oil production in the U.S. had reached its peak in November 1970 and appeared to be headed downward. In five years, as U.S. natural gas consumption reached a record 65 billion cubic feet a day, U.S. proven gas reserves declined from a 289 trillion-cubic-foot high to a 15-year low of 247 trillion cubic feet at the end of 1971.

President Nixon spurred the U.S. energy campaign by dispatching what he described as the first "comprehensive" energy message ever sent to Congress by an American President.

Nixon's message to Congress. President Nixon sent Congress a message June 4 stressing the necessity of developing a program for "an adequate supply of clean energy." Warning that the country faced an increasing shortage of electrical energy, the President told newsmen in releasing the message that industrial societies faced the challenges of finding new sources of energy and finding sources that would not pollute the environment.

In his message, the President chose the "fast breeder" atomic power reactor as the best means to provide cheap, clean energy and asked Congress to increase development funds by $27 million to $130 million for fiscal 1972. He also favored

federal support for the first demonstration breeder plant, which was expected to be in operation before 1980 and would cost at least $500 million. Nixon said he would seek $100 million toward construction of the plant.

The breeder reactor, with its efficient use of nuclear fuel, the President said, "could extend the life of our natural uranium supply from decades to centuries with far less impact on the environment than the power plants which are operating today."

The message also proposed an accelerated program for leasing federal oil lands. New off-shore oil lands were to be released for exploration and oil shale fields in Colorado, Utah and Wyoming were to be opened for the first time to drilling operations and, later, land bids, in a limited way.

The program for leasing off-shore oil regions, currently restricted to as far east as Louisiana in the Gulf of Mexico, was to be extended into waters off Alabama, Mississippi and Florida and possibly within the next five years along the East Coast.

"Drainage" leases adjacent to producing wells were to be offered annually as well as "wildcat" leases in regions new to oil exploration.

Congress also was requested to provide $15 million more to find ways to scrub sulphur dioxide out of coal and oil stack gases and $10 million to augment the program to extract natural gas from coal.

The text of President Nixon's message on energy:

ENERGY RESOURCES

To the Congress of the United States:

For most of our history, a plentiful supply of energy is somethin the American people have taken very much for granted. In the pa twenty years alone, we have been able to double our consumption energy without exhausting the supply. But the assumption that sufficie energy will always be readily available has been brought sharply in question within the last year. The brownouts that have affected som areas of our country, the possible shortages of fuel that were threaten last fall, the sharp increases in certain fuel prices and our growing awar ness of the environmental consequences of energy production have demonstrated that we cannot take our energy supply for granted ar longer.

A sufficient supply of clean energy is essential if we are to susta healthy economic growth and improve the quality of our national lif I am therefore announcing today a broad range of actions to ensure a adequate supply of clean energy for the years ahead. Private industr of course, will still play the major role in providing our energy, but go ernment can do a great deal to help in meeting this challenge.

My program includes the following elements:

To Facilitate Research and Development for Clean Energy:

—A commitment to complete the successful demonstration of th liquid metal fast breeder reactor by 1980.

—More than twice as much Federal support for sulfur oxide co trol demonstration projects in Fiscal Year 1972.

—An expanded program to convert coal into a clean gaseous fue

—Support for a variety of other energy research projects in fiel such as fusion power, magnetohydrodynamic power cycles, and unde ground electric transmission.

To Make Available the Energy Resources on Federal Lands:

—Acceleration of oil and gas lease sales on the Outer Continent Shelf, along with stringent controls to protect the environment.

—A leasing program to develop our vast oil shale resources, provide that environmental questions can be satisfactorily resolved.

—Development of a geothermal leasing program beginning th fall.

To Assure a Timely Supply of Nuclear Fuels:

—Begin work to modernize and expand our uranium enrichme capacity.

Use Our Energy More Wisely:

—A New Federal Housing Administration standard requiring addi-
nal insulation in new federally insured homes.

—Development and publication of additional information on how
nsumers can use energy more efficiently.

—Other efforts to encourage energy conservation.

Balance Environmental and Energy Needs:

—A system of long-range open planning of electric power plant sites
d transmission line routes with approval by a State or regional agency
fore construction.

—An incentive charge to reduce sulfur oxide emissions and to support
rther research.

Organize Federal Efforts More Effectively:

—A single structure within the Department of Natural Resources
iting all important energy resource development programs.

The Nature of the Current Problem

A major cause of our recent energy problems has been the sharp
crease in demand that began about 1967. For decades, energy con-
mption had generally grown at a slower rate than the national output of
ods and services. But in the last four years it has been growing at a
ster pace and forecasts of energy demand a decade from now have been
dergoing significant upward revisions.

This accelerated growth in demand results partly from the fact that
ergy has been relatively inexpensive in this country. During the last
cade, the prices of oil, coal, natural gas and electricity have increased at
much slower rate than consumer prices as a whole. Energy has been
attractive bargain in this country—and demand has responded
cordingly.

In the years ahead, the needs of a growing economy will further stim-
ate this demand. And the new emphasis on environmental protection
eans that the demand for cleaner fuels will be especially acute. The
imary cause of air pollution, for example, is the burning of fossil
els in homes, in cars, in factories and in power plants. If we are to meet
r new national air quality standards, it will be essential for us to use
ack gas cleaning systems in our large power and other industrial plants
d to use cleaner fuels in virtually all of our new residential, com-
ercial and industrial facilities, and in some of our older facilities as well.

Together, these two factors—growing demand for energy and grow-
g emphasis on cleaner fuels—will create an extraordinary pressure on
r fuel supplies.

The task of providing sufficient clean energy is made especially difficult by the long lead times required to increase energy supply. To mo from geological exploration to oil and gas well production now ta from 3 to 7 years. New coal mines typically require 3 to 5 years to re the production stage and it takes 5 to 7 years to complete a large ste power plant. The development of the new technology required to minim environmental damage can further delay the provision of additio energy. If we are to take full advantage of our enormous coal resour for example, we will need mining systems that do not impair the hea and safety of miners or degrade the landscape and combustion syste that do not emit harmful quantities of sulfur oxides, other noxious ga and particulates into the atmosphere. But such systems may take seve years to reach satisfactory performance. That is why our efforts to expa the supply of clean energy in America must immediately be stepped

1. Research and Development Goals for Clean Energy

Our past research in this critical field has produced many prom ing leads. Now we must move quickly to demonstrate the best of th new concepts on a commercial scale. Industry should play the maj role in this area, but government can help by providing technical lead ship and by sharing a portion of the risk for costly demonstration plan The time has now come for government and industry to commit the selves to a joint effort to achieve commercial scale demonstrations in t most crucial and most promising clean energy development areas—t fast breeder reactor, sulfur oxide control technology and coal gasificati

a. Sulfur Oxide Control Technology

A major bottleneck in our clean energy program is the fact th we cannot now burn coal or oil without discharging its sulfur cont into the air. We need new technology which will make it possible remove the sulfur before it is emitted to the air.

Working together, industry and government have developed variety of approaches to this problem. However, the new air qual standards promulgated under the Clean Air Amendments of 1970 quire an even more rapid development of a suitable range of stack g cleaning techniques for removing sulfur oxides. I have therefore quested funds in my 1972 budget to permit the Environmental Protecti Agency to devote an additional $15 million to this area, more th doubling the level of our previous efforts. This expansion means th a total of six different techniques can be demonstrated in partnersh with industry during the next three or four years.

b. *Nuclear Breeder Reactor*

Our best hope today for meeting the Nation's growing demand for economical clean energy lies with the fast breeder reactor. Because of highly efficient use of nuclear fuel, the breeder reactor could extend the life of our natural uranium fuel supply from decades to centuries, with far less impact on the environment than the power plants which are operating today.

For several years, the Atomic Energy Commission has placed the highest priority on developing the liquid metal fast breeder. Now this project is ready to move out of the laboratory and into the demonstration phase with a commercial size plant. But there still are major technical and financial obstacles to the construction of a demonstration plant of some 300 to 500 megawatts. I am therefore requesting an additional $27 million in Fiscal Year 1972 for the Atomic Energy Commission's liquid metal fast breeder reactor program—and for related technological and safety programs—so that the necessary engineering groundwork for demonstration plants can soon be laid.

What about the environmental impact of such plants? It is reassuring to know that the releases of radioactivity from current nuclear reactors are well within the national safety standards. Nevertheless, we will make every effort to see that these new breeder reactors emit even less radioactivity to the environment than the commercial light water reactors which are now in use.

I am therefore directing the Atomic Energy Commission to ensure that the new breeder plants be designed in a way which inherently prevents discharge to the environment from the plant's radioactive effluent systems. The Atomic Energy Commission should also take advantage of the increased efficiency of these breeder plants, designing them to minimize waste heat discharges. Thermal pollution from nuclear power plants can be materially reduced in the more efficient breeder reactors.

We have very high hopes that the breeder reactor will soon become a key element in the national fight against air and water pollution. In order further to inform the interested agencies and the public about the opportunities in this area, I have requested the early preparation and review by all appropriate agencies of a draft environmental impact statement for the breeder demonstration plant in accordance with Section 102 of the National Environmental Policy Act. This procedure will ensure compliance with all environmental quality standards before plant construction begins.

In a related area, it is also pertinent to observe that the safety reco
of civilian power reactors in this country is extraordinary in the histo
of technological advances. For more than a quarter century—since tl
first nuclear chain reaction took place—no member of the public h
been injured by the failure of a reactor or by an accidental release
radioactivity. I am confident that this record can be maintained. Tl
Atomic Energy Commission is giving top priority to safety consideratio
in the basic design of the breeder reactor and this design will also be su
ject to a thorough review by the independent Advisory Committee o
Reactor Safeguards, which will publish the results of its investigatio

I believe it important to the Nation that the commercial demonstr
tion of a breeder reactor be completed by 1980. To help achieve th
goal, I am requesting an additional $50 million in Federal funds for tl
demonstration plant. We expect industry—the utilities and manufa
turers—to contribute the major share of the plant's total cost, since th
have a large and obvious stake in this new technology. But we al
recognize that only if government and industry work closely together ca
we maximize our progress in this vital field and thus introduce a new e
in the production of energy for the people of our land.

c. *Coal Gasification*

As we carry on our search for cleaner fuels, we think immediate
of the cleanest fossil fuel—natural gas. But our reserves of natural gas a
quite limited in comparison with our reserves of coal.

Fortunately, however, it is technically feasible to convert coal in
a clean gas which can be transported through pipelines. The Departme
of the Interior has been working with the natural gas and coal industri
on research to advance our coal gasification efforts and a number of pc
sible methods for accomplishing this conversion are under developmer
A few, in fact, are now in the pilot plant stage.

We are determined to bring greater focus and urgency to this effo
We have therefore initiated a cooperative program with industry to e
pand the number of pilot plants, making it possible to test new metho
more expeditiously so that the appropriate technology can soon be s
lected for a large-scale demonstration plant.

The Federal expenditure for this cooperative program will be e
panded to $20 million a year. Industry has agreed to provide $10 milli
a year for this effort. In general, we expect that the Government will co
tinue to finance the larger share of pilot plants and that industry w
finance the larger share of the demonstration plants. But again, tl

nportant point is that both the Government and industry are now rongly committed to move ahead together as promptly as possible to ake coal gasification a commercial reality.

d. *Other Research and Development Efforts*

The fast breeder reactor, sulfur oxide controls and coal gasification present our highest priority research and development projects in the ean energy field. But they are not our only efforts. Other ongoing ojects include:

—*Coal Mine Health and Safety Research.* In response to a growing ncern for the health and safety of the men who mine the Nation's coal id in accordance with the Federal Coal Mine Health and Safety Act 1969, the Bureau of Mines research effort has been increased from a vel of $2 million in Fiscal Year 1969 to $30 million in Fiscal Year 1972.

—*Controlled Thermonuclear Fusion Research.* For nearly two cades the Government has been funding a sizeable research effort signed to harness the almost limitless energy of nuclear fusion for aceful purposes. Recent progress suggests that the scientific feasibility such projects may be demonstrated in the 1970s and we have therefore quested an additional $2 million to supplement the budget in this field r Fiscal Year 1972. We hope that work in this promising area will con- ue to be expanded as scientific progress justifies larger scale programs.

—*Coal Liquefaction.* In addition to its coal gasification work, the epartment of the Interior has underway a major pilot plant program rected toward converting coal into cleaner liquid fuels.

—*Magnetohydrodynamic Power Cycles.* MHD is a new and more ficient method of converting coal and other fossil fuels into electric ergy by burning the fuel and passing the combustion products through magnetic field at very high temperatures. In partnership with the elec- ic power industry, we have been working to develop this new system electric power generation.

—*Underground Electric Transmission.* Objections have been grow- g to the overhead placement of high voltage power lines, especially in eas of scenic beauty or near centers of population. Again in coopera- on with industry, the Government is funding a research program to velop new and less expensive techniques for burying high voltage ectric transmission lines.

—*Nuclear Reactor Safety and Supporting Technology.* The general search and development work for today's commercial nuclear reactors as completed several years ago, but we must continue to fund safety-

related efforts in order to ensure the continuance of the excellent saf(
record in this field. An additional $3 million has recently been request
for this purpose to supplement the budget in Fiscal Year 1972.

—*Advanced Reactor Concepts.* The liquid metal fast breeder is.t
priority breeder reactor concept under development, but the Aton
Energy Commission is also supporting limited alternate reactor progra
involving gas cooled reactors, molten salt reactors and light wat
breeders.

—*Solar Energy.* The sun offers an almost unlimited supply of ener
if we can learn to use it economically. The National Aeronautics a
Space Administration and the National Science Foundation are curren
re-examining their efforts in this area and we expect to give greater atte
tion to solar energy in the future.

The key to meeting our twin goals of supplying adequate energy a
protecting the environment in the decades ahead will be a balanced a
imaginative research and development program. I have therefore ask
my Science Adviser, with the cooperation of the Council on Environmen
Quality and the interested agencies, to make a detailed assessment
all of the technological opportunities in this area and to recommend ad(
tional projects which should receive priority attention.

2. *Making Available the Energy Resources of Federal Lands*

Over half of our Nation's remaining oil and gas resources, abc
40 percent of our coal and uranium, 80 percent of our oil shale, and so
60 percent of our geothermal energy sources are now located on Fede
lands. Programs to make these resources available to meet the growi
energy requirements of the Nation are therefore essential if shortag
are to be averted. Through appropriate leasing programs, the Gover
ment should be able to recover the fair market value of these resourc
while requiring developers to comply with requirements that will ac
quately protect the environment.

To supplement the efforts already underway to develop the fu
resources of the lower 48 States and Alaska, I am announcing today t
following new programs:

a. *Leasing on the Outer Continental Shelf—An Accelerat Program*

The Outer Continental Shelf has proved to be a prolific source
oil and gas, but it has also been the source of troublesome oil spills
recent years. Our ability to tap the great potential of offshore areas h
been seriously hampered by these environmental problems.

The Department of the Interior has significantly strengthened the nvironmental protection requirements controlling offshore drilling and e will continue to enforce these requirements very strictly. As a pre-quisite to Federal lease sales, environmental assessments will be made accordance with Section 102 of the National Environmental Policy ct of 1969.

Within these clear limits, we will accelerate our efforts to utilize his rich source of fuel. In order to expand productive possibilities as pidly as possible, the accelerated program should include the sale of ew leases not only in the highly productive Gulf of Mexico, but also some her promising areas. I am therefore directing the Secretary of the In-rior to increase the offerings of oil and gas leases and to publish a hedule for lease offerings on the Outer Continental Shelf during the ext five years, beginning with a general lease sale and a drainage sale is year.

b. *Oil Shale—A Program for Orderly Development*

At a time when we are facing possible energy shortages, it is reassur-g to know that there exists in the United States an untapped shale oil source containing some 600 billion barrels in high grade deposits. At urrent consumption rates, this resource represents 150 years supply. bout 80 billion barrels of this shale oil are particularly rich and well tuated for early development. This huge resource of very low sulfur oil located in the Rocky Mountain area, primarily on Federal land.

At present there is no commercial production of shale oil. A mixture problems—environmental, technical and economic—have combined thwart past efforts at development.

I believe the time has come to begin the orderly formulation of a ale oil policy—not by any head-long rush toward development but ther by a well considered program in which both environmental protec-on and the recovery of a fair return to the Government are cardinal inciples under which any leasing takes place. I am therefore requesting e Secretary of the Interior to expedite the development of an oil shale asing program including the preparation of an environmental impact atement. If after reviewing this statement and comments he finds that nvironmental concerns can be satisfied, he shall then proceed with the etailed planning. This work would also involve the States of Wyoming, olorado and Utah and the first test lease would be scheduled for next ar.

c. *Geothermal Energy*

There is a vast quantity of heat stored in the earth itself. Where th
energy source is close to the surface, as it is in the Western States, it ca
readily be tapped to generate electricity, to heat homes, and to meet oth
energy requirements. Again, this resource is located primarily on Feder
lands.

Legislation enacted in recent months permits the Federal Gover
ment, for the first time, to prepare for a leasing program in the field
geothermal energy. Classification of the lands involved is already unde
way in the Department of the Interior. I am requesting the Secretary
the Interior to expedite a final decision on whether the first competiti
lease sale should be scheduled for this fall—taking into account, of cours
his evaluation of the environmental impact statement.

3. *Natural Gas Supply*

For the past 25 years, natural gas has supplied much of the increa
in the energy supply of the United States. Now this relatively clea
form of energy is in even greater demand to help satisfy air quali
standards. Our present supply of natural gas is limited, however, and v
are beginning to face shortages which could intensify as we move to ir
plement the air quality standards. Additional supplies of gas will ther
fore be one of our most urgent energy needs in the next few years.

Federal efforts to augment the available supplies of natural g
include:

—Accelerated leasing on Federal lands to speed discovery and d
velopment of new natural gas fields.

—Moving ahead with a demonstration project to gasify coal.

—Recent actions by the Federal Power Commission providi
greater incentives for industry to increase its search for new sourc
of natural gas and to commit its discoveries to the intersta
market.

—Facilitating imports of both natural and liquefied gas from Ca
ada and from other nations.

—Progress in nuclear stimulation experiments which seek to pr
duce natural gas from tight geologic formations which cann
presently be utilized in ways which are economically and enviro
mentally acceptable.

This administration is keenly aware of the need to take every reasc
able action to enlarge the supply of clean gaseous fuels. We intend to ta
such action and we expect to get good results.

4. Imports from Canada

Over the years, the United States and Canada have steadily increased their trade in energy. The United States exports some coal to Canada, but the major items of trade are oil and gas which are surplus to Canadian needs but which find a ready market in the United States.

The time has come to develop further this mutually advantageous trading relationship. The United States is therefore prepared to move promptly to permit Canadian crude oil to enter this country, free of any quantitative restraints, upon agreement as to measures needed to prevent citizens of both our countries from being subjected to oil shortages, or threats of shortages. We are ready to proceed with negotiations and we look to an early conclusion.

5. Timely Supplies of Nuclear Fuels

The Nation's nuclear fuel supply is in a state of transition. Military needs are now relatively small but civilian needs are growing rapidly and will be our dominant need for nuclear fuel in the future. With the exception of uranium enrichment, the nuclear energy industry is now in private hands.

I expect that private enterprise will eventally assume the responsibility for uranium enrichment as well, but in the meantime the Government must carry out its responsibility to ensure that our enrichment capacity expands at a rate consistent with expected demands.

There is currently no shortage of enriched uranium or enriching capacity. In fact, the Atomic Energy Commission has substantial stocks of enriched uranium which have already been produced for later use. However, plant expansions are required so that we can meet the growing demands for nuclear fuel in the late 1970s—both in the United States and in other nations for which this country is now the principal supplier.

The most economical means presently available for expanding our capacity in this field appears to be the modernization of existing gaseous diffusion plants at Oak Ridge, Tennessee; Portsmouth, Ohio; and Paducah, Kentucky—through a Cascade Improvement Program. This program will take a number of years to complete and we therefore believe that it is prudent to initiate the program at this time rather than run the risk of shortages at a later date. I am therefore releasing $16 million to start the Cascade Improvement Program in Fiscal Year 1972. The pace of the improvement program will be tailored to fit the demands for enriched uranium in the United States and in other countries.

6. *Using Our Energy More Wisely*

We need new sources of energy in this country, but we also need
use existing energy as efficiently as possible. I believe we can achieve t
ends we desire—homes warm in winter and cool in summer, rapid tran
portation, plentiful energy for industrial production and home app
ances—and still place less of a strain on our overtaxed resources.

Historically, we have converted fuels into electricity and have us
other sources of energy with ever increasing efficiency. Recent data su
gest, however, that this trend may be reversing—thus adding to the dra
on available resources. We must get back on the road of increasi
efficiency—both at the point of production and at the point of consum
tion, where the consumer himself can do a great deal to achieve co
siderable savings in his energy bills.

We believe that part of the answer lies in pricing energy on the bas
of its full costs to society. One reason we use energy so lavishly today
that the price of energy does not include all of the social costs of produ
ing it. The costs incurred in protecting the environment and the heal
and safety of workers, for example, are part of the real cost of produ
ing energy—but they are not now all included in the price of the produ
If they were added to that price, we could expect that some of the was
in the use of energy would be eliminated. At the same time, by expandi
clean fuel supplies, we will be working to keep the overall cost of ener
as low as possible.

It is also important that the individual consumer be fully aware
what his energy will cost if he buys a particular home or appliance. Tl
efficiency of home heating or cooling systems and of other energy i
tensive equipment are determined by builders and manufacturers wl
may be concerned more with the initial cost of the equipment than wi
the operating costs which will come afterward. For example, better the
mal insulation in a home or office building may save the consumer larg
sums in the long run—and conserve energy as well—but for the build
it merely represents an added expense.

To help meet one manifestation of this problem, I am directing tl
Secretary of Housing and Urban Development to issue revised standar
for insulation applied in new federally insured homes. The new Fe
eral Housing Administration standards will require sufficient insulatic
to reduce the maximum permissible heat loss by about one-third for
typical 1200 square foot home—and by even more for larger homes.
is estimated that the fuel savings which will result each year from tl

pplication of these new standards will, in an average climate, equal the
ost of the additional insulation required.

While the Federal Government can take some actions to conserve
nergy through such regulations, the consumer who seeks the most for
is energy dollar in the marketplace is the one who can have the most pro-
ound influence. I am therefore asking my Special Assistant for Consumer
ffairs—in cooperation with industry and appropriate Government agen-
ies—to gather and publish additional information in this field to help
onsumers focus on the operating costs as well as the initial cost of energy
ntensive equipment.

In addition, I would note that the Joint Board on Fuel Supply and
'uel Transport chaired by the Director of the Office of Emergency
'reparedness is developing energy conservation measures for industry,
overnment, and the general public to help reduce energy use in times of
articular shortage and during pollution crises.

. *Power Plant Siting*

If we are to meet growing demands for electricity in the years ahead,
ve cannot ignore the need for many new power plants. These plants
nd their associated transmission lines must be located and built so as to
void major damage to the environment, but they must also be com-
leted on time so as to avoid power shortages. These demands are difficult
o reconcile—and often they are not reconciled well. In my judgment the
sson of the recent power shortages and of the continuing disputes over
ower plant siting and transmission line routes is that the existing insti-
tions for making decisions in this area are not adequate for the job.
n my Special Message to the Congress on the Environment last February,
proposed legislation which would help to alleviate these problems
hrough longer range planning by the utilities and through the establish-
ent of State or regional agencies to license new bulk power facilities
rior to their construction.

Hearings are now being held by the Interstate and Foreign Com-
erce Committee of the House of Representatives concerning these pro-
osals and other measures which would provide an open planning and
ecision-making capacity for dealing with these matters. Under the ad-
inistration bill, long-range expansion plans would be presented by the
tilities ten years before construction was scheduled to begin, individual
lternative power plant sites would be identified five years ahead, and
etailed design and location of specific plants and transmission lines would
e considered two years in advance of construction. Public hearings would
e held far enough ahead of construction so that they could influence

the siting decision, helping to avoid environmental problems withou causing undue construction delays. I urge the Congress to take promp and favorable action on this important legislative proposal. At the sam time steps will be taken to ensure that Federal licenses and permits ar handled as expeditiously as possible.

8. *The Role of the Sulfur Oxides Emissions Charge*

In my environmental message last February I also proposed th establishment of a sulfur oxides emissions charge. The emissions charg would have the effect of building the cost of sulfur oxide pollution int the price of energy. It would also provide a strong economic incentiv for achieving the necessary performance to meet sulfur oxide standards

The funds generated by the emissions charge would be used by th Federal Government to expand its programs to improve environmenta quality, with special emphasis on the development of adequate supplie of clean energy.

9. *Government Reorganization—An Energy Administration*

But new programs alone will not be enough. We must also conside how we can make these programs do what we intend them to do. On important way of fostering effective performance is to place responsi bility for energy questions in a single agency which can execute anc modify policies in a comprehensive and unified manner.

The Nation has been without an integrated energy policy in the past One reason for this situation is that energy responsibilities are fragmentec among several agencies. Often authority is divided according to type and uses of energy. Coal, for example, is handled in one place, nuclea energy in another—but responsibility for considering the impact of on on the other is not assigned to any single authority. Nor is there any singl agency responsible for developing new energy sources such as solar energy or new conversion systems such as the fuel cell. New concerns—such a conserving our fossil fuels for non-fuel uses—cannot receive the thorough and thoughtful attention they deserve under present arrangements.

The reason for all these deficiencies is that each existing program was set up to meet a specific problem of the past. As a result, our present structure is not equipped to handle the relationships between these prob lems and the emergence of new concerns.

The need to remedy these problems becomes more pressing every day. For example, the energy industries presently account for some 20 percent of our investment in new plant and equipment. This means that inefficien cies resulting from uncoordinated government programs can be very costly to our economy. It is also true that energy sources are becoming increasingly interchangeable. Coal can be converted to gas, for example, and even to synthetic crude oil. If the Government is to perform adequately in the energy field, then it must act through an agency which has sufficient strength and breadth of responsibility.

Accordingly, I have proposed that all of our important Federal energy resource development programs be consolidated within the new Department of Natural Resources.

The single energy authority which would thus be created would be better able to clarify, express, and execute Federal energy policy than any unit in our present structure. The establishment of this new entity would provide a focal point where energy policy in the executive branch could be harmonized and rationalized.

One of the major advantages of consolidating energy responsibilities would be the broader scope and greater balance this would give to research and development work in the energy field. The Atomic Energy Commission, for instance, has been successful in its mission of advancing civilian nuclear power, but this field is now intimately interrelated with coal, oil and gas, and Federal electric power programs with which the Atomic Energy Commission now has very little to do. We believe that the planning and funding of civilian nuclear energy activities should now be consolidated with other energy efforts in an agency charged with the mission of insuring that the total energy resources of the Nation are effectively utilized. The Atomic Energy Commission would still remain intact, in order to execute the nuclear programs and any related energy research which may be appropriate as part of the overall energy program of the Department of Natural Resources.

Until such time as this new Department comes into being, I will continue to look to the Energy Subcommittee of the Domestic Council for leadership in analyzing and coordinating overall energy policy questions for the executive branch.

CONCLUSION

The program I have set forth today provides the basic ingredients for a new effort to meet our clean energy needs in the years ahead.

The success of this effort will require the cooperation of the Congress and of the State and local governments. It will also depend on the willingness of industry to meet its responsibilities in serving customers and in making necessary capital investments to meet anticipated growth. Consumers, too, will have a key role to play as they learn to conserve energy and as they come to understand that the cost of environmental protection must, to a major extent, be reflected in consumer prices.

I am confident that the various elements of our society will be able to work together to meet our clean energy needs. And I am confident that we can therefore continue to know the blessings of both a high-energy civilization and a beautiful and healthy environment.

RICHARD NIXON

The White House
June 4, 1971

Interior study OKs Alaska pipeline. The Interior Department released a staff study Jan. 13 stating that despite "unavoidable" environmental consequences, a proposed 800-mile oil pipeline across Alaska should be built to meet the crucial oil needs of the nation. The report, described as "tentative" pending public hearings in February, recommended safeguards to keep the "foreseeable environmental costs" of the project at an "acceptable" level.

Construction of the pipeline, held up in 1970 by an injunction, was described in the Interior report as "essential to the strength, growth and security of the United States." The panel noted economic advantages for Alaska but emphasized that U.S. security would be threatened if it became overly dependent on oil from the troubled Middle East.

The pipeline would be built by the Alyeska Pipeline Service Co., a consortium of oil interests. Major stockholders were the Humble Oil and Refining Co., the British Petroleum Co., Ltd. and the Atlantic Richfield Co.

The report was required under an Environmental Policy Act provision for a statement on the possible impact on the environment of government agency proposals. The report admitted that the project would disrupt the area's ecology, noting that: "For those to whom unbroken wilderness is most important, the entire project is adverse because the original character of this corridor in northern Alaska would be lost forever."

The staff said the pipeline would be constructed under the "most stringent environmental and technical stipulations ever imposed upon industry for a project of this nature." Although the oil companies had planned to lay 90%–95% of the pipeline underground, the report said only 52% could definitely go underground without causing serious erosion through melting the permafrost, a situation that could cause breaks in the pipeline and severe oil spills. The staff also recommended crossings for wild animals, protection of fish spawning areas and air and water pollution controls.

Despite safeguards, the report said, "there is a probability that some oil spills will occur even under the most stringent enforcement." The study said oil companies would be required to bear the cost of repairing oil spill damage and that the possibility of major spills was "remote."

Alaska Gov. William A. Egan (D) and former Gov. Walter J. Hickel (R) indicated general approval of the staff study. Egan said Jan. 13 the study was a "sound move." Hickel, who had backed the pipeline before he had been ousted as interior secretary in 1970, said, "This is a great step forward. The public can ask questions now and get the answers." Oil company spokesmen suggested Jan. 14 that construction of the pipeline might cost closer to $2 billion, twice the original estimate.

Although some native spokesmen had accepted the pipeline project if it included guarantees that native groups would share in oil revenues, Joe Upicksoun, president of the Arctic Slope Native Association, said: "My people are not concerned about a phony economy like the Caucasians. We have our land and it has been good to us. The sea has been good to us. The Bureau of Indian Affairs has given us enough education to read about what has happened in the lower 48 [states], what they have done to our land. Now they want to come up here and rape our land."

Interior Department hearings—The Interior Department held a series of public hearings on the pipeline issue, in Washington Feb. 16 and 17 and in Anchorage, Alaska Feb. 24 and 25. Alaska Gov. William A. Egan testified Feb. 16 that his state would face bankruptcy by mid-1976 if the pipeline were not built. He said the project was essential if the state was to remain solvent and help the native Eskimos, Indians and Aleuts, who, he said, "live on a level of poverty below that of any of our other Americans."

Charles Edwardsen Jr., spokesman for the Arctic Slope Native Association, challenged Egan's claim that the pipeline would help the natives. Edwardsen said Feb. 16 the Eskimos depended on hunting, trapping and fishing, and he added, "we don't like welfare." Rep. Les Aspin (D, Wis.), testifying the same day, said the Interior Department report supporting the pipeline had practically ignored "the devastating impact" of the project on Alaskan natives. Aspin also charged that reports from dissenting Interior officials had been suppressed.

Another attack on the staff study came Feb. 17 from former Interior Secretary Stewart L. Udall, who called it a "flawed" document that "brings no credit upon the department." Udall said the study's release in January was "clearly coordinated with a major public relations campaign" by the pipeline company.

Hickel testified in favor of the pipeline at the Feb. 25 Anchorage hearing. Hickel said, "Alaskans deserve the opportunity to enrich the land they love and share its great natural wealth with the rest of the nation and the world."

Canadian developments—Testifying at the Interior Department hearing Feb. 16, David Anderson, a member of the Canadian parliament, called for studies on the potential environmental dangers to Canada's West Coast from tankers moving oil to the U.S. Emphasizing

that he was not speaking for the Canadian government, he suggested that the U.S. failure to consult with Canada about the risks of tanker collisions and oil spills could have serious repercussions on Canada-U.S. relations. Anderson disputed the grounds for the Interior Department study's rejection of an alternate pipeline route through Canada.

Canadian Energy, Mines and Resources Minister J.J. Greene had said Feb. 12 that a Canadian pipeline route for oil and natural gas would ultimately prove more economical for consumers of both countries than the Alaska pipeline. He also suggested that the Canadian route would be more favorable to the environment. Greene said the U.S. oil industry argument that the Alaska route was necessary to U.S. security was only a justification for their decision to build the pipeline. He asked, "Have they not heard of our Norad treaty—the defense agreement between the United States and Canada for mutual defense?"

Greene had met with Nixon Administration officials in Washington March 2 to press for a pipeline route through Canada's Northwest Territories and Prairie provinces. Canadian External Affairs Minister Mitchell Sharp told the Canadian House of Commons March 5 that U.S. Secretary of State William Rogers had agreed to consult with Canada about the pipeline project.

Morton defers decision—Interior Secretary Rogers C. B. Morton said Feb. 19 that he was "a long way" from approving the pipeline. Despite his department staff study, Morton said in testimony before a Senate Appropriations subcommittee that "a great many problems involved" in the project "have not been resolved." He said that Hickel had been under "tremendous pressures" to issue a permit to start the pipeline, but "I'm not under those pressures."

Morton said Alaska had "sold the oil leases too hastily" on the North Slope and "there is a real question whether this was the proper thing to have done." He said the issue was "fraught with emotion and tied up in the native claims" —referring to the pending settlement by

Congress of land claims of Alaska's native population.

Morton said any decisions would not be made on a "profit-loss factor inherent to any economic group. They will be determined on the national need." He added: "It's our responsibility to relate oil reserves on the North Slope to the actual energy requirements of this nation." If the reserves were needed, he said, "we still are going to do everything we can to protect the environment and I'm a long way from deciding that this pipeline is the way to do it."

Morton's stand was attacked Feb. 20 by Alaska House Speaker Gene Guess, a Democrat. Guess charged that "Alaska and Alaskans are being sacrificed on the political altar by the Nixon Administration" and said "Morton is attempting to return Alaska to territorial status."

Army Engineers statement—In a report released by Rep. Aspin March 6, the Army Corps of Engineers said environmental safeguards that the Interior Department said it would require for the pipeline "are too general to support the positive assurances . . . that ecological changes and pollution potential will be eliminated or minimized." Calling for further study, the report said the corps "cannot and will not abandon its regulatory responsibility" and that it would not assure the Interior Department or Alyeska that it would issue the necessary permits.

Pipeline rerouting urged—Nineteen members of Congress urged Aug. 26 that the proposed pipeline to transfer oil from Alaska's North Slope be built across Canada to a point nearer the Atlantic Coast. The 18 Democrats and one Republican, all from the East and Midwest, charged that the proposed pipeline route south to Alaska's port of Valdez would "only serve to increase the already existing disparity" between oil prices in the West and East.

In a letter to Interior Secretary Rogers C. B. Morton, the congressmen said East Coast customers already paid 60¢ more per barrel and Midwest customers 35¢ more for oil than West Coast customers. They said if the proposed route was completed, their constituents "would be paying $3.4 billion per year more."

(Standard Oil of Ohio officials said Sept. 8 that the estimated cost of the Trans-Alaska pipeline had more than doubled, from $900 million to $2.3 billion, since its original estimate.)

Rep. Lee Aspin (D, Wis.) told a committee of Canadian Parliament Dec. 11 that the U.S. Interior Department should await the completion of a $43 million Canadian study on an alternative trans-Canadian oil pipeline before issuing its final report on the trans-Alaska route.

In a series of questions made public in the Interior Department's "apparent failure" to consider oil and gas pipelines together, since a gas pipeline from Alaska across Canada seemed inevitable. He also said that the department should consider in its final report the economic advantages for consumers in the U.S. Midwest and on the east coast, which would result from a trans-Canadian line.

Alaska land bill signed. President Nixon Dec. 18 signed an Alaska native land settlement bill, after the Alaska Federation of Natives voted its approval 511–56. Congress passed the measure Dec. 14, the House 307–60, the Senate by voice vote.

The bill granted a total of $962.5 million, $462.5 million of that amount over 11 years from the federal government, the remainder from state mineral revenues at the rate of 2% a year. Forty million acres of land and mineral rights were to be selected by the natives, mostly surrounding their villages.

The secretary of the interior was authorized to designate 80 million acres for potential recreation and conservation use, subject to Congressional approval within five years, and to set aside an 800-mile corridor should he approve the proposed oil pipeline.

Critics of the bill included the Arctic Slope Native Association, representing 4,500 Eskimos from the North Slope area who were allowed only surface rights to their oil-rich lands, with compensating subsurface rights elsewhere.

Signing of the bill lifted the freeze on state land acquisition set during the previous Administration.

Canadian-U.S. gas search. Four U.S. natural gas distributors signed a $75 million agreement July 9 with Panarctic Oils Ltd. of Canada, under which the latter would receive funds to expand exploration of natural gas reserves in the Canadian Arctic.

Panarctic, a consortium in which the Canadian government held a 45% interest, announced the agreement with Tenneco Oil and Minerals, Ltd., Columbia Gas Systems, Inc., Texas Eastern Transmission Corp. and Northern Natural Gas. Co. The four U.S. companies would get first opportunity to negotiate gas-export agreements with Panarctic. They would also have the right to contract for gas in proportion to their participation in the project.

If substantial gas reserves were discovered, a pipeline would be needed across the barren Northwest Territories to the provinces and the U.S. border.

In September Panarctic confirmed the discovery of a major natural gas field on remote King Christian Island in the Canadian Arctic. Panarctic reported Dec. 15 its fourth discovery of natural gas in the Arctic islands; the new strike was on Ellef Ringnes Island.

Canada bars gas exports to U.S. The Canadian National Energy Board Nov. 19 rejected applications from six companies planning to export 2.66 trillion cubic feet of natural gas valued at approximately $1 billion to the U.S. The board said, "There is at present no surplus of natural gas remaining after due allowance has been made for the reasonably foreseeable requirements for use in Canada." Canadian natural gas currently accounted for about 3% of total U.S. consumption.

Although the applicants argued that untapped gas and oil resources in the Arctic and on the offshore continental shelf were assurance against any future shortage in Canada, their opponents felt that increasing sales of gas to the U.S. would force Canadian consumers to pay unnecessarily high prices for gas in Canada and would hamper efforts to clean up pollution caused by burning dirtier fuels.

Government sources discounted reports that the rejection of the permits was a ploy to improve Canada's bargaining position in current talks with the U.S. over its 10% surcharge.

The ban was the first indication of a reassessment trend undertaken by the government on Canada's energy policies. Only a year before, the energy board had approved long-term gas export contracts to the U.S.

U.S.-Canada energy board urged. In a speech before the American Petroleum Institute Nov. 16, U.S. Sen. Henry M. Jackson (D, Wash.) called for the establishment of a broad North American energy policy that would coordinate the development of fuel sources in the U.S. and Canada.

"I am convinced that our relationship with our Canadian neighbors on energy matters must be strengthened" he said, adding "it may be desirable to establish a joint Canadian-U.S. energy board . . . that would go beyond the traditional arguments about 'peril points,' leasing regulations, tax provisions and the whole gamut of specific industry, environmental and consumer concerns."

Jackson said such a board could deal with problems of mutual concern and exploit opportunities for mutual benefit in the energy field.

Prime Minister Pierre Elliott Trudeau, however, had told the House of Commons Oct. 4 that Canada was preoccupied with other pressing matters and "is not in a position" to continue energy talks with the U.S. The energy talks had begun in 1970.

Home Oil to remain Canadian. The purchase of controlling interest in Home Oil Ltd. of Calgary—the last major Canadian-controlled oil company—by Consumers' Gas Ltd. of Toronto was announced by Energy Minister J. J. Greene in the House of Commons April 22. The agreement achieved the federal government's aim to keep Home Oil in Canadian hands.

The Toronto Globe and Mail reported April 23 that Consumers' Gas would pay $17.6 million for 50.3% of the voting shares of Cygnus Corp., an investment company which controlled Home Oil. R. A. Brown, Jr., president of Cygnus and Home Oil, would continue as chief executive officer of the two companies.

Presssure on the government to intervene in the sale of Home Oil built up in

February after reports that Ashland Oil Inc. of Kentucky was negotiating for purchase of Brown's shares in Cygnus. Critics of foreign ownership of Canada's resources urged the Canadian government to acquire control of Cygnus, if necessary. After exploring that avenue, Greene had announced March 23 that private buyers had re-entered the negotiations. (Consumers' Gas was more than 97% owned by Canadian shareholders.)

Oil import quota raised. President Nixon Dec. 21 authorized an increase in the 1972 quota of oil imports of 100,000 barrels a day, to a total of 1,550,000 barrels

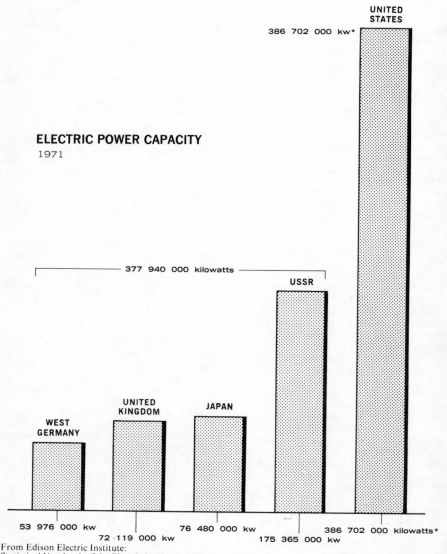

ELECTRIC POWER CAPACITY
1971

UNITED STATES
386 702 000 kw*

377 940 000 kilowatts

USSR

WEST GERMANY

UNITED KINGDOM

JAPAN

53 976 000 kw

72 119 000 kw

76 480 000 kw

175 365 000 kw

386 702 000 kilowatts*

From Edison Electric Institute:
Statistical Yearbook of the Electric Utility Industry for 1972

*Including Alaska and Hawaii.

a day, for states east of the Rocky Mountains.

Two thirds of the increase was allocated to Canada and the announcement said discussions were "proceeding" with Canada for an agreement to permit Canadian crude oil to enter the U.S. without quantitative curbs.

The oil import quota for states west of the Rockies remained unchanged at about 400,000 barrels a day.

Industry view of oil shortage. An assertion that government policy was a major cause of approaching U.S. oil and gas shortages was made by economist Richard Gonzalez, regarded as an oil industry spokesman, in a speech made at the Governors Conference on Meeting Energy Needs. According to the speech's text, inserted in the Congressional Record by Rep. J. J. Pickle (D, Tex.) Nov. 9:

The problems that may cause this nation to find itself short of oil and gas at times during the next decade arise from governmental policies and actions. The principal problems are that federal regulation of the price of gas and oil has discouraged new investments and that environmental regulations will adversely affect the supply and cost of new oil and gas.

Investors have been discouraged from expanding outlays for new oil and gas in the U.S. by the following developments:

1) Regulation of the price of natural gas by the Federal Power Commission at artificially low levels in relation to rising costs and to additional demands created for gas by air pollution regulations.

2) Declining real prices for crude oil under federal "surveillance" of prices under oil import controls adopted for reasons of national security and intended to encourage rather than discourage development of increased domestic capacity.

3) Liberalization of oil imports under continuing political pressures from consuming areas without regard for the consequences on supplies of attractively priced domestic natural gas, for the risks incident to dependence on insecure overseas oil, and for national security.

4) The recommendation of a Cabinet Task Force, fortunately not adopted, that oil import controls be changed in a manner designed to cut the price of crude oil by 30 cents in 1970 and up to 80 cents by 1975.

5) The Tax Reform Act which cut percentage depletion and raised taxes by an amount equal to 5 percent of the gross value of oil and gas, thereby creating fears about further cuts which may be made later. This added risk could have been avoided if the cuts had applied only to production from wells drilled after the Tax Reform Act was adopted.

Environmental regulations have also had an effect on the supply of oil and gas, and will continue to influence costs as well as supply. The most promising areas for development of large new domestic supplies of petroleum at reasonable costs are on the Continental Shelf and in Alaska, where the environmental problems encountered have served to delay drilling and construction of transportation facilities. . . .

Power reserves up. The Federal Power Commission (FPC) Nov. 26 said the nation's electrical power generating capacity was about 27% higher than expected winter needs, above the 20% generally accepted safety margin, but warned that reserves were low in the East Central and Pacific Northwest areas.

Although winter demand had risen 7.4% over 1970, capacity had increased 8.4% to 351,317 megawatts.

The Nixon Administration had announced the previous week a $24.8 million grant for an experimental coal gassification project by the National Coal Association, in the face of reported natural gas shortages.

Oil industry dominates energy? Rep. Robert W. Kastenmeier (D, Wis.) submitted to the House Subcommittee on Special Small Business Problems July 12 a statement amplifying his charges that oil companies were increasingly dominating the coal and uranium industries. According to Kastenmeier's statement:

The oil industry always has dominated natural gas, but the feeding of its insatiable economic appetite through the penetration of other competing fuels, particularly coal, was controlled until the mid-1960's. In September 1963, Gulf Oil acquired Pittsburgh and Midway, a coal company that accounts for about 2 percent of national production. Then, October 13, 1965, Continental Oil Company announced a major breakthrough; an agreement in principle to buy Consolidation Coal, the nation's largest producer which alone accounts for about 12 percent of total coal production. The merger was formally completed in October 1966.

Presently, of the nation's largest 25 oil corporations, at least 12 have holdings in coal and 18 have uranium interests. These include Standard Oil of New Jersey, Texaco, Gulf, Mobil, Standard Oil of Indiana, Shell, Atlantic Richfield, Phillips Petroleum, Continental Oil, Sun Oil, Union Oil of California, Occidental Petroleum, Cities Service,

Getty, Standard Oil of Ohio, Pennzoil United, Inc., Marathon, Amerada-Hess, Ashland, and Kerr-McGee. . . .

It is estimated that there are more than 4,000 producing companies in the coal industry, but many of the most important ones are under the control of the largest producers and virtually all of the remainder are insignificant. The industry has, therefore, become much more highly concentrated and now is clearly dominated by a relatively few large companies. . . .

Four of the nation's largest coal operations now are oil company subsidiaries and these four firms, in 1969, accounted for approximately 23 percent of the country's coal output. A listing of these oil companies, along with their subsidiaries and percentage of the nation's coal production follows:

CONTINENTAL OIL

Consolidation Coal Company, 12.4 percent:
Pittsburgh Coal Co.
Mountaineer Coal Co.
Christopher Coal Co.
Mathies Coal Co.
Hanna Coal Co.
Harmar Coal Co.
Pocahontas Fuel Co.
Tennessee Division.
Bishop Coal Co.
Truax-Traer Coal Co.
Western Division.
Itmann Coal Co.
Ohio Valley Division.
Rowland Division.
Blacksville Division.

OCCIDENTAL PETROLEUM

Island Creek Coal Company, 6.8 percent:
Island Creek Division (W. Va.).
Island Creek Division (E. Ky.).
West Kentucky Division.
Virginia Pocahontas Division (E. Ky.).
Virginia Pocahontas Division (Va.).
Beatrice Pocahontas Co.
National Coal Mining Co.
Northern Division (Pa.).
Northern Division (W. Va.).
Northern Division (Ohio).
Maust Coal and Coke.

STANDARD OIL OF OHIO—2.1 PERCENT

Old Ben Coal Corp. (Ill.).
Coal Processing Corp. (Va.).
Kings Stations Coal Corp. (Ind.).
Old Ben Coal Corp. (Ind.).

GULF OIL CORPORATION—1.9 PERCENT

Pittsburgh and Midway Coal Mining Co.

Studies indicate that the 13 largest coal companies now have about 53 percent of the coal market. Four of these 13 are oil companies, Consolidated Coal (Continental Oil), Island Creek (Occidental Petroleum), Old Ben Coal Corp. (Standard Oil of Ohio) and Pittsburgh and Midway Coal (Gulf Oil). . . .

Attempting to measure coal reserve ownership by individual oil firms is difficult. . . .

However, there is some scattered information that gives us a good picture of the dynamic movement of oil into the coal industry. A 1967 study done for the Federal Trade Commission involving Atlantic-Richfield, Continental Oil, Gulf Oil, Humble Oil and Refining (New Jersey Standard Oil) and Sinclair Oil showed that the combined holdings of these 5 companies amounted to 10,-188.7 million tons of recoverable coal reserves and 2,491,000 acres of coal lands. Compared to 1960, only one of these 5 companies had any holdings, and this amounted to only 7.8 million tons of recoverable coal reserves and 4,524 acres of coal lands. A June 1, 1971 Forbes Magazine interview with Myron Wright, head of Humble Oil, indicated that Humble's coal reserves probably are the largest in the United States. . . .

According to the AEC, the following oil companies are active in uranium exploration: Amerada Petroleum Corp., APCO, Atlantic Richfield, Cabot Corporation, Continental Oil, Earth Resources, Getty Oil Company, Gulf Oil Company, Houston Oil and Mineral Corp., Humble Oil and Refining Company, Kerr-McGee Corp., King Resources, Louisiana Land and Exploration, Magna Oil Company, Mobil Oil Corporation, Orion Oil Company, Pan American Petroleum Corp., Petrotomics Co. (Getty Oil and Continental Oil), Phillips Petroleum Co., and Tenneco Oil Company.

The AEC report on the nuclear industry in 1970 stated that 20 oil companies accounted for 31 percent of the surface drilling in uranium. Kerr-McGee is the largest single producer of uranium in the U.S., accounting for 27 percent of domestic uranium capacity. Other companies are moving aggressively into the uranium field. Atlantic-Richfield owns 544,000 net acres of uranium rights in the U.S., and has rights to explore for uranium on approximately 770,000 acres in Canada. In June 1970, Gulf Oil announced plans for a $50 million uranium development program in Saskatchewan, Canada. . . .

Oil motive in Vietnam? Critics of the Nixon Administration suggested that the U.S. was fighting in Vietnam so that U.S. oil companies could have access to billions of dollars worth of crude oil in Southeast Asia's continental shelf.

Barry Weisberg asserted in the Nation March 8 that "this continental shelf is the largest in the world" and "appears to hold the largest deposits of petroleum."

Prof. Gabriel Kolko of York University (Toronto) reported in the New Republic March 13 that Chairman David Rockefeller of Chase Manhattan Bank had estimated 11 months previously that "the predominantly American international oil firms would spend $35 billion in Asia and the Western Pacific, mainly Southeast Asia, over the next 12 years." According to Kolko, "U.S. explorations off Vietnam continued, and estimates of immense future output in the region became more and more common." Kolko quoted the editor of Petroleum Engineer as predicting that, if the Vietnam war ended favorably, the region might be the site of "one of the biggest booms in the industry's history."

The trade journal World Oil was quoted widely as having estimated in its Aug. 15, 1970 issue that the Asian continental shelf would produce 400 million barrels of oil daily in five years. But World Oil admitted in March 1971 that this was an extreme overestimate that had been picked up from a probably incorrect UPI story carried in the Houston Chronicle March 8, 1970.

Environmental Issue

President's message. Among proposals made by President Nixon in an environmental message to Congress Feb. 8: (a) federal guidelines for state regulation of strip and underground mining; (b) establishment of a single agency in each state or region for supervision of the location and development of power supply and to halt the "growing number' of confrontations" between power suppliers and environmentalists.

EPA sets clean air standards. The Environmental Protection Agency (EPA) announced "tough" national air standards April 30 that would limit permissible amounts of six major air pollutants by 1975. The standards, described by EPA Administrator William D. Ruckelshaus, were issued under provisions of the 1970 Clean Air Act.

To satisfy the standards, Ruckelshaus said, there would have to be "drastic" changes in industrial and electric power practices and in "commuting habits" of persons living near large metropolitan areas. He said many electric generating plants would have to switch from high-sulphur coal to "clean" fuel and warned that the consumer's electric bills might go up.

Under the 1970 Clean Air Act, states had until Jan. 1, 1972 to submit plans for meeting the standards for the six pollutants. The EPA would then either accept the proposals by May 1, 1972 or impose its own plans by July 1 of that year. Then states would have until July 1, 1975 to implement the standards.

The EPA Dec. 21 announced final air pollution standards for new or modified plants in five major industries.

Plants under the new controls included fossil-fueled steam generators (mostly for electric utilities).

In the face of power industry opposition, an EPA spokesman claimed the typical annual family electric bill of $132 would increase by only $1 in affected areas.

Existing plants were to be regulated by state programs designed to meet federal air quality standards.

EPA Administrator William D. Ruckelshaus said that new-plant standards would be set for 18–20 industries in 1972, with a total of 35–40 in the next few years.

Western governors fight dams. The governors of Idaho, Oregon and Washington joined conservationists, according to an Aug. 1 New York Times report, to prevent construction of two hydroelectric dams on the Snake River in Hells Canyon. The three governors opposed the projects in a letter to the Federal Power Commission (FPC).

The letter, drafted by Gov. Cecil D. Andrus (D) of Idaho, said, "It is time we recognized that Hells Canyon is truly a unique and magnificent national treasure." The letter was signed by Govs. Tom McCall of Oregon and Daniel J. Evans of Washington, both Republicans.

The New York Times reported another controversy Aug. 8 over a proposed dam on the Teton River, a tributary of the Snake in Idaho. The project was opposed by the regional office of the Environmental Protection Agency, the Idaho Fish and Game Commission and conservation groups. However, an environmental impact statement by a bureau of the Interior Department concluded that "the benefits outweigh the losses."

Carolina dam controversy—The Conservation Council of North Carolina, ECOS Inc. and three North Carolina residents filed suit Aug. 10 in federal court in Greensboro to halt further construction of the $40 million New Hope Dam in Chatham County. The plaintiffs charged that the project would destroy wildlife and woodland and cause the loss of $1 million in farm and timber revenues.

The suit said the dam would create a permanent lake that would be immediately polluted so as to be unusable for recreation or a water source, which along with flood control were the goals of the project. The plaintiffs cited 29 sources of industrial and muncipal waste discharge in rivers upstream from the dam. The Army Corps of Engineers was scheduled to complete the project in 1973.

Change of venue for power suit—U.S. District Court Judge John H. Pratt granted a motion Aug. 17 for a change of venue from Washington to Arizona in a suit brought by the National Wildlife Federation and the Environmental Defense Fund to halt construction of a power plant complex to serve Utah, Arizona, New Mexico and Colorado. The court accepted the argument of Interior Secretary Rogers C. B. Morton and 11 intervening electric and coal companies that the availability of evidence and convenience of witnesses would be better served by Arizona federal courts.

The environmental groups and American Indian organizations, which filed parallel actions against the Four Corners complex, had opposed the shift because public interest attorneys were available to work on the suit in Washington and because they feared the Arizona courts would be less sympathetic to the standing of citizen group plaintiffs.

Suit dismissed—A federal district court judge in Washington, D.C. dismissed a suit Oct. 6 by the Environmental Defense Fund and other groups trying to prevent construction of the Tellico Dam project in eastern Tennessee by the Tennessee Valley Authority. The suit opposed the project for environmental reasons and because it would flood the ancestral home of the Cherokee Indians. The judge ruled that the suit should not be filed in the District of Columbia.

Thermal pollution suit settled. The Justice Department announced settlement Sept. 1 of the first federal suit involving thermal pollution. The department had filed suit in 1970 charging the Florida Power and Light Co. with disrupting the ecology of Biscayne Bay by discharge of heated water.

Attorney General John N. Mitchell announced a consent decree filed by the government under which the utility agreed to build a $30 million system to insure that discharged water would be compatible with the bay water in salinity and temperature. The company used the bay water to cool its generators.

After a hearing Sept. 9, U.S. District Court Judge C. Clyde Atkins told both sides he agreed with the major points in the settlement and would grant approval with adjustments in the wording of the decree.

Florida to curb thermal pollution. The Florida pollution control board Nov. 22 proposed stiff curbs on hot water discharges from electricity generating plants.

The rules would bar discharges that raise temperatures in surrounding waters more than four degrees in winter and 1.5 degrees in summer.

Oil drilling permits denied. Interior Secretary Rogers C. B. Morton announced Sept. 20 that the U.S. had denied permits for two new oil drilling platforms in the Santa Barbara Channel, scene of a massive oil leak in 1969 that caused the temporary suspension of offshore oil lease grants throughout the country. Morton said the permits were denied for "overriding environmental considerations."

The new oil platforms had been proposed by the Union Oil Co. and the Sun Oil Co. The Union platform would have been on the lease where the 1969 blowout occurred.

In an environmental impact statement reported Sept. 2, the Interior Department had proposed resumption of oil drilling off Santa Barbara, saying that "steps have been taken to minimize the possibility" of major oil spills resulting from accidents. However, the Environmental Protection Agency had said the statement did not adequately assess the risk of spills.

In denying the permits, Morton said the new platforms would be "incompatible" with a 1970 proposal by President Nixon to buy back 35 of the existing 70 leaseholds off the Santa Barbara coast to establish a federal marine sanctuary. However, he indicated that the announcement did not represent a change in policy on selling oil leases. He said the Santa Barbara Channel "is the most unique environmental situation on the entire outer continental shelf."

■ Five oil companies were fined a total of $792,000 in U.S. District Court in New Orleans for failing to install safety valves in Gulf of Mexico oil wells. The following companies were fined after pleading "no contest" to the charges: Tenneco Oil Co., $32,000, Jan. 20; Gulf Oil Corp., $250,000, Jan. 21; Kerr-McGee Corp., $20,000, Jan. 21; Mobil Oil Corp., $150,000, Jan. 28; and Shell Oil Co., $340,000, Feb. 24. Shell had changed its plea from not guilty. None of the charges against Shell involved its oil platform that caught fire Dec. 1, 1970 and was still burning out of control.

■ Interior Secretary Rogers C. B. Morton announced Feb. 16 that the government had refused to grant oil and gas drilling rights in the Los Padres National Forest in California. Morton said he had refused the request of U.S. Royalty Oil Corp. because the area was the nesting ground of about 75% of the remaining 60 to 80 condors, birds with wing spreads of up to nine feet which were considered one of the nation's most endangered species.

Oil spills. Among incidents involving oil tanker spills:

■ Thirty-eight oil companies of the U.S., Japan and Europe signed a voluntary agreement in New York Jan. 14 to provide funds to clean up oil spills. The agreement would provide up to $30 million for any spill involving one of the participating companies. The funds would be used to pay claims beyond the liability of tanker owners, who under a 1969 agreement covering 80% of the world's tonnage undertook to pay up to

$10 million to clean up spills. The oil companies in the new accord received more than 80% of the crude oil and fuel oil transported in the world.

■ Hundreds of thousands of gallons of oil spread for 50 miles along the coast of California following an early-morning collision of two Standard Oil of California tankers Jan. 18. The tankers, maneuvering in a dense fog, collided near the Golden Gate Bridge in San Francisco Bay. A California Standard spokesman said Jan. 19 that the exact amount of oil in the spill had not been determined. He said it could not have exceeded 1,-134,000 gallons.

Half of the oil had drifted out into the ocean, but a high tide Jan. 19 carried much of it back to the coast. Hundreds of birds, covered with oil, were rescued from the bay, but it was believed that many more perished.

California Standard, responsible for cleaning up the spill, hired 250 workers to spread bales of hay to soak up oil on the shore, to man oil-skimming barges and other machinery to pick up oil-soaked birds.

■ A Standard Oil (N.J.) tanker ran aground in fog Jan. 23 at the mouth of the New Haven (Conn.) harbor. The tanker, Esso Gettysburg, hit a rock ledge near the harbor and spilled about 385,000 gallons of light fuel oil into Long Island Sound.

■ Two hundred thirty thousand gallons of oil from a Navy tanker Aug. 20 caused a 65-mile slick along the California coast. President Nixon's beach at the Western White House in San Clemente was one of the beaches smeared by oil in what the Navy acknowledged to be its worst West Coast oil spill. The spill, which was not reported until Aug. 22, occurred while the tanker Manatee was refueling the aircraft carrier Ticonderoga during war games.

■ A Swedish tanker unloading at a Standard Oil Co. wharf spilled 15,000–30,-000 gallons of crude oil into San Francisco Bay Sept. 17. A Richmond, Calif. official said Sept. 19 that probation violation charges were being filed against Standard Oil, which had been found guilty after pleading no contest to misdemeanor charges for two oil spills in July 1970.

TVA sued on strip-mine contracts. Three environmental groups filed suit in federal district court in New York City March 2 charging that the Tennessee Valley Authority (TVA) violated environmental protection legislation in contracting for strip-mine coal in eastern Kentucky and Tennessee. The suit sought to void $111 million worth of contracts with the surface mining industry, which the plaintiffs blamed for "the systematic destruction of mountains and countryside."

The suit was filed by the Natural Resources Defense Council Inc., the Environmental Defense Fund and the Sierra Club. The groups contended that the TVA "blatantly" violated the National Environmental Policy Act by failing to file "environmental impact" statements before contracting for strip-mine coal. About half the agency's coal, 15 million tons annually, came from surface mining.

TVA tightens strip rules. The Tennessee Valley Authority (TVA) Dec. 6 issued new rules tightening environmental requirements on strip (surface mining) coal purchase contracts, as part of its first annual environmental impact statement. But the TVA rejected demands that each of 30–40 annual strip mine contracts include impact statements.

Under the new rules, the TVA would prohibit mining by suppliers on mountain slopes with more than 28° inclines, "scenic" and "wilderness" areas, and areas where stripping would dangerously pollute streams or water supply systems. In addition, 2% of contract payments would be withheld until reclamation work was completed.

The TVA claimed that individual impact statements, which could take up to six months of study each, would make competitive bidding on coal purchasing "practically impossible."

In related developments:

■ Sen. Gaylord Nelson (D, Wis.) Nov. 6 called for a halt to federal leasing of Western coal lands pending environmental impact statements. He charged the Interior Department with "massive leasing" of the lands, rich in low-sulfur coal, with little or no environmental study.

Mining companies had been granted leases or prospecting permits for 2,390 square miles of federally owned lands, according to the Bureau of Land Management and the Bureau of Indian Affairs. The leases were acquired largely in anticipation of a massive gassification program to supply low-pollution fuel.

■ Russell E. Train, chairman of the President's Council on Environmental Quality, and John R. Quarles Jr., general counsel of the Environmental Protection Agency, backed a ban on strip mining that irreparably damaged the environment. They took the stand in Nov. 16 Senate committee testimony.

The Administration had proposed federal strip mine reclamation standards, to be enforced with variations by the states.

The Middle East & North Africa

Oil situation. James E. Akins, director of the Office of Fuels & Energy at the State Department, testified before the House Near East Subcommittee July 15 on the role of Middle East oil in the world. Among his remarks:

The main feature of the world oil picture is that three-fourths of the non-communist world's proven, recoverable reserves lie in the Middle East and North Africa. The main corollary is that Europe and Japan, which now get 90 percent of their supplies from this area, will remain dependent on it for almost all of their hydrocarbons and for much of their energy through 1980, and very likely through the remainder of the century. While the United States is in a more fortunate position with its domestic supplies and with synthetic oils which could be developed, it too could very likely be forced to import one-half of its petroleum needs from the Middle East and North Africa by 1980.

According to recent estimates prepared by various industry sources the world oil reserves are roughly as follows:

[In billions of barrels]

Arab Middle East and North Africa	350
Iran	60
United States (including Alaska)	40
Venezuela	15
Indonesia	15
Canada	10
Other Western Hemisphere	10
Other African	5
Others	5
Total	510

Although this table shows that two-thirds of the non-communist world's reserves are Arab and 80 percent are in the Middle East and North Africa, the picture is in fact understated. The reserves from the Middle East include "only" 150 billion barrels for Saudi Arabia, whereas most authorities are convinced that the figure should be at least twice that high; and the figure for Iraq is only 30 billion whereas the actual figure is more probably closer to 75 billion barrels.

Much has been said about the necessity of looking for oil outside the Middle East and North Africa, with the implication that the size of these reserves reflects the amount of exploration being performed. In fact the opposite is true. Most companies operating in the Middle East and North Africa have already found reserves sufficient to meet their needs for many years to come and almost all exploratory activity today is being carried out in other areas: Indonesia, the North Sea, Canada, and Alaska, off the Coast of Africa—indeed wherever there are sedimentary basins.

Oil states get higher payments. A series of meetings punctuated by threats of a suspension of oil shipments resulted in agreements by Western oil companies to make large increases in payments to the oil countries. *Among the developments of 1971:*

Meetings open—The 10-member Organization of Petroleum Exporting Countries (OPEC) met with representatives of 17 Western oil companies in Teheran, Iran Jan. 12 and 21 to press the oil-country demands.

The OPEC had first aired its new demands for higher revenue at a joint meeting in Caracas, Venezuela Dec. 28, 1970. The conferees issued a manifesto that included a demand that the oil countries' average oil-income tax rate of 55% become standard and that all posted prices be raised to the highest level. The posted prices normally were higher than the actual market prices and were used primarily to compute oil tax and royalty payments to the host country.

The OPEC members were Iran, Iraq, Saudi Arabia, Kuwait, Qatar, Abu Dhabi, Libya, Algeria, Indonesia and Venezuela. They accounted for about 85% of the world's oil production outside the U.S. and the Soviet Union.

The firms represented at the Teheran talks included British Petroleum, Standard Oil Co. (New Jersey), Iranian Oil Participants Ltd. and other American, Dutch and French companies.

In a memorandum submitted to the OPEC Jan. 16, a group of 15 oil firms, including 12 American companies, proposed to negotiate "simultaneously" with the 10 organization members to seek a single five-year "overall endurable settlement" of their demands. Previous oil contracts had been arranged between a company and a single nation. The Persian Gulf producers—Kuwait, Saudi Arabia, Iran, Iraq, Abu Dhabi and Qatar—sought a regional settlement in line with a resolution approved at the Caracas meeting. The memorandum of the 15 firms also expressed opposition to further dealings on an individual basis with Libya, which had attempted to increase its terms beyond those granted to the Persian Gulf producers.

Signing the memorandum were British Petroleum Co., the French Petroleum Co., Gulf Oil Corp., Mobil Oil Corp., the Royal Dutch Shell Group, Standard Oil Co. of California, Standard Oil Co. (New Jersey), Texaco Inc., Continental Oil Co., Marathon Oil Co., Occidental Petroleum Corp., Amerada Hess Corp., Atlantic Richfield Co., W. R. Grace & Co. and Nelson Bunker Hunt, a Dallas oil man.

Shah Mohammed Riza Pahlevi of Iran warned Jan. 24 that if the current talks with the Western companies collapsed in the next nine days "the question of cutting off the flow of oil" to the Western consumer nations "will be definitely considered." The shah promised observance of any agreement between the international firms and the Persian Gulf producers, but he scoffed at proposals for a wider regional settlement as suggested Jan. 16. The differences in geography and marketing factors made an overall global settlement a bad idea, the Iranian leader said.

The oil companies' position had been denounced by members of the Iranian parliament Jan. 23. One deputy charged that "the systematic looting of oil-producing countries has created an explosive situation." Another said the companies were trying to "strangle the owners of the oil."

President Nixon Jan. 16 sent Undersecretary of State John N. Irwin 2d on a mission to Iran, Kuwait and Saudi Arabia to express Nixon's "friendly concern with the situation." Irwin arrived in Teheran Jan. 18 and conferred with the shah and Foreign Minister Ardeshir Zahedi.

Persian Gulf ultimatum—A round of negotiations between representatives of 22 oil companies and the Persian Gulf producing states collapsed in Teheran Feb. 2. The OPEC, to which the Gulf producers belonged, announced Feb. 4 that it had set a Feb. 15 deadline for a settlement; otherwise its members would unilaterally raise prices.

Although details of the talks were not disclosed, the Persian Gulf states were known to be demanding an increase totaling $12.3 billion in revenues over the next five years, while the companies had offered increases amounting to $6.9 billion for the same period. The oil companies had paid the 10 OPEC nations $6.2 billion in 1969.

The OPEC states had threatened to disrupt deliveries to the oil companies unless an agreement was reached by Feb. 3. But a decision to defer such action was taken at a ministerial meeting of the countries in Teheran Feb. 3–4. Major producers backed a suggestion by Shah Mohammed Riza Pahlevi of Iran Feb. 3 that legal or legislative measures be taken to realize their demands for increased revenue. The shah later told a news conference that if the oil companies accepted "our rightful demands," "obviously the OPEC ministers, and especially the six Gulf states, would be prepared to reopen negotiations."

The new president of the OPEC, Iranian Finance Minister Jamshid Amougezar, said Feb. 4, "There is no question of further negotiation. It is simply a question of whether they [the oil companies] accept our terms or not." Amougezar's statement was made at the concluding session of the OPEC ministers' meeting. He said the substance of resolutions adopted at the conference provided for "very strong sanctions" against the companies if they refused to accept legislation raising prices.

A threat to cut off oil shipments to the West was reiterated by OPEC representatives at their European headquarters in Vienna Feb. 7. A resolution warned that "appropriate measures, including a total embargo on shipments of crude oil and petroleum products," would result from a Western company refusal to honor legislation providing for higher oil prices.

Persian Gulf settlement—The 23 Western oil companies signed an agreement in Teheran Feb. 14 to pay the six oil-producing Persian Gulf states an additional $10 billion in revenue in the next five years. This would raise the countries' annual oil income from the current level of $4.4 billion to $7.4 billion in 1975.

The accord climaxed weeks of talks between the firms, 17 of them American-owned, and Iran, Iraq, Saudi Arabia, Kuwait, Abu Dhabi and Qatar. The pact provided increased payments by more than $1.2 billion in 1971, rising to $3 billion in 1975. The Gulf states reportedly had demanded $1.4 billion in increased payments in 1971, rising to $11.8 billion in 1975. The new annual increases were to start June 1, and were to be repeated at the beginning of 1973, 1974 and 1975.

The pact also included a fixed rate of 55% of the companies' net income, even if other oil producing nations gain higher rates, and an immediate boost of 35% a barrel in the posted prices for crude oil at Persian Gulf tanker terminals.

Companies & Arabs resume talks—Oil negotiations between Western companies and four Arab states began Feb. 24 in Tripoli, Libya. The Western companies sent a delegation of more than 20 executives from a dozen firms, headed by George T. Piercy, a senior vice-president of Standard Oil of New Jersey. In a meeting among themselves earlier Feb. 24, the four Arab states—Algeria, Libya, Iraq and Saudi Arabia—agreed to be represented by Libyan Deputy Premier Abdel Salam Jalloud. A communique issued after the meeting of Arab states declared that Libya had been empowered to undertake negotiations with individual oil companies and to report to the other three states in two weeks. The communiqué added: "If these negotiations do not lead to an agreement by the companies to accept the minimum demanded and agreed by the four, then the ministers will meet again in Tripoli to take necessary measures to be applied, including the stoppage of pumping."

The collective negotiations resumed in Libya March 1 despite the apparent Arab insistence on individual negotiations.

Algeria seizes French oil firms — President Houari Boumedienne announced Feb. 24 that his government was taking control of 51% of the shares of all French oil companies in Algeria as well as nationalizing the companies' assets in natural gas and gas pipelines.

Speaking in Algiers at a meeting to mark the 15th anniversary of the General Union of Algerian . Workers, Boumedienne said Algeria would "make compensations" for the assets.

The French government Feb. 25 declared that the nationalization, although done during Franco-Algerian talks, were not a surprise and that France's oil supply "poses no problems." Le Monde of Paris reported Feb. 25: "For several weeks, the tactic of the French authorities has consisted of stalling for time, dragging out negotiations they could not control, hoping to improve their positions through the conference of Teheran, then that of Tripoli, as if the stake was limited to oil taxes or the price of a barrel of oil."

The French statement called on Algeria to specify the "just and equitable" indemnity Boumedienne promised; it also warned that relations between the two countries would undergo "profound change" as a result of the nationalizations.

Algeria Feb. 26 rejected a request made the previous day by Elf-Erap, the French state-owned oil company in Algeria, for 100% nationalization of its interests. The request had reportedly been made in order to insure compensation for the firm's entire investment.

The Algerian government June 30 signed an agreement providing for compensation to one of the nationalized French oil companies.

The accord, resulting from talks begun May 25, was signed for Algeria by Sonatrach, the national oil trust. It gave $50 million to the Compagnie Francaise des Petroles Algerie, a group in which the French government owned 35% of the stock. The company agreed to settle back taxes estimated at $27 million and to invest most of its earnings in Algeria over the next 10 years.

A similar agreement between Algeria and Elf-Erap, the other nationalized French oil company, was signed Dec. 15.

The accord was signed in Algiers by Michel Tenaille, president of Sofretal, an Elf-Erap subsidiary, and Ahmed Ghozali, president of the Algerian national oil company Sonatrach. Tenaille remarked: "The fight is over and I am optimistic about the future." Ghozali said the agreement "definitely settles all differences and clears the way for an association between Elf-Erap and Sonatrach under the new Algerian oil legislation. It provides effective state controls of the oil industry."

The 10-year pact provided that Elf's demands for compensation of about $35 million would be canceled out by taxes owed the Algerian treasury. The government's claims on payment of $80 million in debts would be met by Elf's concession of some of its private holdings, according to the New York Times Dec. 16.

Henceforth two new firms, Elf-Algerie and Total-Algerie, were to have sole management of French oil interests. Disputes were to be settled by Algerian courts.

Algeria raises oil prices — The Algerian government April 13 established the world's highest posted price per barrel of crude oil.

Algeria's oil price increase was announced at an agricultural gathering by President Houari Boumedienne, who set the new cost of a barrel of crude oil at $3.60. The old price was $2.85. The increase was retroactive to March 20.

Boumedienne also declared that French oil companies would be offered $100 million as compensation for the recent nationalization move.

Herve Alphand, secretary general of the French Foreign Ministry, handed Algerian Foreign Minister Abdelaziz Bouteflika the French note April 15 declaring the oil negotiations over. The note halted the special government-to-government talks on "all aspects of cooperation between the two countries" and said that a solution would now be up to "the competent Algerian authorities and the companies operating in Algeria."

French oil companies April 22 suspended the purchase of Algerian crude oil "for the time being." The Algerian cabinet said the same day that the government had given the companies "economic and legal guarantees" for their operation.

Responding to the Algerian statement April 27, French Foreign Minister Maurice Schumann told the French Senate that Algerian nationalization of French oil holdings appeared to leave the companies with "the simple role of money-lender, without any real industrial and commercial responsibility."

Libyan price accord—After nearly six weeks of negotiations in Tripoli, the Libyan government and 25 Western companies April 2 signed a five-year agreement increasing the posted price per barrel of Libyan oil from $2.55 to $3.45.

In announcing the pact, Libyan Deputy Premier Abdel Salam Jalloud said the new price per barrel "includes $3.20 as a permanent price plus 25¢ at variable freight premium." The agreement was expected to increase the Libyan government's annual oil revenue by approximately $620 million.

Jalloud also said the Western companies had "responded to an increase in the tax rate from 50% to 55% with the pledge that they pay the price difference from 1965 to September 1970 on the same basis and manner as agreed upon last September." The companies agreed to an annual inflation allowance of 2.5% of the posted price per barrel plus 7¢. The percentage increase plus 5¢ was to become effective from March 20, the date of the agreement. The remaining 2¢ increase was to be levied from the beginning of 1972 through 1975.

"The companies also pledged to undertake exploration, drilling and other investments in the oil sector in a specific way and satisfactory manner until 1975," Jalloud said.

Within weeks Jalloud specified April 24 that "the only guarantee" for the pact "lies in investment." Jalloud added: "If the companies respond and invest, we shall be prepared to cooperate with them.

If they do not invest and continue only to exhaust present wells, however, the agreement will not go on."

The Libyan government had threatened March 15 and 29 to block the flow of oil supplies unless the negotiating companies submitted to its demands. Col. Muammar el-Qaddafi, leader of the country's military government, said March 29: "The revolution cannot allow the companies to steal even one penny from the people."

The government nationalized all holdings of the British Petroleum Co., Ltd. Dec. 7 because of what it called Britain's "conspiracy" in Iran's occupation of three Persian Gulf islands.

The announcement, broadcast over Libyan radio, said the country was also withdrawing some $1 billion deposited in British banks. BP, owned 49% by the British government, was to be renamed the Arab Gulf Exploration Company.

Compensation for BP operations in Libya, estimated at $150 million, was to be arranged by a panel containing only Libyan representatives, according to a Dec. 8 report.

Iraq—Iraq was to receive about $990 million in oil revenue in 1971 (a rise from $549 million in 1970) under an agreement concluded with the Iraq Petroleum Co. June 7. The firm was owned by British Petroleum, Royal Dutch Shell, Compagnie Francaise des Petroles and Standard Oil of New Jersey. Under the agreement: Iraq Petroleum was to give the Baghdad government a new loan of about $30 million. Government payment to the company of a previous loan of $60 million would be postponed four years. Iraq's posted price was increased 80¢ a barrel, bringing the price to $3.21.

Syria—Under an accord announced by Damascus radio July 6, the Iraq Petroleum Co. was to increase its payments to Syria by more than 50% for the right to pump oil from Iraq across Syria to the Mediterranean. The royalties were to be increased from $54 million annually to $83 million. The company also agreed to pump more oil through the Syrian pipeline, increasing royalty payments to

Damascus by more than $86 million by 1973.

Libyan gas pact. —The government signed an agreement March 4 with Esso Libya for the export to Spain of liquified natural gas from the company's plant at Marsa el Brega.

The agreement with Esso Libya, subsidiary of Standard Oil of New Jersey, ended a dispute over finances that had shut the $350 million plant since June 1970. The accord provided for the export in Esso tankers of 110 million cubic feet of liquified natural gas a day and for transfer for domestic use by the state-owned Libyan National oil company of 200 million cubic feet of the gas a day. Libya was to receive 34¢ per million British thermal units as compared with the old price of 20.6¢ per million BTUs.

Israel-bound tanker attacked. An Israeli-chartered tanker bound for the Israeli port of Elath was shelled from an unmarked launch as it was sailing into the Red Sea June 12. The Popular Front for the Liberation of Palestine (PFLP) claimed credit for the attack.

Three bazooka shells hit the tanker, ripping holes in its side and causing a number of fires which were quickly extinguished. No injuries were reported. The vessel, the Coral Sea, flying the Liberian flag, made its way to Elath June 15 without further incident. It was carrying 70,000 tons of crude oil from Iran that was to be piped from Elath to Ashkelon for transshipment to Europe. The pipeline had been built as an alternative oil route to Western Europe following the closing of the Suez Canal in the 1967 war.

The Coral Sea had entered the 18-mile-long Strait of Bab el Mandeb at the southern end of the Red Sea when one of several boats in the area, with three armed men aboard, approached the tanker and started firing.

A statement issued by the Popular Front for the Liberation of Palestine in Beirut June 13 said the attack on the Coral Sea was only a "warning blow" and part of its campaign to strike at Israeli interests "regardless of where they may be and at the interests of those who help the Zionist enemy." The

statement said the shelling was aimed at the "alliance between the Israeli enemy and the Iranian reactionaries."

Premier Golda Meir reported to the Israeli Cabinet on the incident June 13. She was quoted as saying that Israel expected maritime nations to take measures to prevent attacks on civilian targets in international waters.

Defense Minister Moshe Dayan said June 14 that Israel was taking steps to make certain that oil tankers bound for Elath reached their destination safely.

Other Oil & Gas Producers

Great Britain. Among the year's developments involving Great Britain:

■ The government issued a license to the Soviet-controlled Nafta oil group for the importation of Russian crude oil, according to a London Times report Feb. 3. The license relaxed—but did not end —a 30-year ban on imports of Russian crude oil. British Petroleum had an agreement with Nafta, which served around 400 filling stations in England, to refine 150,000 tons of oil.

■ The Wall Street Journal reported Feb. 22 that Shell-Mex & B.P. Ltd., the marketing group for the British Petroleum and Shell groups, raised prices on gasoline, fuel, diesel and heating oils by 1.2¢ an imperial gallon. The increases followed agreement by 22 international oil companies to pay more revenue and taxes to the six oil-producing Persian Gulf states.

■ The Continental Oil Co. of the U.S., and the Burmah Oil Co., Ltd., of Britain, Feb. 25 announced negotiations for a merger that would create the eighth largest international oil company in terms of assets. Continental had assets of $3 billion and Burmah had assets of $2 billion. Burmah's assets included 23% of the common stock of the British Petroleum Co., Ltd. which in turn was 49% owned by the British government.

■ The House of Commons April 8 approved an emergency amendment authorizing the government to seize or sink any oil tanker threatening to pollute Britain's shores, whether inside or outside its territorial waters. The House of Lords approved the amendment shortly afterward. Approval followed an oil spill in the English Channel attributed to the Liberian-flag tanker Panther, which had gone aground the week before on the Goodwin Sands off southeastern England, located outside Britain's three mile territorial waters. The tanker was refloated and allegedly spilled oil on several beaches in Kent.

North Sea oil & gas—Shell Oil Co., in partnership with Esso, announced April 21 that oil had been found in the group's second well drilled in the British sector of the North Sea, 160 miles off the coast of Scotland. The discovery was the fourth oil strike in the British sector and raised the total number in the North Sea, including those of Norway, Holland and Denmark, to 11 in little over a year.

(Shell announced April 21 an oil strike 13 miles off the coast of Spain, in the River Ebro delta. The Spanish government said the same day that the three wells drilled so far indicated reserves of 40 million tons. Commercial production at the rate of 50,000 barrels a day was expected to begin in 1973.)

■ The Gas Council announced April 13 that it had reached agreement to pay nearly 36¢ for each 1,000 cubic feet of gas produced by a joint venture in the North Sea by the Continental Oil Co. (U.S.) and the National Coal Board. The price was 25% higher than that normally paid to other North Sea gas producers and was thought to serve as an incentive to other producers.)

■ The government announced April 14 a decision to extend by a further 31,-000 square miles the 168,000 square mile area of Britain's continental shelf. The areas involved were off southwest England and northwest Scotland. At the same time the Department of Trade and Industry opened the regions for petroleum exploration licenses.

■ Industry Minister John Eden announced Dec. 22 that 47 companies formed into 11 groupings had been awarded oil exploration licenses for the gas-bearing southern basin of the North Sea. The successful bidders included Shell, Esso, British Petroleum, Continental Oil, Amoco and Norway's Norsk Hydro.

Nigeria. The government reached agreement with Western companies April 22 on increasing the posted price of Nigerian crude oil from $2.36 a barrel to $3.21.

The accord, announced in Lagos and dated from March 20, chiefly affected Shell-BP (British Petroleum). It included a 2½% escalation clause to cover inflation and an annual increase of 5¢ a barrel beginning in January 1973.

In a related development, the government said April 22 that Safrap, a subsidiary of the French state-owned company Elf-Erap, was to resume operations, with Nigeria taking an immediate 35% participating interest.The company had been idle since the beginning of the civil war.

Peru. The government-owned State Petroleum Corp. (Petroperu) and the U.S.-owned Occidental Petroleum Corp. (Oxy) contracted to search for oil in the upper Amazon basin, according to a July 19 report.

The concession, which would encompass an area of three million acres, was to run for 35 years and was technically extended to Petroperu. However, Occidental committed itself to invest $50 million in a seven-year program. Under the unique arrangement, Peru would exempt Occidental from paying royalties or cash taxes. All oil found would go to Petroperu, which would then pay Occidental, in crude oil, for its 50% share of the production of the well-head. The local company would process the oil, while the cost of laying a pipeline would be shared by both partners.

Petroperu contracted with Tenneco Oil Co. and Union Oil Co. of California Sept. 20 for exploration and development of Peru's continental shelf on similar terms. The contract called for the American companies to spend four years in exploration and 26 years in development of any oil deposits found.

Petroperu Nov. 19 signed a 35-year oil exploration and production contract with British Petroleum Co. Ltd.

Petroperu Nov. 17 had announced the discovery of oil in the western reaches of the Amazon basin. Production was estimated at 1,000–3,000 barrels a day.

Venezuela. Bills to nationalize Venezuela's petroleum and natural gas industries became law during 1971.

The hydrocarbons (oil) reversion law was passed by the Chamber of Deputies June 18 and the Senate July 19 and signed by President Rafael Caldera July 30. The measure was to lead to nationalization of the foreign-owned oil industry and legal national ownership of oil installations and equipment when the 40-year concessions expire in 1983–84.

Under terms of the legislation, first introduced March 29, foreign oil companies would be placed under state direction until the concessions expire. It would give the government immediate authority to inspect drilling equipment, pipelines and refineries. The state would have the sole right to grant permission for moving equipment from one oil site to another and could determine when and where new wells would be drilled to develop new oil reserves.

The hydrocarbons law required oil companies to place between $500 million and $1 billion in a central bank fund to guarantee reversion of the properties to the government and to insure the maintenance of the properties. The bill would also require government authorization for any repatriation of profits and dividends by the oil companies to their parent firms.

The bill had the support of the ruling COPEI party and all opposition groups in Congress. Senate and presidental approval were expected, according to a June 20 report. Other pending legislation would give the state oil company (CVP) exclusive domestic marketing rights for petroleum by-products, now shared by foreign oil companies.

The bill had been criticized by members of the foreign-owned oil industry. Robert N. Dolph, president of the Creole Petroleum Co., said in a speech to the Venezuelan Congress, reported June 19, that "the proposed law means de facto nationalization and could lend itself to unworkable absurdities in its interpretation."

Richard Irving, director of Royal Dutch Shell of Venezuela said, according to a June 22 report, that the law would spur the U.S. to find other sources of fuel, inhibit development of new reserves in Venezuela and would mean the loss of present markets to more economical producers in the Middle East.

Caldera indicated optimism over upcoming talks with the U.S. over import quotas. In a press conference June 10, Caldera said "the United States has a greater interest in acquiring Venezuelan oil than ever before." He asserted that "considering the shortage of energy in North America, they need us more than we need them."

In a related development, Reuters reported July 21 that oil workers' unions accused Creole Oil Co., a subsidiary of Standard Oil of New Jersey, and the Shell Oil Co. of closing more than 400 wells in the past weeks in an effort to hinder plans to nationalize oil equipment. The accusations came hours after the Senate approved the oil reversion bill. One union representative said 2,000 men would lose their jobs if the present rate of reduction continued.

Oil company spokesmen said the slowdowns were the result of repairs being made on faulty equipment.

President Rafael Caldera Aug. 26 signed into law a bill creating a state monopoly for the industrialization and export of Venezuela's natural gas reserves to U.S. markets. Caldera signed the law despite major modifications written into the legislation by government opposition.

The legislation was originally sponsored by Caldera and his COPEI party, but the opposition, which controlled both houses of Congress, had introduced modifications that led to COPEI's withdrawal of support in the final Senate vote Aug. 2.

The legislation stipulated that the government could make use of natural gas in concessions to international private enterprises without paying compensation. A Senate committee had eliminated the possible participation of private capital in exporting natural gas. The changes also eliminated compensation for the gas taken from private firms and limited the type available for exportation.

The government had been opposed to the changes, contending that the modifications would limit the government's plans for the investment of $1 billion to export 1.3 billion cubic feet of natural gas daily to the U.S. by 1975.

(The New York Times reported Aug. 2 that Venezuela's Congress was divided over whether to concentrate on selling the gas in liquid form or spending greater amounts to develop a petrochemicals industry.)

The Venezuelan government had agreed July 14 to permit three foreign oil companies to open the first new oil fields granted to private firms in 13 years, according to a Journal of Commerce report July 16. Occidental, Shell and Mobil Oil were to begin drilling the new fields in southern Lake Maracaibo this year. The companies were expected to invest more than $100 million in the initial three-year exploration period.

The service contracts would give the state nearly 90% of profits and allow the state oil companies a major voice in all phases of exploration, development and exportation of oil.

The government of Venezuela had acted March 8 to increase its tax reference price on crude oils by an average 59.5¢ a barrel, effective March 18. In addition, extra charges were imposed for low-gravity crude oil and for the short-haul freight advantage to the U.S. over competitive crude oils from other parts of the world. Venezuela also increased the reference price for heavy industrial fuel oil with high sulphur content by 75¢ a barrel, along with establishing higher prices on low-sulphur fuels.

Venezuela had increased its oil income tax rate from 52% to 60% in December 1970 and had announced at the time that it would unilaterally set tax-reference prices for determination of tax payments.

Venezuela Dec. 21 again raised its tax on oil. The new increase averaged 32¢ a barrel. Heavy high-sulphur fuel oil was reported to have been raised about 30¢ a barrel while low sulphur residual went up an average 75¢ a barrel. The increases would raise taxes paid by foreign oil companies by about $250 million.

In the decree issued jointly by the Ministry of Mines and the Finance Ministry, there was also a provision that U.S. and other foreign oil companies would have to increase their present production of 3.5 million barrels of oil a day to the 1970 rate of 3.7 million daily

or pay increased special taxes. Production drops below 2% of base year 1970 and increases larger than 6% would be penalized by a progressively rising surtax.

Petroleum Minister Hugo Perez la Salvia said Dec. 21 that the move would "avoid substantial decreases in production that could notably alter tax income."

Other Areas

Brazil. The World Bank April 5 announced a loan of $70 million to help finance construction of a 700 megawatt hydroelectric plant and associated transmission facilities at Salto Osorio on the Iguacu River in southern Brazil. The power project was estimated at $152.3 million.

Cuba. The New York Times reported Oct. 3 that Cuba had been experiencing in the past months the worst power shortage since the 1959 revolution.

The Times cited a speech in which Castro reportedly said the situation was particularly serious in four of Cuba's six provinces, including Havana. Castro also said that the power shortage would not ease before the end of 1972.

Egypt. The Aswan High Dam in Egypt was formally dedicated Jan. 15 by U.A.R. President Anwar Sadat and Soviet President Nikolai V. Podgorny.

The dam actually had been in operation for several years, but the last of the 12 turbines had not been installed until July 1970.

The Soviet Union had assisted Egypt in building the project at a cost of more than $1 billion. Construction had begun in 1960. The dam provided Egypt with 10 billion kilowatts of electricity each year, brought 1.3 million acres of new land under cultivation and increased flood control and navigation on the Nile River.

In his dedicatory speech, Podgorny said his country would assist in building a network to transmit power from the dam to the Egyptian countryside, where only 10% of the more than 4,000 villages had electricity.

Rhodesia. It was announced by the government May 13 that gasoline rationing begun in 1965 as a means of countering economic sanctions imposed by Britain, had been abolished. In addition to the abundance of gasoline supplies, the reasons cited for the move were the high cost of administering the rationing system and its inconvenience to the public.

Soviet Union. Finnish Premier Ahti Karjalainen and Soviet Premier Aleksei N. Kosygin signed a new treaty on the expanded development of economic, technical and industrial cooperation between the two nations April 20. Accords were also signed providing for the first Finnish import of Soviet natural gas, beginning in 1974, the sale of a second Russian-built nuclear power plant to Finland, and Finnish development of a forest industry center across the Russian border. The treaty and accords were signed during a visit by Premier Karjalainen to the Soviet Union April 19–23.

The Soviet Union April 7 announced plans for building a major oil pipeline that would connect the Samotlor petroleum pool in West Siberia with outlets in European Russia, a distance of 1,200 miles. The Samotlor field, discovered in 1968, was said to rival in importance the North Slope of Alaska and to contain a total recoverable reserve of two billion tons of petroleum.

Valentin D. Shashin, minister of the oil-extracting industry, revealed June 27, at the opening of the eighth World Petroleum Congress in Moscow, that "work is under way" on a Soviet oil pipeline that would end within 400 miles of Japan. The line was to run 2,600 miles across Siberia from Irkutsk to the Pacific Coast port of Nakhodka.

■ The first plant to process natural gas was in operation at the Orenburg field in the southern Urals, according to a government announcement Oct. 6. The field had reserves estimated at 70 trillion cubic feet of natural gas.

A Soviet government spokesman Dec. 12 announced the completion of the world's largest hydroelectric power station at the new town of Divnogorsk on the Yenisei river in Siberia. The complex, with an estimated generating capacity of six million kilowatts, was to be used primarily to provide power to nearby aluminum plants.

Soviet crude oil production rose from 7,060,000 barrels daily in 1970 to 7,460,000 in 1971.

Atomic Power

Soviet-French uranium pact. France and the Soviet Union March 15 signed a pact worth approximately $7 million by which the Soviet Union agreed to enrich French uranium.

The accord, breaking a monopoly held by the U.S. Atomic Energy Commission in providing fissionable materials to Western countries, obliged the Soviet Union to enrich the uranium (increase its ratio of radioactive isotopes) for a French power plant to be constructed at Fessenheim in Alsace.

■ The EEC Commission took France to the Court of Justice for infraction of the Euratom Treaty, it was reported March 12. France was charged with ignoring the Euratom requirement to negotiate the purchase of nuclear fuels through the organization since it had bought plutonium and other fissile materials directly from outside nations.

The Euratom Council broke an internal deadlock Sept. 20 between France and its five non-nuclear members—West Germany, Italy, Belgium, the Netherlands and Luxembourg—over Euratom inspection of civilian nuclear installations.

The five nations bowed to French demands that only the civilian French nuclear plants using fuel obtained from nations outside France would continue to be subject to Euratom inspection. West Germany had demanded that France open its entire civilian nuclear industry to Euratom checks.

With the internal dispute settled, the five non-nuclear members of the Council of Ministers also approved the terms of inspection for nuclear plants they would seek in negotiations with the International Atomic Energy Agency (IAEA) under the nuclear nonproliferation treaty. The treaty was signed, but not yet ratified, by the five nations. They had opposed IAEA controls over their nuclear facilities. France, which had refused to sign the nonproliferation treaty, would not have been subject to IAEA inspection because it already possessed nuclear weapons.

U.S.S.R. builds atomic converter. The Soviet Union announced March 25 that its scientists had succeeded in constructing a new engineering system that could convert nuclear energy directly into electricity.

In a brief official statement, Tass, the Soviet Press Agency, said "comprehensive tests" had been completed on an advanced converter with an electrical capacity of "several kilowatts." Tass hailed the new converter as "a new major achievement of Soviet atomic science and technology."

(Atomic Energy Commission spokesmen called the Soviet accomplishment "an important technical step." They added, however, that not enough details were available to determine if it was a major breakthrough in the field of direct energy conversion.)

The Tass statement described the new device as a thermionic converter. Scientists in the U.S. and U.S.S.R. had been investigating the thermionic method for directly converting atomic energy into electric power rather than through the heating of steam to drive a turbine generator.

The new development was seen as another step in the attempts by scientists in both the U.S. and the Soviet Union to develop compact generators for on-board power systems in space vehicles.

The Soviet Union did not elaborate on its accomplishment, beyond describing it as "direct [machineless] thermoemission of atomic energy into electricity."

Uranium enrichment talks planned. The U.S. Atomic Energy Commission (AEC) disclosed its readiness to begin "exploratory discussions" with 10 nations on construction of gaseous diffusion plants used in production of fissionable uranium-235 for nuclear reactors, the Washington Post reported July 30.

The nations involved were Britain, Japan, Canada, Australia, France, West Germany, Italy, Belgium, Luxembourg and the Netherlands. The U.S. had previously opposed sharing the formula

for the separation of uranium-235 from ordinary uranium because it was a key element in the production of nuclear arms.

An agreement to help build the plants and share the formula would depend on the conclusion of "appropriate financial and security arrangements."

The AEC currently enriched uranium for the power reactors of France, Belgium, Netherlands, Luxembourg, Italy, West Germany, Switzerland, Spain, Sweden and Japan.

Atoms for Peace conference. The Fourth International Conference on the Peaceful Uses of Atomic Energy was held under U.N. sponsorship in Geneva Sept. 6–16. It was attended by about 4,000 scientists from 79 nations and was presided over by Dr. Glenn Seaborg, former chairman of the U.S. Atomic Energy Commission. The conference focused on the past and future expansion of nuclear power plants to supply electricity and the consequences of this development, including environmental problems.

The desire of the Soviet Union to sell enriched uranium and power reactors to the West was expressed by a Soviet delegate at a press conference Sept. 10. Its selling price for enriched uranium would be slightly more than $27 per kilogram of separative work, almost $5 less than the U.S. price.

Hanford deactivation halted. President Nixon Feb. 4 ordered a suspension of plans to deactivate one of two remaining nuclear reactors at the Atomic Energy Commission's reservation in Hanford, Wash.

The reactor produced plutonium for nuclear weapons and electric power for the Washington Public Power Supply System. The suspension order came after hearings by the Joint Congressional Committee on Atomic Energy Feb. 4 at which Sen. Henry M. Jackson (D, Wash.) testified that the reactor's closure would severely affect power supplies and employment in the Hanford area.

An unidentified Nixon Administration official, quoted by the New York Times Feb. 6, called the reactor "unreliable and a possible safety hazard."

In a related development, the Maryland State Public Service Commission was reported Jan. 23 to have granted a construction permit that week to the Baltimore Gas and Electric Co. to build a nuclear power plant at Calvert Cliffs on the western shore of Chesapeake Bay. At the time the permit was granted, the $387 million plant was more than 30% completed. The permit required plant officials to introduce new anti-pollution controls as technology made such devices available.

Atom plant alterations. The AEC June 7 proposed that nuclear power plants limit their radiation exposure to 1%, or less, of the amount of radiation permitted under current U.S. guidelines for power reactors. AEC spokesmen said only the Humboldt Bay reactor near Eureka, Calif., and the Dresden reactor no. 1 and possibly the Dresden reactor no. 2, both near Chicago, failed to meet the new guidelines.

In another safety measure, the AEC June 19 ordered five atomic power plants to modernize their emergency cooling systems and three plants to lower their peak operating temperatures to 2,300 degrees Fahrenheit. The orders were described as part of an "interim core cooling policy" subject to future changes.

Illinois A-plant delayed. U.S. District Court Judge Barrington D. Parker Dec. 13 enjoined the Atomic Energy Commission (AEC) from issuing a partial operating license for a Cordova, Ill. nuclear power plant, until the agency issued a complete environmental impact statement called for by the 1969 National Environmental Policy Act.

Parker interpreted the act's provision for impact statements as a "mandatory obligation not within the discretion of the agency," which "must be performed in all circumstances prior to the taking of the agency action."

The injunction had been sought by the Illinois attorney general and the Izaak Walton League of America, a conservation group, amid reports that the AEC was planning to grant a permit to the Commonwealth Edison Co. and the Iowa-Illinois Gas and Electric Co. to run the plant at 50% capacity through March 1972.

Judge Parker's decision cited expected temperature rises in river water used as a coolant for the plant, which might have adverse effects on fish reproduction.

Town seeks A-plant—Residents of Midland, Mich. Dec. 2 delivered a petition with 15,000 signatures asking the AEC to approve a construction license for a nuclear power plant for which a local power company had first applied in January 1969. The license had been delayed by hearings and conferences spurred by environmental groups.

The plant had been designed to provide both electrical power and steam used in chemical processes by the Dow Chemical Co., the town's largest employer.

AEC chairman sees new role. Dr. James R. Schlesinger, newly appointed chairman of the AEC, said Oct. 20 that the commission would abandon its role as defender and promoter of the nuclear power industry and concentrate on the public's right to safety, economy and environmental protection.

In a speech to a meeting of the Atomic Industrial Forum and the American Nuclear Society in Bal Harbour, Fla., Schlesinger said it was "not the responsibility of the AEC to solve industry's problems," or oppose all industry critics. He said the agency would work to prevent "small accidents, small spills, unplanned shutdowns, power interruptions and associated higher construction and maintenance costs," which he considered the industry's greatest potential problems. The agency would continue to conduct research in atomic power generating techniques up "to the point of commercial application."

Schlesinger also defended the AEC's new rules for nuclear power plant construction, issued after a federal court denounced the agency's environmental policies, against industry hopes that the decision would be appealed.

1972

National leaders were under increasing pressure
during 1972 to find ways to close the growing energy
gap. 1972 was a Presidential election year, and both
political parties adopted platforms that included
planks proposing solutions to the energy problem. A
survey predicted that the U.S. energy demand would
quadruple by 1990. Controversy continued over the
issue of how to bring Alaskan oil to the Southern 48
states. The U.S. continued to relax curbs on oil
imports; it resumed the acceptance of bids for oil-
drilling in the Gulf of Mexico, and proposed curbs
on coal strip-mining were shelved. Some utility rates
and oil and gas prices rose. Oil-producing countries
raised prices again, and Iraq nationalized the
Western-owned Iraq Petroleum Co. A 47-day coal-
miner strike slowed British factories to a four-day
week and caused other disruptions by cutting power
output. North Sea drilling produced promising gas
and oil finds. U.S. atomic energy planners placed
their major hope on "fast breeder" reactor development, and the U.S.S.R. put the world's first commercial fast-breeder plant into operation.

Mounting Pressure to Solve Energy Problem

U.S. Plans & Actions

During the Presidential election year of 1972, government officials and political leaders of both parties responded to the challenge to find ways to cope with the U.S.'s growing energy problems.

The installed generating capacity of the U.S.'s electric utilities rose from 367,396,-000 kilowatts at the end of 1971 to a record 399,606,000 Dec. 31, 1972. Utility output rose from 1,613,936,000,000 kilowatt hours in 1971 to a record 1,747,323,000,000 in 1972. The purchase of 6,267,000,000 kilowatt-hours of electricity from industrial sources and the import of 10,373,000,000 from Canada and Mexico during 1972 increased the year's input to a record 1,763,-960,000,000 kilowatt-hours.

Motor vehicles in the U.S. (military use excluded) used a record 110.1 billion gallons of motor vehicle fuel during 1972.

Crude oil and natural gas liquids provided 46% (5.96 billion barrels, producing 32,812 trillion BTUs of energy) of the fuel used in the U.S. during 1972. The other major fuels used were: natural gas 32% (22.607 trillion cubic feet, 23,308 trillion BTUs); coal 17% (517,053,000 short tons, 12,428 trillion BTUs); waterpower 4% (280.2 billion kilowatt-hours, 2,937 trillion BTUs); nuclear power 1% (56.9 billion kilowatt-hours, 606 trillion BTUs).

U.S. output of crude oil and natural gas liquids declined about 5% from the November 1970 high to an average of some 11.6 million barrels a day in 1972; imports during 1972 averaged 4.7 million barrels a day, some 1.4 million of which came from the Eastern Hemisphere (primarily the Middle East). Bituminous coal and lignite production in the U.S. dropped from 603 million tons in 1970 to about 570 million in 1972.

Democrats outline energy policy. The Democratic National Convention, meeting in Miami Beach, Fla., July 11–12 adopted a platform containing the following planks on energy resources:

The earth's natural resources, once in abundant and seemingly unlimited supply, can no longer be taken for granted. In particular, the United States is facing major changes in the pattern of energy supply that will force us to reassess traditional policies. By 1980, we may well have to depend on imports from the Eastern Hemisphere for as much as 30 to 50 percent of our oil supplies. At the same time new forms of energy supply—such as nuclear, solar or geothermal power—lag far behind in research and development.

In view of these concerns, it is shocking that the Nixon Administration still steadfastly refuses to develop a national energy policy.

The Democratic Party would remedy that glaring oversight. To begin with, we should:

■ Promote greater research and development, both by government and by private industry, of unconventional energy sources, such as solar power, geothermal power, energy from water and a variety of nuclear power possibilities to design clean breeder fission and fusion techniques. Public funding in this area needs to be expanded, while retaining the principle of public administration of public funds;

■ Re-examine our traditional view of national security requirements in energy to reconcile them with our need for long-term abundant supplies of clean energy at reasonable cost;

■ Expand research on coal technology to minimize pollution, while making it possible to expand the efficiency of coal in meeting our energy needs;

■ Establish a national power plant siting procedure to examine and protect environmental values;

■ Reconcile the demand for energy with the demand to protect the environment;

■ Redistribute the cost of power among consumers, so that all, especially the poor, may be guaranteed adequate power at reasonable costs;

■ Develop a national power grid to improve the reliability and efficiency of our electricity system;

■ End the practice of allowing promotional utility advertising as an expense when rates are set; and

■ Find new techniques to encourage the conservation of energy. We must also require full disclosure of the energy needs of consumer products and home heating to enable consumers to make informed decisions on their use of energy.

GOP states energy policy.
The Republican National Convention, meeting in Miami Beach, Fla., Aug. 22 adopted a platform containing the following planks on energy resources:

No modern nation can thrive without meeting its energy needs, and our needs are vast and growing. Last year we proposed a broad range of actions to facilitate research and development for clean energy, provide energy resources on Federal lands, assure a timely supply of nuclear fuels, use energy more efficiently, balance environmental and energy needs, and better organize Federal efforts.

The National Minerals Policy Act of 1970 encourages development of domestic resources by private enterprise. A program to tap our vast shale resources has been initiated consistent with the National Environmental Policy Act of 1969.

We need a Department of Natural Resources to continue to develop a national, integrated energy policy and to administer and implement that policy as the United States approaches the 21st Century. Energy sources so vitally important to the welfare of our nation are becoming increasingly interchangeable. There is nothing inherently incompatible between an adequate energy supply and a healthy environment.

Indeed, vast quantities of energy are needed to do the work necessary to clean up our air and streams. Without sufficient supplies of power we will not be able to attain our goals of reducing unemployment and poverty and enhancing the American standard of living.

Responsible government must consider both the short-term and the long-run aspects of our energy supplies. Avoidance of brown-outs and power disruptions now and in the future call for sound policies supporting incentives that will encourage the exploration for, and development of, our fossil fuels. Such policies will buy us time to develop the sophisticated and complex technologies needed to utilize the exotic energy sources of the future.

National security and the importance of a favorable balance of trade and balance of payments dictate that we must not permit our nation to become overly dependent on foreign sources of energy. Since more than half our nation's domestic fossil resources now lie under Federal lands, high priority must be given to the governmental steps necessary to the development of these resources by private industry.

A liquid metal fast breeder reactor demonstration plant will be built with the financial support of the Atomic Energy Commission, the electric power industry, and the Tennessee Valley Authority.

We will accelerate research on harnessing thermonuclear energy and continue to provide leadership in the production of energy from the sun and geothermal steam. We recognize the serious problem of assuring adequate electric generating capacity in the nation, and pledge to meet this need without doing violence to our environment.

Power demand to quadruple.
The Federal Power Commission (FPC) released the results of a six-year survey of the nation's electric power resources and needs April 15, concluding that demand would quadruple between 1970 and 1990.

The FPC estimated that power generating capacity would have to rise from 340,000 megawatts in 1970 to 1,-260,000 megawatts by 1990 to meet the demand, at an investment cost of $400 billion, and with increased research to handle environmental and transmission problems. Atomic energy plants would generate 47% of total power output in 1990, compared with less than 2% in 1970.

The consumer price per kilowatt-hour, which the report said had declined from 2.7¢ to 1.53¢ between 1926 and 1968, was expected to rise to 1.83¢ in 1968 dollars by 1990, or 3.51¢ taking inflation into account. Rising fuel and construction costs would cause the increase.

Generating equipment.
The electric utility industry of the contiguous 48 states had a capability of 394,100,000 kilowatts by the end of 1972, up 27,400,000 kilowatts during the year. This total capability was supplied in these proportions by the following types of generators (data from Edison Electric Institute's electric power survey):

Conventional Steam Turbine-Generators	73.7%
Nuclear Steam Turbine-Generators	2.6
Conventional Hydro Generators	14.3

Pumped-Storage Generators	1.0
Gas Turbine-Generators	7.5
Combined-Cycle Generators	0.3
Diesel Engine-Generators and others	0.6

High court backs FPC. The U.S. Supreme Court June 7 agreed that the Federal Power Commission (FPC) had the authority to allocate natural gas supplies among its users during times of gas shortages.

In the decision written by Justice William Brennan, the court said the FPC authority extended to supplies of industrial users of natural gas as well as the amount of gas that could be channeled to non-industrial consumers.

The case had come to the court on an appeal of a decision by the 5th U.S. Circuit Court of Appeals involving a suit by the Louisiana Power & Light Co. against the United Gas Pipe Line Co. In that ruling, the appeals court held that Louisiana Power & Light could not be denied delivery of all the gas provided in its contract because the Federal Power Act did not apply to dealings with industrial gas consumers.

Plains states resource study. Secretary of the Interior Rogers C. B. Morton reported Oct. 3 that a federal-state task force would be established to set policy for developing the huge coal deposits and other resources in Montana, Wyoming, South Dakota, North Dakota and Nebraska.

The study group would include representatives of the Interior, Agriculture and Commerce Departments, the Environmental Protection Agency and the governors of the five states. The region was thought to contain 40% of national coal reserves, as well as uranium, oil, gas, forests and other resources. About 80% of the mineral deposits were on federal land, according to Morton.

U.S. acts in fuel shortage. The Office of Emergency Preparedness (OEP) reported Dec. 21 and Dec. 22 the measures it was taking to meet shortages of fuel oil and propane gas in the Midwest. OEP said it was running an emergency coordinating center with representatives from other government agencies, had requested additional production of heating oil by refineries, and had asked railroads to expedite shipments of propane gas.

The shortage had been partly caused by cautiously low inventories after a mild 1971-72 winter, an early cold wave, and wet weather, leading to a huge demand for propane gas to help dry harvested crops. A spokesman for Shell Oil Co. said Dec. 21 that a switch by industries from coal to oil, to meet air pollution controls, had exacerbated the crisis. The company said it had begun to allocate its supplies among customers.

The Nebraska Petroleum Managers Association said Nov. 19 that heating oil supplies would be cut off to at least 8,000 Omaha area homes because of the shortage.

Shortage called deliberate—Maine Gov. Kenneth M. Curtis (D) said Dec. 22 that the Nixon Administration had deliberately developed an oil shortage to gain support for such policies as fuel price increases and continuation of tax loopholes, and to lessen environmentalist opposition to the proposed trans-Alaska oil pipeline, Atlantic Ocean offshore oil drilling and construction of deep water oil ports.

Interior Secretary Rogers C. B. Morton had announced Nov. 11 creation of an energy board to develop and coordinate "energy resource development objectives."

California energy curb asked. A Rand Corporation report financed by the California legislature, the National Science Foundation and the Rockefeller Foundation said power shortages, fuel depletion and environmental deterioration threatened California and the nation, and recommended that the demand for electricity be reduced by 60% of the projected figure for the year 2000.

To achieve this goal, the report, announced Oct. 16, called for large-scale substitution of gas for electricity in homes and development of solar energy and better insulation for buildings, and suggested that California encourage these practices through tax incentives, educational programs and legislation.

Projecting annual demand increases of 6% for natural gas and 4% for petroleum,

as in the past decade, the report said domestic supplies would be exhausted in 30
years, while coal reserves would be depleted within 70 years. Atomic power
could not appreciably alter the situation
by the year 2000, since fossil fuels would
still have to provide 75% of energy needs,
according to the highest current predictions of nuclear plant construction.

MIT study sees resource crisis. "The
Limits of Growth," a study of current
world production and population trends
financed by the Club of Rome, predicted
a catastrophic world economic and
population decline in the next century,
unless production and consumption habits were put under deliberate international
control.

The study, conducted at the Massachusetts Institute of Technology (MIT)
and planned for publication in March,
centered on a computer model of the
interrelations between such economic
factors as population, natural resources,
industrial and food production, and pollution.

The study was directed by MIT professor Dennis L. Meadows, at the request of the Club of Rome, an international group of business and economic
leaders.

Alaska oil pipeline potential threat. The
Interior Department, abiding by a 1970
federal court order barring a permit for
construction of the trans-Alaska oil
pipeline without an extensive environmental impact statement, issued a six-
volume report March 20, conceding
that any method of transporting North
Shore oil would pose a potential threat
to the environment.

But in a three-volume economic and
national security analysis accompanying
the report, the department called recovery of the oil "an important national
security objective," and called the proposed 789 mile Prudhoe-Valdez pipeline
the earliest feasible route.

In a news conference March 20, Interior Undersecretary William T.
Pecora said the department would hold
no further public hearings, as demanded
the same day in a telegram to President
Nixon by the Wilderness Society, the
Friends of the Earth and the Environ-

mental Defense Fund, the groups which
filed the initial suit against the project.

Oil spillage would be the major environmental threat from the pipeline,
according to the report. It was considered almost certain that at least one
large magnitude earthquake would
strike the region traversed by the 48-inch
pipe during its lifetime. Once the oil
would be loaded on tankers in Valdez,
"persistent low-level discharge from the
ballast treatment facility and tank cleaning operations" might endanger salmon
and other fishing industries even more
than major one-shot spills.

A proposed route across northwest
Canada to the Midwest would incur less
danger from earthquakes and, if it followed the same route as a proposed
natural gas pipeline, would avoid doubling the environmental damage that
two separate routes would entail. But
engineering studies for a Canadian route
had not advanced, and Canadian environmental legislation would require
still further delays for detailed impact
consideration, the report said.

According to the economic analysis,
trans-Alaska oil deliveries could reach
500,000 barrels a day by 1975 and two
million barrels by 1980, lessening U.S.
dependence on oil imports. Construction cost estimates had risen from $1
billion in 1969 to $5 billion currently.

U.S. District Court Judge George L.
Hart had refused Jan. 14 to order further Interior Department hearings, and
refused once again to allow a Canadian
parliament member to enter the case.

Alaska Gov. William A. Egan, it was
reported March 4, had expressed concern that soaring cost estimates might
make it impossible for the state "to
derive logical and sensible financial
benefits." Over-optimistic royalty predictions for the late 1970's could cause
a state budgetary crisis, according to
Alaska officials.

Permit for pipeline scheduled—Interior
Secretary Rogers C. B. Morton announced May 11 that he would grant a
permit for construction of the controversial trans-Alaska oil pipeline.

In a five-page statement, Morton conceded that the 789-mile Prudhoe-Valdez
pipeline would involve "some environmental costs," but said it represented the

fastest way to transport the North Slope oil which "the U.S. vitally needs."

Conservationists immediately accused Morton of bowing to oil interests—which had reportedly invested $2 billion or more in the project—and vowed to continue a court challenge to the pipeline on grounds that it would cause irreparable damage to the environment.

In approving the Alaska route, Morton rejected construction of a pipeline across Canada to the Midwest, which he said would take much longer. However, three U.S. environmental groups—the Wilderness Society, Friends of the Earth, and the Environmental Defense Fund—had claimed in a report a week earlier that Morton did not have enough information to make "a rational decision" against the Canadian route.

Canada Energy Minister Donald S. Macdonald expressed disappointment at the U.S. Interior Department permit approval for the pipeline. A statement said "the government of Canada remains convinced that the alternative Canadian route through the Mackenzie Valley is the best from all aspects."

Rival Canadian route urged—At a hearing June 9 before the Congressional Joint Economic Committee, Canadian and U.S. environment and energy experts testified that the Interior Department had not given adequate consideration to a cross-Canada oil route in approving a permit for the trans-Alaska pipeline.

The Canadians had argued that even if only one tanker a year ran aground in the waters off British Columbia, between the pipe's southern terminal at Valdez, Alaska and northwest U.S. markets, the province's $20 million fishing industry could be threatened, as well as $1 billion in recreation property investments.

David Anderson, Canadian parliamentary environment expert, reported that Canada had spent $43 million on 30 economic and environmental studies of a possible oil and gas pipeline route from the North Slope in Alaska through the Mackenzie River Valley to the U.S. Midwest. Anderson claimed that U.S. Interior Secretary Rogers C. B. Morton had ignored this information, and warned that the Canadian government might not

approve a natural gas pipeline, which the U.S. was expected to favor, if the oil were not moved through the same corridor.

Economist Richard D. Nehring, who resigned from the Interior Department to protest the Alaska pipe decision, told the committee that the draft environmental impact statement on the issue had concluded that the Canadian route would cause less ecological damage. He said the conclusion was dropped from the published version. Nehring also maintained that most Interior Department economists believed the Canada route would be economically more practical, since the need for oil was expected to be greater in the East and Midwest than on the West Coast.

Interior Secretary Morton appeared before the committee June 22 to explain his opposition to Interior staff reports backing the trans-Canada route.

Morton claimed that the Canadian route "will produce more environmental damage in the long run," largely because it would pass through longer sections of permafrost and over 12 major rivers as opposed to one river on the Alaska route. Morton claimed that far more gravel would be needed as a bed for the Canadian route, requiring 1,000 excavation pits.

As for the possibility of laying the oil pipeline in the same corridor as a proposed Canadian natural gas pipeline, Morton claimed that gas transported underground at sub-freezing temperatures and hot oil transported through an elevated pipe required different terrains "a considerable distance apart." He added that Canadian money markets could not handle financing of 51% of both lines, which he assumed would be required by the Canadian government.

Morton claimed that the Canada route would cost $6 billion, as opposed to the $4.5 billion combined cost of the Alaska pipe and tanker fleet, and he said preliminary work needed on the Canada route would cause a delay of "as much as seven years" in oil deliveries.

Morton also announced new rules being considered to bar any oil discharge from tankers carrying oil from Valdez, the southern terminal of the

Alaska route. The rules concerned construction and inspection of the tankers, Coast Guard control of shipping lanes and a ban on ship movement through fishing zones or "ecologically sensitive areas."

Consortiums merge—Two of three major international consortiums investigating development of a $5 billion trans-Canada natural gas pipeline announced a merger and said they would apply with U.S. and Canadian authorities early in 1973 to begin construction work, it was reported June 15. The groups, Gas Arctic Systems and Northwest Project Study Group, said a 48-inch pipe could deliver 3.5 billion cubic feet of North Slope gas daily to Canadian and U.S. markets.

Intervention sought—Attorneys for 28 natural gas companies eager to exploit North Slope deposits sought June 7 to intervene in U.S. district court in Washington, on the side of the government and oil companies, in the suit blocking final approval of a trans-Alaska permit. Although the Alaska route, which would entail sea-borne transport on the last leg, could not be used by the gas companies, attorney Tilford A. Jones said "when the oil starts flowing, the gas will start flowing."

Alaska pipeline stay lifted. U.S. District Court Judge George L. Hart Aug. 15 lifted an injunction he had issued in April 1970 against construction of the trans-Alaska oil pipeline.

Hart ruled in Washington that the Interior Department's environmental impact statement had adequately considered alternatives, and that all right-of-way and landuse permits had been legally issued. He said he had issued a final ruling to expedite the appeals process, since "it can be confidently anticipated that the final decision in this matter rests with the Supreme Court."

A Justice Department spokesman said the Interior Department was not yet ready to issue a construction permit. The Alyeska Pipeline Service Co., the consortium of oil companies that would build the pipeline, announced

that it would not begin work until all legal issues were resolved by the court of appeals, and not without 30 days notice to environmental groups.

The Alaska Federation of Natives had been reported Aug. 13 to have charged that any further delay in construction could "undermine and violate" the land claims settlement act enacted in 1971. The act mandated a long term payment of $500 million to the natives, to be paid from mineral royalties, mostly from North Slope oil.

In a related development, nine oil companies filed suit in Alaska superior court in Anchorage Sept. 12 against new laws that would, they contended, sharply increase taxes on North Slope oil, and require state right-of-way permits for the pipeline. The companies said the tax increase would violate agreements under which they had already paid $910 million to the state for oil and gas leases, and that transport of oil through the pipeline, largely crossing federal lands, constituted interstate commerce not subject to Alaska law.

Alaska gas pipe studied. The El Paso Natural Gas Co. said in Washington and Anchorage Dec. 4 it was considering building a natural gas pipeline across Alaska in conjunction with the proposed oil pipeline. The $3 billion project would include liquefaction facilities and a fleet of tankers. The company said it was also considering an all-land transport route across Canada.

El Paso had filed with the Federal Power Commission to build the world's first commercial coal gassification plant in New Mexico, at a cost of $420 million, it was reported Nov. 16.

Canadian Arctic developments. An agreement between three major U.S. pipeline companies and a Canadian drilling firm to drill for gas on Canada's Arctic islands was announced in Houston, Texas and Calgary, Alberta Feb. 1.

The three U.S. firms—the Consolidated Gas Supply Corp., a subsidiary of the Consolidated Natural Gas Co.; the Panhandle Eastern Pipeline Co.; and the Texas Gas Exploration, Ltd., an affiliate of the Texas Gas Transmission Corp.—pledged $30 million to finance drilling by the Dome Petroleum, Ltd. The agreement commited Dome to sell 75% of any gas discoveries to the companies provided they paid as much as any other bidder.

Panarctic Oils, Ltd., a government-industry consortium, announced Feb. 24 an oil strike on Ellesmere Island, 700 miles below the North Pole.

Charles Hetherington, president of the firm, said a sample taken from the Fosheim Peninsula of the island and reaching Calgary that day was "oil, clear and sweet." Hetherington added that it could not yet be determined whether the strike had commercial implications but said Panarctic was "continuing to drill into the sand and to drill stem test in order to evaluate the magnitude of the discovery."

The low-sulphur oil had been located at a depth of 3,500 feet. Panarctic had already made four commercial gas finds in the area.

It was revealed Nov. 30 that Panarctic Oils had made a natural gas discovery on Melville Island. The newest find was on the west side of the Sabine Peninsula.

In another oil development, Peter Usher, an official in the Department of Indian Affairs and Northern Development, charged that the federal government appeared to favor oil companies and the economy of southern Canada at the expense of Eskimos and the preservation of the northern environment, the Toronto Globe and Mail reported Feb. 1.

His charge was contained in a report on the impact of oil exploration on the Eskimo community of Banks Island. The report urged a moratorium on all oil and mineral development in areas occupied by native peoples unless the interests of the natives were more adequately taken into consideration.

Some nine months later, Jean Chretien, minister for Indian Affairs and Northern Development, was reported Nov. 18 to have ordered a French-owned oil company to remove its exploration equipment from the traditional hunting grounds of Eskimos in the Arctic region at Tuktoyaktuk, near Cape Bathurst on the Beaufort Sea.

The firm, Klf Oil Exploration and Production of Canada Ltd., was enjoined for one year from conducting seismic operations which the Eskimos felt would disturb the habitat of caribou, foxes and bears. According to a spokesman for Chretien, a decision would be made at the end of a year whether to allow the operations to go on.

U.S. utility to buy Quebec power. Consolidated Edison Co. of New York had agreed in principle to buy some $500 million worth of electric power from the Quebec Hydro-Electric Commission over a 20-year period, the Wall Street Journal reported July 27.

Quebec Premier Robert Bourassa said the accord would open to Hydro-Quebec the important U.S. market for the excess power the Quebec energy network would have at its disposal in future years. He added that the contract and a mutual assistance agreement would provide a permanent working relationship between Hydro-Quebec and the East Coast Power Pool, of which Con Edison was a member.

U.S. to get Canadian power. The federal Cabinet Aug. 28 adopted a report by the National Energy Board recommending construction of a generating plant in New Brunswick which would export electrical power to the U.S.

"It is the first time the board has been asked to license the export of power from generating facilities which would be constructed partially to serve the U.S. market," the report said, concluding that benefits from employment and savings through large-scale construction would offset the disadvantages of possible environmental pollution.

The $184 million plant was to be built at Lorneville, a few miles west of Saint John on the Bay of Fundy, where a deep water port was to be constructed to supply the generators with cheap foreign oil from supertankers. The New Brunswick Electric Power Commission was given authority to build the plant and export to the U.S. 400,000 kilowatts of power between 1976 and 1986. NBEPC could also export 876 million kilowatt hours a year of interruptible power between 1976 and 1985.

Oil import curbs lifted. The Office of Emergency Preparedness announced Dec. 8 that it was suspending import curbs on No. 2 crude oil used in home heating. The fuel had been in short supply in the East and Midwest.

The suspension, which would be in effect from January to May 1973, would permit the import of any foreign petroleum sources.

"During the next three to five years, a further deterioration of the domestic energy supply position is anticipated, and as a result fuel imports will have to be increased sharply," according to a report released Dec. 11 by the National Petroleum Council.

"The nation's dependence on imports of oil and gas increased to 12% of total energy requirements in 1970 and is likely to be 20%–25% by 1975," the industry report added.

Nixon had moved twice previously in 1972 to ease oil import curbs. The President May 11 had ordered a 15% increase in the crude oil import quota, Canadian imports increasing from 540,000 to 570,000 tons a day, Venezuelan and Middle East imports rising from 965,000 to 1,165,000 tons a day. George A. Lincoln, director of the Office of Emergency Preparedness, said May 11 the "supply-demand balance . . . showed that under present conditions, domestic production would not meet the anticipated demand throughout the year without added imports." In a May 11 speech in New York, Assistant Secretary of the Interior Hollis M. Dole reported that in the past the oil import quotas were designed to guarantee a surplus of oil east of the Rockies. However, he said since that surplus stock has now been exhausted, the "new emphasis on the program will be to permit imports to fill the gap between demand and production." The present increase affected only Interior Department Districts I–IV, all east of the Rocky Mountains.

Nixon acted Sept. 18 to increase oil imports by 35% for the remainder of 1972 in the area east of the Rockies.

The action, taken on the recommendation of the Office of Emergency Preparedness (OEP), whose study had found that oil production was leveling off

while demand was rising, would double the daily import rates of No. 2 fuel oil.

Authorities had feared that a severe winter could lead to fuel shortages in New England, which used No. 2 oil, according to the New York Times Sept. 18.

Sen. Edward M. Kennedy (D, Mass.) termed the Administration action a "piecemeal and short-sighted solution" and called for the termination of oil import quotas, "whether or not the major oil companies like it."

Elmer Bennett, assistant director of OEP, acknowledged that the loosened import restrictions "highlights the fact that we are becoming more and more dependent on Middle East sources of crude oil."

Arguments against the oil import quotas had been presented by Barbara Heller in the May issue of the Friends of the Earth publication Not Man Apart:

New England pays the highest prices in the nation for electricity because it pays the highest prices for oil. Why? The answer is simple. The import quota system restricts the free market mechanisms which should govern the world oil market. We simply cannot obtain anything over a certain specified amount of foreign oil. . . .

What would happen if the import quota system were suddenly lifted? Chances are reasonably good that New England's oil prices would decrease; . . .

Arguments supporting the quotas had been presented May 4 by Sen. John Tower (R, Tex.). As recorded in the Congressional Record, Tower said of "the benefits of the mandatory oil import program":

First, a great deal of natural gas is discovered in the search for oil. It is estimated that 25 percent of our natural gas production comes in association with domestic oil and gas production. This represents a saving to the consumer of approximately $1.7 billion, assuming that such gas, in the absence of domestic production, would have had to be imported at a 65 cent per million cubic feet greater than that of domestic gas.

Second, it is important to note that during the years 1959–70, the years in which the program was in existence and for which we now have the necessary data, the reserves of natural gas increased at an average annual rate of 19 trillion feet of natural gas. For oil, the comparable figure was 3.3 billion barrels. Furthermore, this has enabled us to steadily increase domestic production of both oil and gas. . . .

In addition to the above, there are further reasons which justify the existence of an oil import program.

A large increase in imports would damage the U.S. international trade balance at a time when this Nation is working hard to improve that balance.

Second, any substantial increase in the importation of oil and gas would have to come from the Eastern Hemisphere—a less than reliable source. Since 1948, in fact, there have been at least 10 disruptions in the movement of oil from these sources. Presently, the east coast is more than 95-percent dependent on imports of residual fuel oil and is approximately 50 percent dependent on imported oil. With such oil-producing States as Texas and Louisiana producing at their maximum, the east coast could be in serious trouble in case of a long interruption of oil imports.

The price we pay for imported oil will increase also. Much of the cause for this increase can be attributed to the unified bargaining stance of the Organization of Petroleum Exporting Countries. . . .

The OPEC has been quick to threaten sanctions in various forms upon purchasers who are slow to abide by OPEC edicts. . . .

Third, heavy reliance on imported oil involves risk to national security and freedom of action in international affairs.

Finally, the import program insures the continuation of competition in the domestic petroleum industry. In the absence of oil quotas, only the large international companies could survive.

Superports recommended. The Maritime Administration recommended Nov. 27 that two floating deep water oil ports be built off the Atlantic and Gulf coasts, to handle the next generation of supertankers.

The agency recommended a site off Delaware, to be developed at a cost of $499 million, and a Louisiana site with a $191 million estimated cost. Tankers already built weighing as much as 350,000 dead-weight tons and ships under construction to weigh up to 500,000 tons required waters deeper than those in most conventional ports. The cost of dredging, the agency said, would be prohibitive and entry of supertankers into existing ports would entail serious environmental risks.

The Army Corps of Engineers released a report by a private consulting firm Nov. 9 providing additional support for a superport program. The firm said savings in ocean transport costs by use of the tankers would justify building the deepwater ports, but conceded that it had not been able to calculate possible environmental costs of the project. A group of environmentalist organizations had filed suit Oct. 1 to bar any Commerce Department subsidies for supertanker construction until an environmental impact statement were issued.

A group of 10 major oil companies announced Nov. 22 it had signed an option to lease 1,450 acres of land off the Louisiana coast to build a $500 million oil port.

Off-Shore Oil Developments. Interior Secretary Rogers C. B. Morton promised governors and officials of several Eastern Seaboard states Jan. 11 to establish a "meaningful communications system" prior to granting leases for oil exploration on the continental shelf.

Morton said drilling was 7–10 years away, even if commercial deposits were found, and if the Supreme Court ruled on conflicting state-federal rights in off-shore areas. In 1971, the Interior Department had reported plans to begin leasing before 1976.

Although Morton said the final decision on leasing would be up to the White House, Gov. Marvin Mandel (D, Md.) said after the meeting that consultation with state agencies was promised in preparing the required environmental impact statement.

Oil companies submitted offers totaling more than $586 million for 74 tracts of land beneath the Gulf of Mexico off Louisiana Sept. 12, after a nine-month leasing delay caused by environmentalist suits.

The Sierra Club had announced Aug. 16 that it would no longer contest the sale of the leases, after the Interior Department filed a new environmental impact statement that considered alternative energy sources and the possibility of reducing demand.

The department Jan. 20 had returned unopened bids for leases to over 300,000 potentially oil-rich acres on the outer continental shelf off eastern Louisiana, after U.S. District Court Judge Charles R. Richey refused Jan. 19 to lift his temporary injunction against the lease sale. The Court of Appeals for the District of Columbia Jan. 13 had upheld Richey's Dec. 16, 1971 injunction by a 2–1 vote. Richey had acted after environmental groups charged that the government's environmental impact statement for the sale was inadequate because it did not discuss alternate sources of oil to avoid the danger of offshore spills. The government later wrote an addendum to its impact statement, but Richey said he needed more time for a final ruling.

Oil companies Dec. 16 bid $1.6 billion more for leases to about 600,000 acres of submerged land in the Gulf of Mexico.

The Bureau of Land Management had said in a study issued Dec. 15 that some $2 trillion in oil reserves lay beneath the Atlantic continental shelf, in addition to other mineral deposits. More than half the oil lay between Sandy Hook, N.J. and Cape Hatteras, N.C.

A U.S. district court judge had ruled Dec. 14 in favor of Alaska in a dispute with the federal government over title to 4,000 square miles in Cook Inlet near Anchorage. At least $2 billion in petroleum reserves were believed to lie beneath the ocean floor, and Alaska expected to derive some $235 million in oil royalties.

Puerto Rico Gov. Luis A. Ferre had announced Aug. 3 that oil companies conducting explorations had found indications of petroleum deposits in Puerto Rico. An aide later said the deposits were off the island's coast near Aguirre.

U.S. District Court Judge Francis C. Whelan ruled in Los Angeles June 21 that Interior Secretary Rogers Morton had exceeded his powers when he suspended oil drilling operations in April 1971 on 35 leases in the Santa Barbara Channel off California.

Morton had suspended operations to give Congress time to consider measures canceling the leases and setting up a national energy reserve. The channel had been the scene of a massive oil well blowout in 1969 that caused considerable damage to wildlife and property. Whelan ruled that the Outer Continental Shelf Lands Act gave the interior secretary power to suspend operations only "when necessary to prevent waste or damage to personal property."

The judge ordered new permits granted to Gulf Oil Corp., Mobil Oil Corp., Texaco, Inc. and Union Oil Co. of California, and extended the exploratory drilling period 32 months.

Santa Barbara charges dismissed— Judge Morton L. Barker dismissed 342 counts of criminal pollution against four oil companies Jan. 10, after Union Oil Co., Mobil Oil Corp., Texaco, Inc. and Gulf Oil Corp pleaded guilty to one charge each in connection with the 1969 oil spill in the Santa Barbara Channel.

The companies were fined only $500 each, since the judge ruled they had "suffered sufficiently" from earlier civil damage settlements.

Environmental Issue

North Central plan opposed. The Environmental Defense Fund called on the federal government to set up an independent committee to evaluate the proposed mammoth North Central Power Project, under which the Bureau of Reclamation and 35 utility companies would mine coal and generate power in Wyoming to supply Midwest cities via continental transmission wires.

The group, it was reported Sept. 2, said the project would entail strip mining an area half the size of Rhode Island, emitting more nitrogen oxides, sulfur dioxide and particulates than all present sources in New York City and the Los Angeles area combined, consume more water than New York City, thereby reducing the flow of the Yellowstone River by 81%, to produce more electricity than currently produced in Japan.

2-country oil-spills committee. External Affairs Minister Mitchell Sharp announced June 19 that the U.S. and Canada had agreed to form a committee to try to prevent oil spills off Canada's west coast which might occur if the two governments went ahead with the proposed Alaskan oil pipeline.

The group was to be headed for Canada by Environment Minister Jack Davis and for the U.S. by Russell Train, chairman of the Council on Environmental Quality.

Tanker control bill signed. President Nixon signed a bill July 10 to regulate construction and operation of tankers to prevent spills of oil or other harmful cargoes.

The new Port & Waterways Safety Act authorized the Transportation Department, in consultation with the Environmental Protection Agency (EPA) and other federal agencies, to issue rules regulating tanker design, construction and operation, to be effected no later than Jan. 1, 1976. The government would have increased inspection powers, and could set manning requirements and tougher qualifications for officers and seamen.

The transportation secretary would control ship traffic in U.S. inland and coastal waters, regulate handling of dangerous cargo on piers and set standards for waterfront equipment.

Foreign flag ships would have to comply with the new regulations in American waters, and ships already in use would be covered by rules to be set by the Transportation Department.

EPA orders Delaware curb. The Environmental Protection Agency (EPA), in the first regular legal action under the 1970 Clean Air Act, March 8 ordered the Delmarva Power and Light Co. of Delaware City, Del. to cut back on sulphur dioxide emissions that violated the EPA-approved state pollution control plan.

The power company had obtained a court order enjoining the state Water and Air Resources Commission from enforcing the state plan, which had called for a 40% sulphur dioxide emission cut for New Castle County by Jan. 1. The EPA gave the company 30 days to comply, by purchasing low-sulphur fuel, or face penalties of up to $25,000 a day.

House delays Tocks dam. The House Appropriations Committee voted June 19 to delay construction work on the proposed Tocks Island dam project on the Delaware River between New York, New Jersey and Pennsylvania.

The committee included, in a $5.4 billion public works appropriations bill, the entire $14.8 million requested for the project by the Army Corps of Engineers, but it ordered that all the money be used for land acquisition pending solution of environmental problems.

The dam was designed to create a 14,800 acre, 37-mile reservoir for flood control, water supply and recreation. Conservationists charged that the $380 million project would cause oxygen depletion and excessive algae growth, and seasonally expose acres of lifeless mudflats, endangering fish life.

The disputed Tocks Island proposal suffered a further setback Sept. 13 when New Jersey Gov. William T. Cahill told a meeting of the Delaware River Basin Commission that he could not support the project unless major modifications and reductions in size were made.

Although Cahill lacked veto authority, his reservations, supported by Delaware Gov. Russell W. Peterson and by Henry L. Diamond, representing New York Gov. Nelson A. Rockefeller, threatened to delay work on the project for at least two years, according to commission officials.

Cahill said New Jersey faced "adverse environmental and financial impact" unless recreational facilities were redesigned to accommodate 4 million rather than 10 million visitors annually, a massive sewage plant were replaced by a series of far smaller plants, federal funds were guaranteed to pay for the planned road system, and the danger of eutrophication controlled. The governor conceded that the dam would aid flood control, but said state regulation of flood plains was the key to the problem. He said he still favored federal acquisition of the entire 72,000-acre tract for a national park.

Storm King challenge dismissed. By an 8–1 vote, the Supreme Court June 19 dismissed the appeals of conservation groups and the City of New York to block construction of one of the world's largest electrical power plants on Storm King Mountain near Cornwall, N.Y.

The court, with only Justice Douglas dissenting, rejected the appeals by refusing to review a decision by the Federal Power Commission that approved construction of the Storm King plant by the Consolidated Edison Co.

Still blocking construction were appeals by the same plaintiffs before a New York state court.

Four Corners plants criticized. The Senate Interior Committee, in a study reported Aug. 13, charged that piece-meal decision-making" and inadequate planning had resulted in avoidable environmental damage by a complex of coal-burning power plants in the Four Corners region of Arizona, New Mexico, Colorado and Utah.

The Interior Department, which had issued mining, road and plant leases and permits, had not adequately taken into account the recreational potential of the region, alternate plant sites, or the possibility that the plants would keep out other industries "by allocating scarce natural resources and preempting for that purpose [energy production] the capacity of the region to absorb pollutants," according to the report.

However, the committee said the plants were necessary to meet current needs, and that more would be required by 1980, although these should not be built without region-wide study.

FPC sees plant delays. The Federal Power Commission (FPC) said Nov. 2 that approval of over 100 hydroelectric power plants as well as natural gas pipelines would be delayed six months or more to meet a recent court order requiring the agency to issue its own environmental impact statements before approving license applications. The new rules were published Oct. 30.

The Supreme Court had refused Oct. 10 to review the lower court ruling in a suit brought by the Greene County, N.Y. Planning Board against a high-voltage transmission line. The agency had previously accepted environmental statements submitted by the license applicants.

The new rules, with a similar court-ordered policy at the Atomic Energy Commission would affect plants designed to supply more than half of new U.S. power needs in the 1970s.

Among the projects delayed, the FPC said Nov. 2, was the Appalachian Power Co. "Blue Ridge" two-dam complex on the New River in Virginia and North Carolina, to generate electricity and store water to flush pollution from the Kanawha River at Charleston, W.Va. The $350 million project had been initially proposed in 1965.

TVA dam delayed. The 6th U.S. Circuit Court of Appeals Dec. 13 upheld a lower court order barring construction by the Tennessee Valley Authority (TVA) of the Tellico Dam project on the Little Tennessee River in eastern Tennessee, until an environmental impact statement was completed.

The TVA had decided not to proceed with a 14-dam project on the French Broad River in western North Carolina, it was reported Nov. 16. The agency said local support for the $125 million earthen dams had eroded. The project had been intended for flood control, recreation and industrial purposes.

Interior scored on stripping. The General Accounting Office, in a report released Aug. 24, charged that the Interior Department had failed to abide by its own rules and to enforce the Environmental Policy Act in regulating strip coal mining on federal and Indian lands in the West.

The report, prepared at the request of the House Conservation and Natural Resources Subcommittee, said Interior had frequently issued prospecting and production permits and leases to mining companies without preparing adequate environmental impact statements or securing exploration plans. Compliance bonds, to cover reclamation costs, had not always been adequate, and had not been supplied at all in some cases. Coal companies had usually failed to supply data on reclamation methods and acreage reclaimed.

Most of the regulations had been set by Interior in 1969. Permits or leases had since been issued for 2.3 million acres of public or Indian lands, in response to the growing demand for energy in the Western states.

Subcommittee chairman Henry S. Reuss (D, Wis.) and ranking Republican Guy Vander Jagt (Mich.), in a letter to Russell E. Train, chairman of the President's Council on Environmental Quality, suggested that regulation and restoration of strip mines be turned over to the Army Corps of Engineers or the Environmental Protection Agency, to avoid a conflict of interest within the Interior Department, which was responsible for promoting exploitation of natural resources.

Strip mine curb dies. Senate Majority Leader Mike Mansfield (Mont.) said Oct. 14 he was removing from the calendar both the House and Senate Interior Committee versions of a bill to control strip coal mining, since he could not obtain agreement on a time limit for debate.

The House had passed a bill, by a 265–75 vote Oct. 11, that would have banned stripping on all inclines above 20 degrees, unless the miner proved the land could be reclaimed. No stripping could have taken place without permits issued by the Interior Department, which could have banned any mining it considered unsafe or likely to cause "irrevocable and lasting injury" to the environment.

Miners would have had to prepare reclaiming plans, and residents near proposed stripping sites would have had the right to demand Interior Department hearings.

Rep. Wayne Hays (D, Ohio), a sponsor of the measure, conceded in debate Oct. 11 that the bill would increase consumer electric bills by about 15¢ a month.

Supporters of the bill, heartened by the substantial House majority, predicted eventual passage of a strong measure. Rep. Ken Hechler (D, W. Va.), leader of the House anti-stripping forces, said Oct. 14 "there will be a pell-mell rush to strip huge tonnages of coal," which would "spur public outrage and result in vastly increased support for abolition."

Shale hearings begin. The Interior Department began a series of environmental impact hearings Oct. 10 in Denver in preparation for offering six tracts of federal land in Colorado, Utah and Wyoming for oil shale extraction.

The leases, each covering about 5,000 acres, would be designed to encourage development of commercial oil extraction methods. Royalties would be paid on future production.

EPA curbs leaded gas. The Environmental Protection Agency (EPA) ordered Dec. 27 that most gasoline stations offer customers lead-free fuel by July 1974 but delayed for 60 days a final ruling on its proposal to eliminate lead from all gasoline gradually beginning in 1975.

A coalition of 13 environmental groups filed suit Dec. 27 in U.S. Court of Appeals in Washington to force implementation of the lead ban program.

Prices & Other Developments

Utility rate guidelines. New guidelines for regulating utility rate increases were announced by the Price Commission March 17. The commission's chairman, C. Jackson Grayson Jr., said it had been established in recent hearings that "many rate increases, some of them substantial, would clearly be necessary in order to supply continuing adequate, safe and pollution-free services."

Under the new guidelines, supervision of utility rate increases would remain with state or federal regulatory agencies. But the agencies were to develop specific stabilization regulations, based on rates of return and cost factors, that would require approval by the commission, which would then issue certificates of compliance to permit the agencies to approve the increases without review. The commission would monitor the performance of the regulator agencies and review, within a 60-day period, noncertificated rate increases granted.

The commission March 8 had extended the freeze on utility rates until March 25, but said it would issue new regulations before that time. The freeze was first imposed Feb. 10.

Oil price rise OKed. The Price Commission April 5 approved average price increases of 2% for four major oil companies, with the maximum single-product increase held to 8%. The companies: Standard Oil Co. of California, Standard Oil Co. (Ohio), Continental Oil Co. and Humble Oil & Refining Co.

Standard Oil (Calif.) won approval of the Price Commission Nov. 20 to raise the price of low sulphur fuel used by electrical utilities by up to 9.2% in California and by up to 23.9% in Hawaii.

Texaco Inc. also was granted permission for an 8.3% price increase for fuel oil.

Natural gas price rise set. The Federal Power Commission announced April 6 that it would permit natural-gas producers to increase prices, particularly on new supplies, to spur exploration and offset the current short supply. An increase in the natural gas price was endorsed April 10 by Interior Secretary Rogers C. B. Morton, who told the House Committee on Interior and Insular Affairs "we are facing a fuel and power crisis." Undersecretary of State John N. Irwin 2d told the panel that the U.S. faced the prospect of importing half the oil it used by 1980.

To assure "an adequate and reliable supply of natural gas," the FPC had established an "optional procedure" regulating the sale of natural gas at the wellhead to interstate pipelines. Under the new rule, the producer could sell at over-ceiling prices on proof that the sale would serve the "public convenience and necessity."

In a formal complaint filed with the FPC May 15, 12 Democratic and two Republican senators charged the agency with an "anti-consumer bias" and "an arrogant usurpation of the legislative function" in its April 6 decision. The charges followed hearings before the Senate Antitrust & Monopoly Subcommittee April 28.

The senators charged that the natural gas industry could benefit by as much as $1 trillion over a period of the life of known gas reserves as a result of the proposed ruling. The cost to consumers was estimated at an additional $15 billion annually.

Tenneco gas price rollback. The FPC May 19 approved a voluntary settlement by the Tennessee Gas Pipeline Co., a division of Tenneco, Inc., and its customers and state regulatory commissions, rolling back a wholesale natural gas rate increase by nearly $46 million a year. The FPC staff and the General Services Administration had contested the settlement.

A proposed rate increase to $140 million a year had been requested in August 1970 and put in effect March 1971. The approved rate increase to $94.3 million represented a 19.8% hike and was opposed by two FPC members as a violation of Price Commission guidelines.

FPC approval resulted in part from a company promise to spend $100 million a year for five years for expansion of its delivery system, conditional on the availability of additional gas.

New gas policy. The FPC Aug. 3 approved a new natural gas pricing policy which would raise rates for household and industrial users.

The policy, which would permit producers to deliver new reserves of gas at prices above present ceilings, was taken "to spur domestic exploration and development," the FPC said. The agency said this course would be cheaper than relying on other fuels, foreign or domestic.

The FPC said its policy of increasing the supply of "the cleanest-burning and least polluting of all the fossil fuels" had the support of the Environmental Protection Agency and the Commerce and Interior Departments.

FPC bars new gas ceiling rates. The Federal Power Commission (FPC) ruled unanimously Dec. 12 that it would not raise the ceiling rate for natural gas sold on the interstate market. The commission instead urged producers to utilize the "optional pricing procedure" which the FPC had promulgated in August.

The FPC's policy of allowing free market forces to drive up the price of natural gas in an attempt to foster exploration and development of the fuel appeared to parallel an Administration position.

Interior Secretary Rogers C. B. Morton said Dec. 11 that the White House was considering asking Congress to end the regulation of natural gas prices.

The FPC had reported Nov. 29 that this winter's record natural gas shortage would be twice as large as the gap in supply during the winter of 1971.

The net shortage was estimated at 700 billion cubic feet (3% of the total national gas consumption).

An earlier FPC decision rendered Dec. 7 limited the commission's jurisdiction over synthetic gas. The agency declared that its regulatory power over the interstate price of synthetic gas extended only to cases where natural gas had been mixed with a synthetic fuel, such as liquid naphtha.

Electricity cost up in 1970. In its annual report issued Sept. 17, the FPC found that electricity costs for the average homeowner were higher in 1970 than at any time before 1940 and were 6% higher than 1969 costs.

Although actual costs per kilowatt hour declined from 2.7¢ to 1.53¢ from 1926–1968, heavy electrical usage caused consumers' bills to rise.

Geothermal power studied. Former Interior Secretary Walter J. Hickel asked in a study reported Dec. 10 that the U.S. begin a 10-year $685 million study of the economic potential of geothermal energy, the use of underground heat to produce electrical power. The study, prepared at the University of Alaska and sponsored by a National Science Foundation agency, predicted that geothermal power output by the year 2000 could exceed the amount of power currently produced by all sources in the U.S., if technical problems were solved.

Oilman new Transportation secretary. A California oil-company executive was named Dec. 7 as President Nixon's choice to be transportation secretary. Claude S. Brinegar, a senior vice president of Union Oil Co. of California, was chosen to succeed John A. Volpe.

Brinegar, 45, had been with Union Oil for 20 years. He was a member of its board of directors, president of its gasoline division, Union 76, and a director of International Speedway Corp., operator of the Daytona Speedway in Florida.

Brinegar's appointment was denounced later Dec. 7 by the Highway Action Coalition, an antihighway lobby, which cited his association with Union Oil. The lobby said it was "outraged" by the selection because his company had been involved in the Santa Barbara, Calif. oil spill in 1969 and other pollution incidents.

(William P. Clements Jr. was appointed Dec. 12 as deputy secretary of defense. Clements was chairman of the board of Sedco, Inc. of Dallas, an oil-drilling firm.)

Middle East & North Africa

Persian Gulf oil agreements. The six Persian Gulf members of the Organization of Petroleum Exporting Countries (OPEC) signed an agreement in Geneva Jan. 20 with Western oil companies to compensate them for devaluation of the U.S. dollar in December 1971.

The accord—affecting Abu Dhabi, Iran, Iraq, Kuwait, Qatar and Saudi Arabia—was to go into operation immediately and would give the six countries an 8.49% increase in the posted price of oil, which was $2.23 a barrel before negotiations began. (OPEC had requested an 8.57% increase.)

Annual revenue paid by Western companies to the Gulf states was expected to increase at least $700 million to approximately $8.9 billion. The agreement would expire in 1975, along with a pact signed in February 1971 between the Western companies and the six states.

Five of the Persian Gulf oil producing countries and the major Western oil companies reached initial agreement in New York Oct. 5 on sharing ownership of petroleum production. Details of the accord were not made public pending final approval of individual agreements that were to be worked out between the firms and the governments of Saudi Arabia, Qatar, Kuwait, Iraq and Abu Dhabi.

The companies involved were the Standard Oil Co. (New Jersey); the

Standard Oil Co. of Calif.; Texaco, Inc.; the Mobil Oil Corp.; the Royal Dutch Shell Group; the British Petroleum Co., Ltd.; and Compagnie Francaise des Petroles.

The chief negotiator for the Gulf states was Sheik Ahmed Zaki Yamani, Saudi Arabia's minister of oil and minerals. The interm agreement climaxed nine months of negotiations.

Algerian-U.S. oil-gas deals. Standard Oil Co. of New Jersey and the Algerian government Jan. 3 signed an agreement for crude oil deliveries valued at $400 million.

The agreement, extending until 1975 and calling for the purchase by Standard Oil of 100,000 barrels of oil a day, was facilitated by the Algerian-French petroleum agreement of December 1971, which ended a French boycott of Algerian oil and removed from non-French companies the threat of legal action for entering the disputed market.

President Houari Boumedienne conferred March 27 with David D. Newsom, U.S. assistant secretary of state for African affairs. The two men reportedly discussed prospects for the purchase of Algerian natural gas by U.S. firms. It was believed to be the first time Boumedienne had received a U.S. official of this rank since the two states broke diplomatic relations in 1967.

The U.S. Federal Power Commission (FPC) June 28 approved plans for a U.S. firm to import $1.5 billion worth of liquefied natural gas from Algeria, a move the FPC's presiding examiner, Max L. Kane, had tentatively authorized in May.

The FPC ordered El Paso Algeria—a company formed for the deal by the El Paso Natural Gas Co.—to place itself under FPC jurisdiction as a gas importer and not to use current U.S. price controls as a means of determining future gas prices.

The natural gas was to be liquefied at Arzew, Algeria in a plant constructed with the help of a $250 million Export-Import Bank loan. (In New York June 28, Henry Kearns, the bank's president, said his organization had a "preliminary pact" with Sonatrach, the Algerian state-owned petroleum corporation, to help in the project.) The liquefied gas was to be

sold by Sonatrach to El Paso Algeria at the price of 30.5¢ per million BTUs (British thermal units). El Paso would load the gas into nine tankers, to be purchased for an estimated $742 million, and when the tankers reached the high seas the gas would become the property of three U.S. utilities firms which had received FPC authorization. The Columbia LNG Corp. of Wilmington, Del. and the Consolidated System LNG Co. of Pittsburgh were to pay El Paso Algeria 77¢ per million BTUs and the Southern Energy Co. of Birmingham was to pay El Paso 83¢ for the same quantity.

The three firms planned to spend some $270 million to construct pipeline facilities and a terminal at Cove Point, Md., on Chesapeake Bay, to store the liquefied gas and convert it to its natural state. The terminal was expected to supply as many as 12 states and the District of Columbia.

El Paso Natural Gas and the Algerian government announced May 30 that the U.S. firm had contracted to purchase quantities of liquified natural gas in addition to the amount whose import had been tentatively approved by the Federal Power Commission earlier in the month.

The new agreement doubled the volume of gas to be imported, raising the total to two billion cubic feet per day over a 25-year period.

Eascogas LNG (liquified natural gas) Co. had reported May 25 the signing of a contract with the Algerian government for the import to the U.S. of 4.2 trillion cubic feet of the gas over a 22-year period. Eascogas consisted of the Public Service Electric & Gas Co. of New Jersey and the Algonquin Gas Transmission Co. of Boston. The latter two companies announced June 28 that they had signed an agreement for Burmah Oil Tankers, Ltd. of Scotland to transport LNG from Algeria to New York harbor and Narragansett Bay, R.I.

In order to make the project "economically viable," the FPC Oct. 5 modified restrictions it had attached to its June decision authorizing the import of Algerian LNG. The FPC removed the ceiling on escalation of prices El Paso Algeria would be allowed to pay the Algerian government for the gas. It thus validated a contract formula that tied this price to costs and wages in the U.S. oil and steel industries.

If costs of the project, "reasonably and prudently incurred, exceed present estimates," El Paso Algeria would be allowed in the future to apply for increases in the prices it charged the three pipeline companies involved in the deal.

The FPC also ruled that distributing firms—those buying from the pipeline companies—would not be required, as in the June decision, to sell the gas separately from existing supplies and that they could do so on the basis of incremental pricing.

European group orders Algerian gas. A European consortium, composed of France's Gaz de France and West German, Dutch and Belgian distributors, signed a 20-year contract in Algiers Dec. 15 with Algeria's state-owned oil and gas agency, Sonatrach, for delivery of 858 billion cubic feet of liquefied natural gas annually. The contract required the approval of the governments concerned. The European firms were also obliged to grant Sonatrach, by March 15, 1973, a major loan to cover its planned investments for the project.

In another major sale, a consortium that included British Petroleum and Mitsui & Co., the latter a Japanese firm, won a 20-year contract for delivery of an estimated $3 billion worth of liquefied natural gas from the Persian Gulf state of Abu Dhabi to Tokyo Electric Power Co., the Wall Street Journal reported Dec. 18. The contract, would require approval from the Abu Dhabi and Japanese governments.

Soviet-Libyan deal. The U.S.S.R. March 4 announced the signing of an agreement with a Libyan delegation in Moscow.

The Soviet news agency Tass said that the pact "provides for cooperation in prospecting, extracting and refinishing oil, in developing power generation and other branches of Libya's economy, as well as prospecting for mineral deposits and gas, and in training Libyan national cadres."

The Libyan delegates had included Abdel Salam Jalloud, minister of industry and economy and the second-ranking figure in his country's government.

Following Libya's nationalization of British oil holdings in December 1971, production had dropped by nearly half to 215,000 barrels a day, the London Times had reported Feb. 4.

The Times said the country had been unable to sell any of its oil because of British threats of legal action against foreign purchasers.

Soviet-Iraqi pact signed. The Soviet Union and Iraq signed a 15-year treaty of friendship and cooperation in Baghdad April 9. The 14-article document was initialed by President Ahmed Hassan al-Bakr and Soviet Premier Aleksei N. Kosygin, who had arrived in the Iraqi capital April 6.

The pact's Article 9 indicated that the Soviet Union was prepared to provide Iraq with more arms

The treaty also provided for more economic and technical cooperation between the Soviet Union and Iraq, including the development of oil production.

Kosygin had attended the dedication April 7 of the Soviet-financed nationalized North Rumaila oil field. The Iraq Petroleum Co., which had owned North Rumaila before it was seized by the Baghdad government in 1961, had warned April 6 that it would not give up its claims to the field. The consortium of Western oil firms said it would take legal action with regard to oil being produced and about to be shipped for the first time from North Rumaila.

The Soviet government newspaper Izvestia disclosed July 21 that both nations also had signed a trade and economic accord June 7. Under the agreement, the Soviet Union was to aid Iraq in further development of the North Rumaila oilfield and to build an oil refinery at Mosul and a 370-mile pipeline between Baghdad and Basra.

(The Iraqi news agency reported Aug. 7 that Brazil's state oil concern Petrobras had signed a contract for prospecting, production and marketing of oil in Iraq.)

Iraq seizes Western oil firm. Iraq June 1 seized the Iraq Petroleum Co., a Western-owned consortium of U.S., British, Dutch and French firms. Syria the same day followed suit and nationalized all IPC's assets on its territory, which included a section of a pipeline that passed from Iraq through Syria to Lebanon.

Iraq's nationalization move followed the collapse of negotiations between the government and IPC, which produced 10% of Middle East oil. A final offer by Iraq Petroleum had been rejected by Baghdad May 31, the last day of a two-week ultimatum the government had given May 17. During the negotiations, the government had demanded an increase in the company's lagging oil production.

In announcing the nationalization action, President Ahmed Hassan al Bakr accused IPC of "cutting production to a point which was unprecedented in Iraq or in any other oil-producing country." The drop in output was aimed at plunging Iraq into an "economic ordeal" and to "force the revolution to retreat from its objectives," Bakr said. According to the president, the cutback since the start of 1972 had thus far cost his country $86 million in tax and royalty revenue.

The Baghdad government had also accused IPC of cutting production to force Iraq to settle a number of disputes with the firm. The most controversial argument involved the company's long-standing demand for compensation for the government's 1961 seizure of the North Rumaila oil fields.

The British government June 2 called on its allies in the Central Treaty Organization (CENTO) to pressure Iraq to give fair compensation for IPC's seized assets. Foreign Secretary Sir Alec Douglas-Home made the request at the conclusion of a two-day CENTO conference in London.

The Soviet Union June 2 lauded the nationalization of IPC's assets by Iraq and Syria. The government newspaper Izvestia described the action as "a great victory for the Arab peoples, who are now struggling against the forces of imperialism and for a strengthening of their national economies."

Iraq June 3 offered to sell its nationalized petroleum at "reduced and competitive" prices. The government said crude oil from IPC's former Kirkuk oil fields would be sold in the Syrian terminal at Baniyas and the Lebanese terminal port of Tripoli "on long-term contracts which may be signed immediately."

France assured of Iraqi oil—Iraq guaranteed France a 10-year supply of oil from the northern Kirkuk fields under conditions that prevailed before the IPC's seizure. The accord, announced June 18, was negotiated during a five-day visit to Paris by Saddam Hussein Takriti, vice chairman of the Iraq Revolutionary Command Council.

Under the agreement, the Cie. Francaise des Petroles (CFP), partly owned by the French government, could continue to buy 23.75% of the oil produced at Kirkuk for 10 years under prenationalization conditions. The figure represented the CFP's ownership in the IPC, a U.S.-dominated consortium, and reportedly amounted to 14–18 billion barrels of crude oil annually.

French sources June 18 emphasized that the accord did not mean France would pull out of the IPC, which the Iraqi government reportedly sought to break up. France, which was heavily dependent on Iraqi oil, reportedly was mainly interested in assuring its supplies would not be cut off.

Italian accord set—Iraq found another market for its nationalized oil June 19, signing an agreement with Italy's National Hydrocarbons Agency (ENI) for the expansion of oil sales to Italy in exchange for increased technical aid.

The amount of additional oil to be shipped to Italy was not disclosed. A 10-year agreement had been signed in March under which ENI would supply equipment and services, while Iraq would sell Italy 20 million tons of crude oil.

(Iraq also agreed to supply 50,000 tons of crude per month to India's Barauni refinery, the Journal of Commerce reported June 20.)

Arab oil-bloc loan—The 10 nations that formed the Organization of Arab Petroleum Exporting Countries (OA-PEC) agreed June 20 to lend Iraq and Syria $169 million to help meet exchange shortages arising from their nationalization of the IPC.

The loan, which would cover the three months beginning June 1, would give Iraq $151 million and Syria $18 million. The money would come mainly from Kuwait, Abu Dhabi and Libya.

The OAPEC gave its full support to Iraq and Syria June 20, warning the IPC's parent companies against applying sanctions to stop Iraq from selling its nationalized crude oil. The Organization of Petroleum Exporting Countries had expressed its support for Iraq and Syria June 19.

In a related development June 22, the U.S. Treasury Department announced that the U.S. had voted against a $13.4 million World Bank education loan to Iraq, approved by the bank the same day. The U.S., which had never voted formally against a World Bank loan, had unsuccessfully sought to delay the grant "pending evidence of reasonable progress toward compensation" of the IPC companies. Iraq had promised compensation, but with the amount to be offset by unspecified back taxes and other charges, according to the New York Times June 23.

El Paso ruling left unchanged. The Supreme Court refused April 17 to disturb an order it handed down in 1964 requiring the El Paso Natural Gas Co. to divest itself of its Pacific Northwest pipeline properties.

The El Paso case had come to the court again after the Pacific Gas and Electric Co. of California asked the justices to rescind the 1964 order and a 1967 ruling that called for divestiture, "without delay."

The company asked the court to reconsider in light of "changed circumstances," foremost among them a shortage of natural gas. Pacific Gas asserted that the breaking up of El Paso would harm the public interest.

California, Wyoming, Idaho, Washington State, and Arizona backed the appeal, but Utah and the U.S. Justice Department opposed it.

Service stations purchase challenged. Texaco Inc.'s acquisition of 222 Douglas Oil Co. service stations was challenged March 27 by the Justice Department in an antitrust suit filed in federal district court in Los Angeles. The suit alleged that Texaco's purchase of almost half the service stations owned or leased by Douglas eliminated actual and potential competition for retail gasoline sales between the two companies in the western U.S.

Esso becomes Exxon. The Standard Oil Co. of New Jersey became the Exxon Corp., effective Nov. 1, as a result of a shareholders' vote Oct. 24.

The name change was made for efficiency, spokesmen said. Esso, the company's tradename, was introduced after the 1911 breakup of the Standard Oil group was ordered as a result of antitrust violations.

Former members of the Standard Oil group objected to the use of Esso in their marketing areas and a court ruling eventually restricted the use of the name to only 18 states.

Coal firms indicted. Five Midwest coal companies were indicted by a federal grand jury in Milwaukee April 11 on charges of conspiring to fix prices and allocate customers for coal sold through Lake Michigan and Lake Superior dock facilities.

The companies were the major sellers of dock coal in Minnesota and parts of Wisconsin, North Dakota and South Dakota, with about $42 million in 1968 sales. They were Great Lakes Coal & Dock Co., C. Reiss Coal Co., Pickands Mather & Co. (a subsidiary of Diamond Shamrock Corp.), Hometown, Inc. and Youghiogheny & Ohio Coal Co.

Latin America

Energy meeting. Ministers of power and petroleum from 20 countries concluded an informal meeting in Caracas Aug. 24, calling on Latin American importers and exporters to deal directly with one another, bypassing the large international companies. They also proposed the formation of a joint Latin American organization dealing with power, and creation of a unified power market in the region. Their final communique struck a nationalist tone, supporting the right of all Latin American countries to defend their natural resources.

Argentina. The Inter-American Development Bank agreed to lend Argentina $50 million for the construction of a hydroelectric plant on the Futaleufu river in the province of Chubut.

Bolivia. President Hugo Banzer Suarez and Brazilian President Emilio Garrastazu Medici met in the Brazilian town of Corumba, on the Bolivian border, April 4 and signed agreements that included a Brazilian commitment to purchase Bolivian crude oil from fields near Santa Cruz for two years beginning in July.

Bolivia began to export natural gas to Argentina April 29 after President Banzer opened the 328-mile Colpa-Yacuiba gas pipeline. The sale of natural gas to Argentina would bring Bolivia $15 million a year for the next 20 years.

Brazil. The state oil concern Petrobras signed a contract with Tennessee Colombia, a subsidiary of the U.S. firm Tenneco Oil Co., for joint exploration of concessions in Colombia (reported May 5). Petrobras announced in November that it would begin to operate in Colombia in 1976 on completion of the first Brazilian refinery to be built abroad. The plant would refine 15,000 barrels of oil daily.

Petrobras would build a refinery in Recife with a capacity of 300,000 barrels a day, it was reported Nov. 17. Along with the new Apucarana refinery in Parana, it would increase Petrobras' refining capacity by 60% from 600,000 barrels daily.

The Inter-American Development Bank Oct. 27 approved three loans totalling $57 million to help expand electric power production in northeast Brazil.

Colombia. The state-owned oil company Ecopetrol announced May 19 that it would launch an intensive exploration and exploitation campaign in seven areas in eastern, central and southern Colombia, representing an investment of $100 million. Much of the capital would be foreign, including, for the first time, investments by a French company, Petrol D'Aquitaine.

In a related development, the Mobil Oil Corp. sold its 49.9% interest in Colpet, a local subsidiary, to Ecopetrol.

The manager of the government's Industrial Development Institute (IFI) announced Sept. 18 that IFI and the Peabody Coal Co., a subsidiary of the Kennecott Copper Corp., had signed a fifty-fifty $120 million "association contract" for exploration and exploitation of El Cerrejon coal field in the northeastern state of Guajira. Proven reserves at the field were 80 million tons, with possible reserves of 200 million tons and a likely extraction rate of 5 million tons yearly, Uribe said.

Cuba. Annual petroleum output fell from 1970's 200,000-metric-ton peak to a peak of 200,000 metric tons in 1970 to 125,000 tons, increasing the island's already heavy dependence on the Soviet Union for oil supply, the Miami Herald reported Sept. 11. Russian oil shipments to Cuba had reportedly reached 3.9 million metric tons annually, up by about 75,000 metric tons from 1970.

The decrease in domestic oil production was widely attributed to the failure of the Guanabacoa oil fields east of Havana, discovered in 1968.

Ecuador. Brig. Gen. Guillermo Rodriguez Lara, head of a military junta that had seized control in Ecuador Feb. 15, announced Feb. 18 that he would revise oil contracts "if necessary" to follow his policy of changing the country's economic structure.

More than 30 foreign companies had oil exploration or exploitation contracts in Ecuador. A Texaco-Gulf consortium, which was scheduled to begin exportation in August, alone had investments estimated at $330 million.

The armed forces, because of provisions made by ex-President Jose Maria Velasco Ibarra, already shared substantially in oil tax revenues.

As had been planned, a first export of 314,000 barrels of oil from the new Texaco-Gulf fields in Oriente province left the country Aug. 15, reportedly making Ecuador Latin America's second largest petroleum exporter after Venezuela. The oil, bound for a Texaco refinery in Panama, had been pumped through a recently opened pipeline from the eastern Amazonian fields to the new petroleum terminal at Balao, on the northern coast.

The first shipment aboard a tanker from Ecuador's new petroleum fleet left Balao Aug. 26. The fleet had been formed by the navy and the Japanese firm Kawasaki Kisen Kaisha to handle 50% of the total production at the Amazonian fields.

Meanwhile, Gulf and Texaco were locked in negotiations with the Ecuadorean government over crucial phases of their operations, the Times of the Americas reported Sept. 6. The talks reportedly covered royalties to be paid by the companies, taxes for which they would be liable, the extent of their surface rights, and the size of social security contributions they would pay for their Ecuadorean employes.

Sources from another U.S. company, American Oil Co. (Amoco), disclosed Sept. 29 that the firm would suspend oil explorations in Oriente and return its concessions to the government. Observers noted the move followed announcement by the state oil concern CEPE that all contracts between the government and oil firms would be reviewed.

Oil shipments averaged 130,000 barrels a day during the first 1½ months of exportation, and 230,000 barrels daily during the first 10 days of October. (Oil revenues enabled the government to reduce the fiscal deficit from $106.6 million to $6.8 million, and to increase foreign reserves to $72 million, the London newsletter Latin America reported Nov. 10.)

Ecuador had sharply increased rentals on concessions to foreign companies exploring for oil in the eastern Amazon region, retroactive to Oct. 1, 1971, the New York Times reported Nov. 19. The companies reportedly responded by abandoning exploration work or sharply reducing operations.

Ecuador Nov. 29 announced the cancellation of the drilling concessions of a number of U.S. oil companies in the Gulf of Guayaquil, citing "irregularities" in the granting of the concessions in 1968. The ruling covered two million acres, where an estimated four trillion cubic feet of natural gas reserves had been discovered by a consortium headed by Ada Oil Co. of Houston.

The consortium had drilled nine wells at a cost of about $30 million, but production was not yet under way. Company officials were reported confused about the status of the concessions, not being certain they had actually been canceled.

Announcement of the cancellation followed the resignations of two civilian Cabinet officials, Finance Minister Nestor Vega Moreno and Production Minister Felipe Orellana. Vega Moreno, who played a major role in granting the concessions, resigned Nov. 28; Orellana quit Nov. 30. Vega Moreno's resignation was linked by most observers to internal disputes over oil policy. The government attributed Orellana's resignation to a dispute with the agrarian reform director, the Miami Herald reported Dec. 1.

The cancellation of the concessions and the dismissal of Vega Moreno were seen as a further tightening of the government's grip on the making of oil policy and a move to break all continuity with the policies of past governments

Guatemala. The Guatemalan government bought controlling interest in the country's electric power company from the Boise Cascade Corp. for $18 million cash, free of taxes, the Miami Herald reported May 20. The corporation, whose 50-year contract to supply Guatemala with power expired in May, had reportedly been accused by the left of making "excessive profits."

Peru. State-owned Petroperu announced that it had received government approval to award contracts to four groups containing seven U.S. oil companies to search for oil in the Amazon jungles of eastern Peru.

The companies named in the Jan. 17 Wall Street Journal report were Atlantic Richfield Co., Standard Oil Co. (Indiana), Shell Oil Co., Phillips Petroleum Co., Getty Oil Co., Panocean Oil Corp. and Transworld Oil Corp.

Two wells of high-grade, low-sulphur oil had been struck in the eastern jungle area near the Amazon river town of Iquites, the Miami Herald reported May 1.

The Marubeni Corp., a Japanese consortium, had offered to finance construction of a pipeline from the Amazon oil region to Peru's Pacific coast, El Nacional of Caracas reported May 8. The government reportedly planned to build the pipeline, which would require an investment of more than $400 million, as soon as Peruvian production reached 100,000 barrels a day.

Japan's government petroleum concern, Petroleum Development Co., disclosed Aug. 24 that it had obtained a concession from Petroperu to explore and develop oil resources in southeastern Peru.

President Juan Velasco Alvarado asserted in Lima that Peru would pay "not one cent more" for the expropriated assets of the International Petroleum Corp. (IPC), a U.S. enterprise seized without compensation in 1968, the Miami Herald reported July 7.

Velasco claimed Peru had paid $100 million for the company but the money had reverted to the state as partial payment of a $690 million government claim against IPC for 40 years of allegedly illegal oil extraction. He said he had discussed the matter with former U.S. Treasury Secretary John B. Connally and had been assured President Nixon would seek a solution to the problem.

Earlier, an official of the Rand Corporation had accused the U.S. June 14 of cutting off all economic aid to Peru in retaliation for the nationalization of IPC. The official, Luigi Einaudi, accused IPC of pressuring other U.S. companies not to invest in Peru.

In response to similar charges by the Inter-American Committee on the Alliance for Progress, which had accused the U.S. of opposing development loans to Peru in international lending agencies, an official from the U.S. Agency for International Development, Richard Weber, stated June 14 that Peru did not have "an absolute right to

development assistance, and we are under no obligation to provide it."

Petroperu Sept. 6 signed exploration and exploitation contracts with three foreign companies—Champlin Petroleum Co., Sun Oil Co. and Continental Oil Co.—authorizing them to operate over 2,470,000 acres in the Amazon jungle region northeast of Lima. The companies would be required to give 54% of any oil produced to Petroperu, and to extend to the government loans totaling $7.5–$8 million. Remaining reserves and equipment in the contract area would revert to the state after 35 years.

Petroperu announced Sept. 30 that it had struck oil for the fourth time in the Amazon region. The find was made near the site of the company's first discovery at Trompeteros on the Corrientes River, 600 miles northeast of Lima.

Uruguay. The Uruguayan government gave its blessing Aug. 3 to an immediate search for oil on the continental shelf under national waters in the Atlantic. The decision followed rumors that a report by a French firm, Compagnie Generale de Geophysique, had forecast rich oil reserves there. However, the state concern ANCAP warned the report was only "moderately optimistic."

Venezuela. It was reported Jan. 3 that the Venezuelan government had received two loans from the Inter-American Development Bank (IDB) totalling $33.1 million for development and expansion of its electric power facilities.

The Miami Herald reported March 5 that Venezuela had agreed to create a special fund to guarantee that oil company assets to be handed over when oil leases expired in the 1980s would be delivered in good condition. The assets would go to the state under the terms of the 1971 oil reversion act.

The government was charging that U.S. oil companies operating in Venezuela were keeping output at a minimum in protest against the oil reversion law.

Citing government statistics, Mines and Hydrocarbons Minister Hugo Perez La Salvia told newsmen that 1971 production had increased at a normal rate until April 13, when the law appeared in Congress. After that, he said, apart from a brief increase in October, production had steadily decreased, with the end of the year figure showing a 4.3% drop in comparison with 1970.

The oil companies and U.S. Ambassador Robert McClintock maintained the decline was due to the mild U.S. and European winter and the consequent drop in demand for crude oil. However, recent statements by McClintock and U.S. Secretary of State William P. Rogers had reinforced Venezuela's conviction that its oil production was being manipulated by external forces.

In a paper to the U.S. Congress, Rogers had said Venezuela was no longer reliable as an oil producer and warned that the U.S. would have to depend more on the Middle East.

In a recent message to the Venezuelan Congress, President Rafael Caldera had echoed Perez La Salvia's views on the decline in production and vowed that Venezuela would "not retreat or vacillate before any threat or tactic."

Shell-Venezuela claimed May 11 that during the first 10 days of the month it had increased production by 140,000 barrels a day—reaching 875,000 barrels —and the Creole Petroleum Corp. reported an increase of 155,000 barrels a day since the end of February. Shell said the increase was due to a greater demand for crude oil on the internationaal market.

Production was given a further boost May 11 when President Nixon authorized U.S. import of an additional 230,000 barrels a day to ward off anticipated shortages.

■ Perez La Salvia inaugurated a new Venezuelan Petroleum Corp. oil camp in the central state of Guarico April 13. Initial production at the camp would be about 17,000 barrels a day.

■ The second oil well completed under the new system of oil service contracts had been abandoned by the Mobil Oil Corp. because it was dry, El Nacional of Caracas reported May 13. The first well had also been dry.

The Congress of Venezuela had approved a bill guaranteeing state control of the domestic petroleum market by 1975, the London newsletter Latin America reported July 28.

The action followed a decision by the U.S. June 26 not to raise existing crude oil and oil product tariffs despite the expiration June 30 of a 20-year bilateral trade agreement with Venezuela. State Department sources said June 26 that the two countries would keep open discussions to explore the "development of petroleum and other economic relationships."

Two international oil companies, Mobil Oil Corp. and the Royal Dutch-Shell Group, announced Aug. 23 that their Venezuelan subsidiaries had suspended exploration in the southern part of Lake Maracaibo after striking several dry holes. The companies had received service contracts in 1971 from the state-owned oil company CVP, which would own any oil or gas discovered in the area.

The announcement followed disclosure by the Mines and Hydrocarbons Ministry Aug. 15 that Venezuela had produced only 3,086,000 barrels of oil daily in January-April, a drop of 712,000 barrels daily from the corresponding period in 1971. However, reports to the ministry by the oil companies included plans for an increase in production during the last four months of 1972 to balance the year's output with that of 1971, El Nacional of Caracas reported Sept. 8.

A Venezuelan government spokesman announced Oct. 19 that prices for 1973 oil exports would rise by about 12¢ a barrel. The increase, the third in as many years, compared with more than 60¢ in 1971 and more than 30¢ in 1972. It was accompanied by a new scale of shipping charges designed to profit from the expected rise in world tanker rates.

Former Mines Minister Juan Pablo Perez Alfonso attacked the increase Nov. 4, charging it did not "even reach 12¢ and is closer to 10¢." He had previously called for a price rise of 24¢ a barrel and a cut in production from 3.4 to 2.9 million barrels a day, saying the moves were necessary to protect dwindling reserves. He further warned that unless government spending was not sharply reduced, the country would face an "outrageous" budgetary deficit.

Mines Minister Hugo Perez La Salvia charged Nov. 6 that Perez Alfonso was spreading "falsehoods" in the guise of rational arguments and intimating that the government was lying about oil income. Finance Minister Luis Enrique Oberto said the price increase would raise Venezuela's oil income by more than $136 million in 1973, to a total of $2.2 billion, the Times of the Americas reported Nov. 1.

About 90 new oil wells had been drilled by various firms during August and September in the southern region of the eastern state of Monagas, El Nacional of Caracas reported Nov. 12. The majority had been drilled by the U.S.-owned Creole Petroleum Corp. and the Venezuelan state oil concern CVP. Creole said Nov. 13 that it had drilled 229 new wells in Venezuela during 1972.

Venezuela and the U.S. had been holding talks on possible joint exploitation of the Orinoco heavy oil belt, which contained oil reserves estimated at 700 billion barrels, it was reported Dec. 9. President Rafael Caldera reportedly had proposed an agreement providing for a guaranteed market in exchange for a guaranteed source of supply.

Caldera's proposal was in keeping with recent Venezuelan reservations about a U.S. decision, announced Dec. 8, to suspend for the first four months of 1973 hemispheric preferential treatment in imports of No. 2 oil, used for heating homes. The treatment had given Venezuela and Caribbean suppliers exclusivity in supplying the U.S. east coast.

The first section of the $924 million El Tablazo petrochemicals complex, on the northeastern shore of Lake Maracaibo, was inaugurated by President Caldera Dec. 16.

Talks between Venezuela and Colombia over national sovereignty in the potentially oil-rich Gulf of Venezuela were suspended Dec. 12 until February 1973.

Other Areas

Australia. The Mitsubishi Corp. of Japan announced Sept. 19 that it

had received an order from the Australian Gas Light Co., the major gas authority in New South Wales, for a 34-inch pipeline worth about $A54 million ($US65.3 million) to carry natural gas from Moomba in South Australia to Sydney.

The announcement drew strong protests from Australian firms and a threat of a trade union ban on work on the pipeline if a portion of the order was not placed with Australian companies. The Australian Gas Light Co. had said Australian steelmakers were not sufficiently equipped or experienced to produce the pipeline.

The Senate decided Oct. 18 to appoint a committee to investigate why the pipeline order had been awarded to Mitsubishi.

A New South Wales government minister announced Oct. 10 that the Santos-Delhi-Vamgas consortium, exploring for natural gas in the Cooper Basin in South Australia, had advised the Australian Gas Light Co. it had found the two trillion cubic feet of natural gas required to justify construction of a pipeline to Sydney. Construction of the pipeline had been held up because of the consortium's failure to prove sufficient reserves.

An accord establishing a permanent boundary in the potentially oil- and gas-rich seabed in the Timor Sea between northern Australia and Indonesia was signed in Djakarta Oct. 9 by Australian Foreign Minister Nigel Bowen and Indonesian Mining Minister Sumantri.

Under the agreement, the new boundary would extend to the southern edge of the Timor Sea trough and give Australia exploration rights to about 200 miles off the northwest Australian coast. Australia had originally claimed seabed rights to about 300 miles off its coast, while Indonesia had demanded an equal division of the rights.

The seabed border between Australia and Portuguese Timor would have to be negotiated separately with the Portuguese government.

Ghana. Ghana April 12 signed an agreement with the U.S. Offshore Exploration Oil Co. giving the firm a license to explore some 1,200 square miles off the Ghanaian coast.

Great Britain—*Coal strike.* The British government declared a state of emergency Feb. 9 to conserve power during the first nationwide coal strike in 46 years, in effect since Jan. 9. A ban on the use of electricity for advertising and display went into effect Feb. 11 as the first emergency measure.

The state of emergency was deemed necessary because picketing miners had blocked delivery of coal to power stations in many areas.

Layoffs were begun Feb. 14 when government emergency regulations went into effect closing down 20,000 medium-sized factories for four days a week and imposing 50% cuts in electricity consumption by the nation's biggest companies.

Other power reductions caused blackouts of up to nine hours daily on a staggered basis for homes, offices and stores. The government imposed a total ban on the use of electric heat in offices, shops, public halls, restaurants and buildings used for recreation or entertainment. The cuts were designed to conserve dwindling coal supplies at electric power stations.

The emergency measures won approval in the House of Commons Feb. 14 by a 315–276 vote.

As public criticism increased over Prime Minister Edward Heath's refusal to intervene in the dispute, Heath met Feb. 15 with business leaders' and with the general secretary of the Trades Union Congress (TUC), Victor Feather, in an attempt to deal with the growing crisis. Heath vainly urged Feather to press striking miners to end their picketing and allow coal supplies to reach power stations. He reiterated an earlier appeal to miners to return to work and accept a temporary increase of about $8 a week pending the outcome of a three-man court of inquiry set up to make non-binding recommendations for a settlement.

The miners voted Feb. 25 to end the strike, and they returned to the mines Feb. 28.

Acting on the recommendation of the National Union of Mineworkers, the miners had voted by a 96% majority to accept the recommendations of a special court of inquiry, headed by Lord Wilberforce, for a strike settlement.

The settlement meant an average pay increase of nearly 20%, considerably higher than the government's unofficial wage increase ceiling of 8%.

The Wilberforce report, released Feb. 18, had acknowledged the government's concern in avoiding inflationary pay increases, but concluded the miners deserved "special treatment" because of the "combination of danger, health hazard, discomfort in working conditions, social inconvenience and community isolation." It found that the "fall in the ranking of coal mining pay has been quite unwarranted" and ruled that the government should provide the funds if the National Coal Board could not afford the pay raises.

In line with another Wilberforce recommendation, leaders of the mine union and the coal board agreed Feb. 21 to set up a working party to seek ways to improve the coal industry's productivity.

Coal prices increased—A 7½% increase in coal prices, effective March 26, was announced by John Davies, secretary of state for trade and industry, March 6. The increase was designed to provide for part of the miners' additional wages recommended by the Wilberforce committee to end the 47-day coal strike.

To meet the rest of the cost, the government would increase the National Coal Board's accumulated deficit and make an emergency £100 million ($260 million) grant to the board.

The government Dec. 11 announced the grant of £720 million in coal subsidies over the next five years and a write-off of the £475 million accumulated deficit of the National Coal Board. The aid would slow the closure of marginally uneconomic pits and ease the social costs of closing hopelessly uneconomic pits.

Offshore oil & gas developments—The government announced March 15 that 66 groups of companies had been allocated oil exploration licenses covering 246 blocks in the North Sea, Irish Sea and Celtic Sea north and west of Britain. The successful applicants, chosen according to the companies' exploration and exploitation plans, were committed to drilling at least 224 wells over six years.

The biggest winner of the new exploration licenses was British Petroleum Co., awarded about 28 licenses either singly or in groups. Other large winners included the Royal Dutch-Shell Group and Standard Oil (New Jersey) and various groups involving Phillips Petroleum Co., Italy's state-controlled Agip and Belgium's Petrofina S.A.

A consortium composed of the U.S. Esso Petroleum Co. and Shell U.K. Ltd. confirmed Sept. 13 that a major new oil field, called Brent, had reserves of up to one billion barrels of oil and a production rate capacity of 300,000 barrels a day. The Brent field, located 100 miles northeast of the Shetland Islands in the North Sea, was apparently second only to the British Petroleum's Forties field off Aberdeen, expected to yield 400,000 barrels a day.

The Hamilton Brothers Petroleum Corp. (U.K.) had announced June 20 that a test well, located 200 miles east of Scotland, had yielded a crude oil flow of 3,000–3,400 barrels a day, with indications that the yield from a single platform could exceed 100,000 barrels a day.

The Mobil Oil North Sea, operators for an oil exploration consortium, announced Sept. 24 the discovery of a "sizable" oil and gas field 100 miles east of the Shetland Islands. Initial flow rates registered 3,400 barrels daily of low sulphur crude oil. The consortium was composed of Mobil Oil, the U.K. Gas Council, Amoco Petroleum (U.K.), Amerada Exploration and Texas Eastern (U.K.).

An international consortium headed by Phillips Petroleum Co. announced Aug. 18 an oil strike in a field named Eldfisk in the Norwegian sector of the North Sea. The field was estimated to have a production capacity of 10,000 barrels of oil a day. It was located about 12.5 miles south of the Ekofisk field, discovered by the consortium in 1970 and currently producing more than 40,000 barrels a day. The consortium was composed of Phillips, Petrofina S.A. of Belgium, the French-Norwegian Petronord Group, and Italy's state-controlled Agip.

The Phillips consortium announced Sept. 6 it had discovered another oil field, named Edda, in Norway's North Sea sector. The field, located eight miles northwest of Eldfisk, had estimated flow rates of about 10,000 barrels daily of low sulphur crude.

The North Sea would produce three million barrels of oil a day and attract $12.5 billion in investment by 1980, according to a study released Sept. 19 by Cazenove & Co., a London brokerage firm.

The report estimated North Sea reserves as of Jan. 1 at more than 12 billion barrels of oil and 50 trillion cubic feet of natural gas. In contrast, reserves of Alaska's Arctic North Slope were estimated at 10.5 billion barrels of oil and 30.5 trillion cubic feet of gas.

A separate estimate of about three million barrels a day for North Sea's oil output by the early 1980s was advanced at a London conference on North Sea oil Sept. 19 by Jack Birks, president of BP Trading Ltd., a unit of the British Petroleum Co. He said two-thirds of the total would come from British waters.

Greece. A lucrative oil-refinery concession was awarded as two separate projects July 26 to Stratis Andreadis, a banker and industrialist, and Ioannis Latsis, a tanker operator.

Deputy Premier Nikolaos Makarezos, who announced the contracts at a press conference, said Andreadis would construct at Pachi, near Athens, a $76 million oil refinery with a yearly capacity of six million tons. The contract called for the supply, transport and refining of 68 million tons of crude oil. Latsis would expand his present export refinery at Eleusis, near Athens, from a capacity of 1.2 million tons to 3.5 million tons yearly at a cost of $45 million. His contract called for the supply, transport and refining of 38 million tons of crude oil.

The concession had previously been awarded to Aristotle Onassis, the shipping magnate, who then had backed down because of what he contended was the project's economic unfeasibility.

An agreement under which the Soviet state engineering agency would build an $89 million electric power plant in northern Greece was signed by the Greek and Soviet governments in Moscow Nov. 17. The plant, to comprise three 125 megawatt generators, would be built at Philippi to profit from the area's rich peat deposits.

Japan. Japan announced May 23 that it would use $3 billion of its $16.5 billion of foreign-exchange reserves as loans for Japanese firms developing oil and other natural resources. The loans would be provided through financial institutions at 3% interest.

The fifth Japan-Soviet Joint Economic Committee completed a four-day session Feb. 24 in Tokyo with a Soviet request for Japanese economic assistance to build a trans-Siberian oil pipeline.

The request was for $151 million in loans to help finance a line paralleling the one joining the Tyumen oil fields and Irkutsk. The new line would also be extended to Nakhodka, a port on the Sea of Japan. Smaller requests were made for projects to study the development of coal mines in Yakutsk and natural gas deposits on the continental shelf near Sakhalin.

Ivan F. Semichastnov, the Soviet first deputy foreign trade minister, headed the Soviet delegation. Japan's was led by Kogoro Uemura, head of the Federation of Economic Organizations.

Japan and the U.S.S.R. in Tokyo Nov. 24 signed an agreement to explore jointly for oil and natural gas off Sakhalin island. The accord was reached after a week of talks between a Soviet delegation headed by Deputy Foreign Trade Minister Nikolai G. Osipov and 14 Japanese businessmen.

Under the accord, Japan was to provide Moscow with credits to purchase the necessary equipment, machinery and materials. The Soviets were said to have asked for $200 million in credits and an additional $30 million for the purchase of consumer goods from Japan.

The Soviet Union also was seeking an additional $3 billion in bank loans from Japan and the U.S. to finance part of the Sakhalin project and two other development programs—natural gas deposits at Yakutsk in East Siberia and oil fields at Tyumen in West Siberia.

North Vietnam. A second fuel link between China and North Vietnam was nearing completion, the U.S. Defense Department reported Aug. 4. The pipeline between the Chinese border town of Pingsiang and Kep, 30 miles northeast of Hanoi, was expected to increase North Vietnam's fuel supplies from China to 1,000 tons a day when completed later in August, according to the department. It ran parallel to another pipeline from Pingsiang to Hanoi, which had been completed in July and carried 400 tons of fuel a day.

Soviet Union. The government newspaper Izvestia announced April 22 the opening of a huge natural gas field at Medyezhye, in Siberia near the Arctic Circle. By 1975 the field was expected to produce more than 1,200 billion cubic feet of gas a year. The field's estimated reserves were 50 trillion cubic feet.

Soviet oil production rose from 7,460,000 barrels a day in 1971 to 7,900,000 in 1972.

U.S. firm gets Soviet deal—Dr. Armand Hammer, chairman of the Occidental Petroleum Corp., announced in London July 18 his firm had agreed to provide scientific and technical information in five fields in return for Soviet oil, natural gas and metals. Newspaper accounts indicated that the deal might also involve construction in the Soviet Union of several motels by Holiday Inns.

Hammer refused to estimate the dollar value of the agreement, which was concluded July 14 with the State Committee for Science and Technology. The areas covered by the pact were: exploration and production of oil and gas, production of agricultural fertilizers and chemicals, metal treating and plating, design and construction of hotels and conversion of garbage into fuels.

U.S. Commerce Secretary Peter G. Peterson, reporting Aug. 14 on a Soviet visit, said three U.S. firms were engaged in talks with the Soviet Union aimed at tapping that country's natural gas reserves in western Siberia. The El Paso Natural Gas Co. was discussing cooperation with the Japanese to export liquefied gas from the Soviet Far East to the U.S. West Coast. Tenneco, Inc. and the Texas Eastern Transmission Corp. were exploring plans to ship the gas from Murmansk to the U.S. East Coast. Peterson declared that other Soviet materials being considered for export to the U.S. were platinum, oil, copper, chrome, zinc and timber.

Franco-Soviet gas contract—The French Finance Ministry confirmed July 31 a contract for the purchase of 2.5 billion cubic meters of Soviet natural gas annually for 20 years, effective in 1976. The purchases would be worth 250 million francs (about $50 million) annually.

Arrangements for delivery of the gas were agreed on during the visit of French Premier Pompidou to Italy, when the Italian government approved the exchange of Soviet natural gas sold to France for Dutch natural gas sold to Italy. The Soviet gas would be supplied to Italy through a pipeline being built to deliver the gas Italy contracted to buy from the Soviet Union in 1969. In return Italy would cede to France 2.5 billion cubic meters of the six billion cubic meters of gas it would get under a contract with the Netherlands. The transport problem had stemmed from the absence of a pipeline connecting France to the Soviet gas fields.

Asian power projects—The official press agency Tass reported June 18 that final approval had been given for construction of an Asian power-generating project that would supply European Russia with electricity by the end of the decade.

The complex, located near strip mines in the Ekibastuz coal district in northeast Kazakhstan, would consist of four plants each with a capacity of four million kilowatts. A power line would run from Ekibastuz west to the central Russian town of Gryazi, where it would feed into a larger grid.

One of nine 300,000-kilowatt turbines was set in motion Nov. 15 at the Nurek hydroelectric station on the Vakhsh River in the Tadzhikistan Republic. The project was expected to be the world's largest rock-fill dam.

The U.S.S.R. opened its first major electric plant in the West Siberian oil fields Dec. 29. The new unit was located at Surgut, on the middle reaches of the Ob River, and was expected to achieve a generating capacity of 2.4 million kilowatts.

Trinidad-Tobago. The Miami Herald reported Jan. 23 that oil had begun to flow through a 23-mile submarine pipeline from the first of four offshore wells being tapped by the Amoco Trinidad Oil Co., a subsidiary of the Amoco International Oil Co. An Amoco spokesman said production from the well was expected to reach 70,000 barrels a day by the end of 1972.

Finance Minister George Chambers had urged that prices for Trinidad oil be increased.

Chambers cited prices received by other oil producers among developing countries. Libya, Chambers said, obtained $2.03 a barrel, Qatar received $1.25, and eight other oil producers, including Venezuela ($1.56), received prices in between. Trinidad received 69¢ a barrel, far below such countries.

Yugoslavia. President Tito of Yugoslavia and Rumanian President Nicolae Ceausescu opened the Iron Gates Dam, which the two countries had built accross the Danube River, May 16.

The project, reportedly the largest hydroelectric power plant in Europe, was completed at a cost of $450 million and was expected to have an annual production of 12 billion kilowatt hours.

In speeches at the dedication ceremonies, both leaders praised the Soviet Union for giving the project technical assistance.

Atomic Power in the U.S.

Suit against AEC's conflicting functions. Six environmental groups filed suit Jan. 6 to force the Atomic Energy Commission to divest itself of one of two competing functions—the promotion or the regulation of the nuclear power industry. The plaintiffs also asked for a temporary injunction barring the AEC from issuing permits for construction or operation of any new nuclear power plants.

The suit, filed in the U.S. District Court for the District of Columbia, charged that the law granting the commission the conflicting powers violated constitutional due process rights, since "there is no unbiased legal forum before which a citizen or group can secure a fair hearing in matters relating to atomic energy." Earlier attempts by the AEC to separate its promotional and regulatory activities were called "a fiction, not a fact."

To demonstrate their claim that "the commission is biased in favor of its promotional activity," the environmental groups said the AEC sells uranium fuel to utilities before fuel permits are granted, and allows purchases of "millions of dollars" of equipment before granting construction permits.

As a result of AEC bias, the suit charged, the public was deprived of a voice in protecting its health and safety, and in the matters of plant financing and eventual consumer power rates.

Filing the suit were the Conservation Society of Southern Vermont, the Chesapeake Bay Foundation, the Lloyd Harbor Study Group of Long Island, Businessmen for the Public Interest in Chicago, the Colorado Student Lobby, and the Cortlandt (N.Y.) Conservation Association.

U.S. to build breeder reactor. Atomic Energy Commission (AEC) Chairman James R. Schlesinger announced Jan. 14 that the Tennessee Valley Authority (TVA) and the Commonwealth Edison Co. of Chicago would construct a breeder nuclear power plant in Tennessee, the first of a type that was expected to become the mainstay of future power needs.

The facility, which would provide 350,000–500,000 kilowatts to the TVA, when completed in 1980, would cost $500 million. The federal government and the TVA would each supply $100 million, with public and private utilities and other corporations supplying the rest. A construction start was scheduled early in 1973.

Schlesinger said breeder reactors, which would create more fuel than they consume, would reduce thermal pollution of the environment, a problem with current nuclear reactors, because of a more efficient heat exchange system.

"We are reaching the point that supplies of fossil fuel—coal, gas and oil—are recognized to be limited," he said. "Furthermore, the availability of low-grade uranium is not too substantial,

with estimates being that at its present rate of use in non-breeding reactors it will be exhausted in several decades."

Possible drawbacks to the breeder reactors were the extreme combustibility of liquid sodium used in the process, and the dangerously long radioactive life of plutonium, the by-product fuel.

Schlesinger claimed that the proposed design incorporated more safety features than a Soviet breeder reactor plant already constructed.

Breeder reactor opposed. A group of 31 scientists and other experts issued a statement April 25 asking Congress to deny an Administration request for development funds for an experimental fast breeder reactor power plant because of safety and environmental hazards.

The group, including Dr. Linus Pauling, Dr. Harold C. Urey, Barry Commoner, Dr. Paul Ehrlich and Dr. George Weil, warned that the heat generated in a breeder reactor could lead, in cases of human or mechanical error, to a substantial atomic explosion. Problems of plutonium transport and disposal were also raised, as well as the danger of black market plutonium diversion for weapons manufacture.

Weil, the group's spokesman, urged that coal be reconsidered as an alternative energy source, since "it can be cleaned up, hopefully."

The AEC had defended the breeder reactor April 14 in an environmental impact statement for the proposed demonstration plant, to be built at an eventual cost of $2 billion–$3 billion as part of the Tennessee Valley Authority power grid. (The AEC had refused to prepare an impact statement for an entire generation of plants, as demanded by environmental groups.) The agency claimed that coal mining disrupted as much as 4,000 times as much land as equivalent nuclear power systems. The breeder reactor would produce additional fuels as it operated.

Doubts about the reactor were expressed by the Interior Department, which said April 14 that the danger of a "hydrogen explosion" had been insufficiently explored.

Breeder impact statement barred. U.S. District Court Judge George L. Hart Jr. dismissed March 24 in Washington a suit brought by the Scientists' Institute for Public Information, to force the AEC to issue a general environmental impact statement on the breeder nuclear reactor program.

The institute attorney, J. Gus Speth, had said that the type of reactor planned by the government would entail greater risk of explosion, thermal pollution and plutonium emission, and produce more radioactive wastes, than other types, and argued that all the problems should be discussed in public before the program's momentum and budget commitments made it inevitable.

Speth said annual federal expenditures of $100 million and the AEC plan to give the breeder reactor top priority together constituted a major federal act and a legislative proposal, thereby invoking the impact statement requirement.

Judge Hart ruled that the breeder program would have no tangible impact for some time, and that "there is no certainty any kind of system will ever be economically feasible."

Breeder reactor site set. The site of the first U.S. commerical-type fast breeder nuclear reactor power plant would be a 360-acre tract between Oak Ridge and Kingston, Tenn., AEC Chairman James R. Schlesinger and Aubrey J. Wagner, chairman of the Tennessee Valley Authority, announced Aug. 7. Wagner said construction would probably not start for two years.

Westinghouse to build breeder. The Westinghouse Electric Corp. was selected as prime contractor for the nation's first fast breeder reactor plant, it was reported Nov. 22.

The choice was made by the Project Management Corp., the plant's operating consortium led by the Tennessee Valley Authority and the Commonwealth Edison Co. of Chicago, and approved by the AEC. The losing bidders for the estimated $300 million job, General Electric Co. and Atomics International, a division of North American Rockwell, were expected to become subcontractors.

Michigan breeder to close. The only U.S. operating nuclear breeder reactor power plant, at Lagoona Beach, Mich., was to be closed, it was reported Nov. 29. The Enrico Fermi plant had run into continuous operating difficulties in its nine-year history as a demonstration plant. It was considered obsolete.

Pennsylvania vs A-plant. The state of Pennsylvania petitioned the AEC to deny a construction permit to the Public Service Electric & Gas Co., to build a nuclear power plant on a Delaware River island 11 miles from Philadelphia, it was reported Jan. 17.

Pennsylvania based its objections on the danger, in the event of an accident or release of water-borne radioactivity, "of a substantial number of deaths and injuries." The state cited a report by three AEC safety consultants, who asked consideration of additional safety features for the 2 million kilowatt plant, considering its location in a high population area.

Reactor safety disputed. Concern within the AEC that standard nuclear power plant safety equipment might be inadequate surfaced during hearings begun by the agency Feb. 2, in response to criticism by environmental and scientist groups.

Philip L. Rittenhouse of the AEC's Oak Ridge (Tenn.) National Laboratory testified March 10 that he and 28 other nuclear safety experts, many of them AEC employes, questioned the efficacy of the emergency core cooling system. The system was designed to flood nuclear reactors with cool water if mechanical failure caused reactor temperatures to rise dangerously. A cooling failure could theoretically cause an atomic explosion. The AEC had ordered modifications in eight existing plants in 1971 as part of an "interim core cooling policy," after laboratory models of the cooling system had failed to function properly, despite computer predictions of success.

A coalition of 60 environmental groups, the National Intervenors, had asked the AEC Feb. 1 to suspend all nuclear power plant licensing pending full scale tests of the cooling system, scheduled for 1975. The hearings were expected to continue into the summer.

A-power safety left to U.S. The Supreme Court ruled April 3, by a vote of 7-2, that the Atomic Energy Commission (AEC) alone had the authority to regulate radiation control standards for nuclear power plants.

The decision dealt a serious setback to efforts by local governments to seek tighter safety controls.

The court affirmed a lower court ruling involving the state of Minnesota and a nuclear power plant built by the Northern States Power Co. 30 miles north of Minneapolis.

Northern States had successfully sought an AEC operating license that would have permitted a daily discharge of 41,400 curies of radioactive debris. A curie was a unit of radioactive disintegration.

That discharge, however, would have violated a limit of 860 curies a day set earlier by Minnesota's State Pollution Control Agency.

Two lower U.S. courts had sustained Northern States' contention that it would have been either impossible or prohibitively expensive to meet Minnesota's limits. More than a dozen states had sided with Minnesota's position during the litigation.

Offshore plant planned. The Public Service Electric & Gas Co. announced March 22 it had chosen a site three miles off the New Jersey coast for construction by 1980 of a 1.15 million kilowatt nuclear power plant, to isolate possible thermal or radioactive dangers.

The New Jersey State Assembly had passed a resolution March 20 asking the AEC and the State Environmental Protection Department to oppose the project, the first of two envisioned by the firm, until further studies were made.

Florida offshore A-plant set. A 1,150 megawatt floating atomic power plant will be constructed near Jacksonville, Fla. by Westinghouse Electric Corp. and Tenneco, Inc., it was announced May 25. The plant was planned for 1978 or 1979 operation. A project spokesman

said the companies saw "the offshore nuclear power generating station as the only logical answer to power shortages in coastal urban areas."

Surface waste storage planned. In the face of opposition to its plans to bury radioactive wastes underground, the U.S. Atomic Energy Commission (AEC) announced May 19 it would build a huge steel and concrete waste storage facility above ground or just below the surface.

The facility would be ready by 1979 or 1980, at an estimated cost of $100 million. In the meantime, the AEC said it would continue research on burying the waste products of power plant fission reactions in abandoned salt mines or other underground formations.

A 1966 report by the National Academy of Sciences had advised the underground alternative. About $8 million had already been spent to prepare a salt mine near Lyons, Kans. to receive the wastes, before opposition from the state of Kansas and the Sierra Club blocked the project in 1971.

EPA offers model state law—The Environmental Protection Agency offered a model state law to control radioactive tailings (wastes) from Western uranium processing mills, it was reported Feb. 18.

The law, presented at a state-federal Southwestern water pollution conference in Las Vegas, would hold uranium mills responsible for the tailings, and require contouring and planting of grass. The tailings were a problem in nine Western states—Wyoming, Colorado, Utah, South Dakota, New Mexico, Texas, Arizona, Washington and Oregon, where 15 operating and 20 defunct uranium mills would be joined in the next 10 years by 10 new plants.

The AEC had refused legal responsibility for plants built before the 1969 Environmental Policy Act, but said waste control would be required before any new mills were licensed.

Temporary permits OKd. The Senate approved a House-passed measure 80-0 to grant temporary operating permits for nuclear power plants whose permanent licenses were under challenge, it was reported May 19.

The bill, designed to prevent possible power blackouts or brownouts in 1972, had been passed by the House May 3. The AEC would be empowered to grant the permits if it determined that safe operation were possible, until Oct. 30, 1973. About 12–15 plants nearing completion would be affected, whose permanent licenses were in some cases being challenged on environmental grounds.

The bill had been revised by the Joint Atomic Energy Committee, after criticism by environmentalists, to require a public hearing and a detailed environmental impact statement before even the temporary permit could be granted.

A-plant amendment dies. An attempt to amend the National Environmental Policy Act to permit temporary emergency licensing of nuclear power plants before environmental impact statements were completed was tabled July 19 by the Senate Interior Committee, killing chances for 1972 passage.

Some senators, including Republican-Conservative James L. Buckley (N.Y.) feared that the amendment would set a precedent endangering the entire environment act, while others said the Atomic Energy Commission (AEC) had not specified any region which would suffer a power shortage this summer from environmental delays in licensing completed plants.

Congress had already passed an amendment to the atomic energy act permitting interim licenses, and an AEC spokesman said July 19 that the agency would consider that authority adequate.

AEC Commissioner James R. Schlesinger had told the Joint Committee on Atomic Energy March 16 that without emergency licenses, five completed nuclear plants would be unable to meet possible 1972 and 1973 power shortages in New York City, Illinois, Iowa, Michigan and Wisconsin.

Under the proposed amendment, the AEC would have the power through June 1973 to issue licenses without public hearings and without submitting statements to the Council on Environ-

mental Quality. After July 1, 1973 no new interim licenses would be issued, but any license could be continued if the AEC declared the emergency still in effect.

Russell E. Train, chairman of the Council on Environmental Quality, and William D. Ruckelshaus, administrator of the Environmental Protection Agency, testified in favor of the amendment, leading some congressmen to charge that the Administration was planning to cripple the environmental act.

Illinois license granted—The most critical energy situation that would be covered by the amendment, Schlesinger said, concerned the Quad Cities dual-reactor plant at Cordova, Ill., whose license had been blocked by a federal judge at the request of the State of Illinois, the Izaak Walton League and the United Auto Workers.

The plaintiffs agreed March 30 to drop their objections when the companies, the Commonwealth Edison Co. and the Iowa-Illinois Gas & Electric Co. agreed to build a four-mile $30 million cooling system to prevent thermal pollution of the Mississippi River. Testing licenses for the plant were issued March 31.

Schlesinger said March 30 that the accord "alleviates the immediate pressure" for new legislation, but he still hoped for Congressional action to facilitate licensing of at least 13 other nuclear plants under construction.

AEC curbs NY A-plant. The Atomic Energy Commission (AEC) announced July 27 that it had ordered a nuclear power plant in Rochester, N.Y. to reduce power to 83% of capacity in June, after damage to some fuel rods was found.

The agency reported that about 4% of the rods had been flattened in sections, raising the danger of cracks and "higher levels of gaseous radioactivity in the reactor coolant system." Although three other operating nuclear power plants in the U.S. used the same type of fuel rod, the AEC said none had developed problems.

AEC orders cooling change. In its first regulatory act concerning non-nuclear problems at any nuclear power plant, the AEC ordered the Consolidated Edison Co. (Con Ed) Oct. 2 to install a "closed cycle" cooling system at its Indian Point, N.Y. complex by 1978, in order to end thermal pollution and massive fish kills in the Hudson River.

The AEC had been ordered in 1971 by the U.S. Court of Appeals for the District of Columbia to consider a plant's environmental impact before granting plant permits. The AEC told Con Ed to replace its current cooling system, which drew over a million gallons of water per minute from the Hudson and returned it 20 degrees warmer, with two 400-foot-high cooling chimneys, which would recycle the water. The chimneys were expected to cost as much as $150 million to build and $2.5 million a year to operate.

The utility was granted a five-year delay to alleviate the scarcity of power in the New York area, and to allow the company to retire air polluting plants in New York City.

A-safety probe asked. A coalition of 60 environmental groups Sept. 13 called the Consolidated National Intervenors asked the U.S. Congressional Joint Atomic Energy Committee to begin a "full-fledged, in-depth" study of safety problems in atomic energy plants that had been debated in eight months of still uncompleted hearings before the Atomic Energy Commission.

The problems centered on the reliability of the emergency cooling system. The Intervenors criticized the AEC for issuing two new operating licenses while the controversy continued, although the commission had ordered reductions in maximum permissible operating temperatures for the 25 previously licensed plants. A spokesman for the environment group claimed that a West German panel of scientists equivalent to the AEC Safety and Licensing Board had asked for a moratorium on new licenses in that country.

Dr. Edward E. David Jr., presidential science adviser, told the Congressional Committee Sept. 12 that atomic plants, and in particular fast breeder reactors, were the most promising source of pollution-free energy for the next 20

years. He said in addition that research in nuclear fusion as an energy source had progressed far enough so that the program would be expanded beyond this year's $61 million level, but he reported no important gains in coal gassification research.

The AEC had reported to the committee Sept. 7 that it had committed itself to pay all costs of developing the fast breeder reactor above the $250 million it expected private industry to contribute. Costs, including development and five years of operation, were expected to be about $700 million.

New safety rules suggested. L. Manning Muntzing, the Atomic Energy Commission's (AEC) director of regulation, proposed new atomic power plant operating procedures Oct. 26 to reduce the risk of accidental radiation leakage. The rules could lead to increasing energy costs.

Under the new rules, the 26 existing water-cooled plants would be required to reduce maximum temperatures of safety blankets around fuel rods from 2,300 to 2,200 degrees Farenheit, and adopt a more conservative formula for calculating temperatures. While the rules would not take effect until ordered by the five-member commission after further hearings, Muntzing said "prudent people would start to get ready first."

The rules would require power output reductions of up to 20% at some plants, and could lead to extensive modifications in plants now operating or under construction. The Washington Post cited an industry source Oct. 27 as estimating that the rules would cost the industry $50–$100 million over the next five years.

Buried plants studied. A study released by the California Institute of Technology and reported Sept. 23 found that underground nuclear power plants could be built with current technology, at a 10% higher cost.

The underground alternative would provide greater insurance against radiation leaks in case of accidents, and avoid defacing the landscape. Underground construction technology had already been perfected to house military stores and headquarters, the study stated.

Con Ed scores A-plants.

Louis H. Roddis Jr., president of Consolidated Edison Co. of New York, told an international conference of the Atomic Industrial Forum in Washington Nov. 11 that maintenance problems in nuclear power plants were the chief cause of their failure to meet projected output levels.

Roddis said 18 nuclear power plants in operation had recorded an average energy deliverability of 60.9%, although early studies had predicted an 80% rate. This meant that the plants had to be shut down for repairs and maintenance nearly 40% of the time they should have been in operation. Conventional fossil fuel plants had 75% deliverability.

One reason for the increased repair time and cost, Roddis said, was the need to limit radiation exposure among workers. A broken cooling pipe at the company's Indian Point plant had required the use of 700 rotating repairmen over seven months, compared with 25 men and two weeks needed for a comparable conventional plant problem.

Uranium plant need seen.

The Atomic Industrial Forum, the association of nuclear power producers, charged Oct. 16 that the Atomic Energy Commission (AEC) had seriously underestimated the amount of enriched uranium that would be required by the power industries of the non-Communist world by the 1980s, and called for an immediate start on planning for new enrichment plants.

Forum Secretary Edwin A. Wiggin said the AEC, which supplied most of the enriched uranium used in non-Communist countries, was not committed to begin planning a new plant until fiscal 1976. Wiggins said 8–11 years were needed to plan, finance and build a plant. He asked the AEC to start planning work on four new plants between 1972 and 1977.

The report estimated that 300 million kilowatts of electricity would be produced outside the U.S. by 1985, a figure 17% higher than the AEC estimate, in addition to expected domestic production increases.

Private uranium processing backed.

In a major policy turnabout, Atomic Energy Commission (AEC) Chairman James R. Schlesinger called on private industry Dec. 8 to build the uranium enrichment facilities that would be needed to supply the nation's growing electric power needs. The AEC published proposed regulations for corporate access to secret government technology involving the enrichment process, until now a carefully guarded government monopoly because of its use in the production of atomic bombs.

In a related move, Schlesinger announced the same day that the commission would draw up new, more businesslike contracts for the agency's sales of enriched uranium to the electric utilities using nuclear power reactors. The new contracts would be more specific about the quantity and delivery dates of the needed uranium and would carry a penalty for cancellation.

Schlesinger said private industry would also be needed to maintain the U.S.'s lead in the export of nuclear power technology. He predicted that export sales of U.S. enriched uranium and reactors would total $3.5 billion annually by the mid-1980's, compared with the present $900 million.

Fusion breakthrough.

The AEC Dec. 1 reported a major advance toward the goal of a controlled nuclear fusion reaction, which might eventually lead to a clean, almost inexhaustible supply of energy.

By using a process which Soviet scientists had pioneered, scientists at the Plasma Physics Laboratory at Princeton University were able to raise hydrogen plasma to densities and temperatures several times those achieved in a controlled environment, although still well below temperatures needed for fusion reactions.

Gas project radiation cited.

In a Denver briefing reported Oct. 1, AEC scientists estimated that natural gas that might be recovered through underground nuclear explosions would expose Los Angeles residents who would use the gas in unvented heaters to only 2 millirems of radiation a year and expose San Francisco users to 2.5 millirems. The accepted civilian exposure limit was 500 millirems a year, background radiation in Colorado constituted 200 millirems a year and one-time X-ray exposure constituted 50 millirems a year.

The AEC planned two more tests in its program to tap 300 trillion cubic feet of natural gas buried in Western rock deposits.

Nuclear Power in Other Areas

Breeder reactor built in U.S.S.R. The Soviet government newspaper Izvestia announced Jan. 4 that construction of the world's first commercial "fast breeder" nuclear power reactor had been completed. The new plant, located in the town of Shevchenko on the eastern shore of the Caspian Sea, was capable of producing about 350,000 kilowatts of energy.

The plant, which began operation Nov. 30, would yield 150,000 kilowatts of electrical power and would convert 30 million gallons of salt water into fresh water daily.

(The U.S.S.R. announced Dec. 26 it had started up a fourth conventional nuclear reactor at its biggest nuclear power station near Voronezh, in central European Russia. The new reactor would increase the station's total electrical capacity to 1.5 million kilowatts.)

U.S., U.S.S.R. set joint effort. The AEC announced Sept. 29 that the U.S. and Soviet Union had signed an agreement earlier in the week to expand their cooperative efforts in the peaceful uses of atomic energy to cover controlled thermonuclear fusion reactions and "fast breeder" reactors. Other exchanges covered in the U.S.-Soviet memorandum—the fifth since 1959—included atom-smasher machine studies, radiation chemistry, energy conversion and disposal of radioactive wastes.

Brazilian plant. The U.S. Export-Import Bank announced Feb. 1 that it would lend Brazil $138 million for construction of a nuclear power plant in Angra dos Reis, 90 miles southwest of Rio de Janeiro.

3 European nations in A-project. West Germany, Belgium and the Netherlands would cooperate in construction of a prototype natrium-cooled, fast-breeder nuclear power station, the West German Education and Science Ministry announced March 16.

Cost of the project was initially estimated at DM 1.07 billion (about $331 million). The plant would be located at Kalkar, on the West German side of the Dutch border.

Australia, Japan sign accord. Australia and Japan signed an agreement Feb. 21 on the peaceful uses of atomic energy. The accord would facilitate commercial agreements for the sale of Australian uranium to Japan and would provide for future collaboration between the two nations in research and commercial ventures.

Australian uranium exports. Contracts worth more than $A44 million ($US52.4 million) for the export to Europe and Japan of uranium oxide were negotiated by Queensland Mines Ltd. and Mary Kathleen Uranium Ltd., the latter 35% owned by Kathleen Investments Ltd., for the sale of 3,340 short tons over ten years from 1975.

National Development Minister Sir Reginald Swartz said the contracts had been won despite a world over-supply of uranium.

Second Indian plant starts. India's second atomic power plant began operation Aug. 11. The 400 megawatt facility, constructed with Canadian aid, was located at Rana Pratap Sagar in Rajasthan State, 250 miles south of New Delhi.

India's first nuclear plant had been opened in 1969 at Tarapur near Bombay. Under the Indian Atomic Energy Commission's 10-year program, four more reactor complexes were to be built by 1980 and were to produce 2,700 megawatts of power. The next plant was to be completed in Madras State in 1973. The atomic power plants were to replace India's more expensive traditional methods of power production with water and coal.

U.S., South Korea sign pact. The U.S. signed a 30-year nuclear agreement Nov.

24 to supply more than 25,000 pounds of enriched uranium and to permit export of a 600-megawatt nuclear reactor to South Korea. The pact was signed by South Korean Ambassador Kim Dong Jo and U.S. Assistant Secretary of State Marshall Green.

Finland to buy nuclear reactor. The Finnish government Dec. 22 approved plans by private industry to purchase a 660 megawatt nuclear power plant from a Swedish company. The plant would be built near Pori on Finland's west coast at an estimated cost of $600 million.

1973

The outbreak of war in the Middle East triggered a
sharp deterioration in energy supplies as Arab oil
nations cut petroleum output and embargoed oil to the
U.S. and Holland. Other oil producers joined the
Arabs in posting runaway price increases. Earlier in
1973 the U.S. had again relaxed curbs on oil imports.
President Nixon, in his second energy message, outlined his plans for making the U.S. self-sufficient in
energy. The U.S. again called on other oil-importing
nations to join together in solving the problems of
assuring adequate energy supplies to all. Spot gasoline
and diesel oil shortages developed in the U.S. Congressional probers investigated charges that major oil
and gas companies had deliberately provoked the
"energy crisis" in order to monopolize the energy
field. Nixon revamped his energy policy organization
and chose Colorado Gov. John A. Love as his "energy
czar." But Love later resigned on learning that Nixon
had decided to make William E. Simon the U.S.' top
energy official. Mandatory allocations were imposed
on home heating, jet, diesel and other distillate fuels.
Pressure for easing environmental standards increased
as the energy shortage grew. Libya nationalized U.S.
oil companies in rataliation for U.S. support of Israel.
Serious disruptions were taking place in the automobile, trucking, airline and other industries dependent
on petroleum. Great Britain went on a three-day week
as a coal-mine slowdown caused power shortages.

Before the Arab Oil Embargo

U.S. Developments

A dramatic worsening of the already serious worldwide energy situation took place in October 1973 after war erupted in the Middle East.

Egypt and Syria had attacked Israel on two fronts Oct. 6, and other Arab nations quickly entered the war against Israel. The U.S.S.R., which had armed the Arab belligerents before the fighting, began a heavy resupply operation after the attack to replace Arab equipment lost in combat, and the U.S. soon began emergency airlifts of weapons to Israel to replace Israeli losses.

The Arab oil nations decided Oct. 17 to use what was described as their "oil weapon." In an effort to get the U.S. to change its policy toward Israel, the Arabs first cut oil production and then imposed an embargo on oil shipments to the U.S.

The Arab oil action added a new and troubling problem to what was increasingly acknowledged to be the deteriorating energy predicament of the U.S. and the other developed nations.

These were among the developments in the energy situation prior to the Middle East conflict.

Shortages increase. A widening of the energy gap was reported early in the year.

In the week ending Jan. 5, the American Petroleum Institute reported that less than a six-week supply of home and industrial oil remained.

The energy crisis was exacerbated by severe cold weather throughout the Northeast, Midwest and South, by the chronic shortage of natural gas and by oil refineries operating below capacity, according to the Washington Post Jan. 13.

In addition, more stringent air pollution laws had caused factories to utilize greater quantities of No. 6 oil and No. 2 oil, the principal home heating fuel which was low in sulphur content. And in order to meet the growing demand for electricity, utilities were forced to use oil-powered turbine engines to reach peak electrical capacity, the Post reported.

Three major oil companies—Texaco, Inc., Shell Oil Co. and Mobil Oil Corp. —had begun to allocate supplies of home heating oil, kerosene, diesel fuel and aviation jet fuel, according to the Wall Street Journal Jan. 11.

On Jan. 26, Texaco announced that it had begun to ration heating oil in New Jersey, Delaware, Maryland and in the Philadelphia area of Pennsylvania.

The company also reported it would "allocate" deliveries to Massachusetts, where it supplied 35% of the state's No. 6 residual oil. After Jan. 27, Massachusetts' schools and industries would receive deliveries sufficient only to prevent pipes from bursting in freezing weather, according to Texaco.

"It's difficult to comprehend how a major oil company with the resources of Texaco can find itself in such a predicament," Massachusetts Gov. Francis W. Sargent declared Jan. 26.

As cold weather and short supplies began to affect the East, schools and factories were closing down in an 11-state region from Colorado to Ohio, according to the Post Jan. 13. Homes in Michigan and Illinois were without heat and families were being evacuated from homes in Minnesota.

Trucking companies were forced to curtail shipments of heating oil in Michigan because of the diesel fuel shortage and buses, barges and railroads were also affected by the dwindling supplies of the fuel.

Shutdowns were narrowly averted in the three major New York City airports when the Interior Department released imported jet fuel held in bond, the Journal reported Jan. 8.

The Pentagon confirmed Jan. 12 that it would purchase an additional 315 million gallons of jet fuel to cover supplies for the next six months. Its first fuel order, placed in October 1972, proved inadequate because of the "unanticipated demands for jet fuel that came from the increased air traffic in Southeast Asia," a spokesman said.

In a letter released Feb. 16, Commerce Secretary Frederick B. Dent urged the presidents of 45,000 companies to conserve energy and suggested 19 measures to deal with the "very tight fuel situation."

The actions included reducing window drafts, recycling waste as fuel, installing insulation and turning off lights.

Crude oil, gas, gasoline statistics—The American Petroleum Institute reported March 18 that proven reserves of crude oil (those known to exist and subject to an economical recovery process) fell by 1.7 billion barrels during 1972 from the previous year. Including the north slope of Alaska, oil reserves were put at 36.3 billion barrels, or 4.5% below demand.

A report issued the same day by the American Gas Association indicated proven reserves of natural gas declined 4.6% in 1972 to 266.1 trillion cubic feet. During 1972, the U.S. consumed 22.5 trillion cubic feet of gas—all that was produced in the country—as well as an additional one trillion cubic feet from Canada.

Gas reserves had been declining for the past five years, the association said.

The Petroleum Industry Research Foundation Inc. predicted April 2 that there would be a gasoline shortage of 125,-000 barrels a day, equaling 2% of demand, for the part of the U.S. east of the Rockies from March–September.

It was not expected that the supply problem could be easily alleviated because current refinery output was near maximum capacity and because a world shortage of crude oil was causing most foreign refineries to operate below capacity.

According to the report, gasoline supplies would be 3%–5% below demand by 1975.

A British Petroleum Co. report, released May 8, said total world consumption of oil totaled 2.59 billion tons in 1972 (up 192 million tons from 1971), and world production rose by 134 million tons to 2.61 billion; however, total demand grew by 8% while output increased at a rate of 5.4%

Estimates of proven reserves rose by 3.9 billion tons. Half of the total reserves of 90.9 billion tons were located in the Middle East. (The U.S. possessed 6.2% of the world's reserves.) Saudi Arabia accounted for 46% of total production in 1972 when the entire Middle East produced nearly 70% of the world's oil.

Nation lags, Patman says—Rep. Wright Patman (D, Tex.), chairman of the Congressional Joint Committee on Defense Production, Feb. 4 cited evidence, presented in the committee's annual report, that the nation was failing to deal with the worsening energy crisis.

Patman charged that 49 of 56 civilian nuclear power plants were behind schedule an average of 14.3 months each. He was also critical of a 1980 deadline for the commercial demonstration of a liquid metal fast breeder reactor, which the Atomic Energy Commission believed was essential for meeting nuclear energy needs.

While coal constituted 73% of fossil fuel reserves, no commercially feasible process for desulphurizing stack gases had been developed, Patman declared.

Despite these delays in federal action, government experts put U.S. dependence on foreign supplies of oil by 1985 at one-third to one-half of the nation's energy requirements.

FPC asks gas priority plan. The Federal Power Commission (FPC) Jan. 8 proposed a priority system of natural gas deliveries that would put major industrial gas consumers, especially electric power companies, on an interruptible basis.

The proposal, as described, would put residential and small commercial customers in a first priority category, and provide lower rates for the interruptible users. The FPC said large users could take advantage of pollution control equipment if forced to switch to alternative, more polluting energy sources.

While the proposal was being debated, the agency put into effect a priority list of shortage deliveries, with residential and small commercial users in the highest of eight categories.

The shortage had closed hundreds of schools, civil buildings and factories in Iowa, Illinois, Colorado, Nebraska, Indiana and Ohio.

The FPC reported Jan. 17 that natural gas reserves under the interstate system had declined in 1971 for the fourth straight year. The drop was 7.1%. Production showed no significant increase for the first time since 1963.

The "reserve to production ratio," a measure of industry's ability to meet future demands, dropped from 12.3 years of deliverability in 1970 to 11.5 years in 1971.

More than 2 trillion cubic feet of gas were added as new supplies in 1971 but were immediately used, with the effect that net ground reserves fell by 12.3 trillion cubic feet to a total of 161.3 trillion.

Gas price raised. The Federal Power Commission (FPC) Feb. 21 allowed a natural gas producer, George Mitchell and Associates Inc. of Texas, to raise interstate rates above FPC ceilings on the condition that the firm invest $16.5 million in a five-year gas exploration program.

Oil import quotas relaxed. President Nixon Jan. 17 authorized the unlimited purchase of home heating oil and diesel oil from foreign sources for the next four months. Quotas were also relaxed for the import of crude oil during 1973 with states east of the Rockies to receive 2.7 million barrels a day (the 1972 level was 1.78 million barrels a day) and with the West Coast area to receive 800,000 barrels a day (up from 717,000 in 1972).

Despite the Administration action, White House spokesman George A. Lincoln, director of the Office of Emergency Preparedness (OEP), warned, "We're going to have a tight fuel situation all winter."

Lincoln said oil refineries had increased production by 5% in the last two months under government pressure, but that another 5% gain of "about a million barrels a day" was still needed.

Higher prices for home heating oil, permitted under Phase 3 regulations, could provide a likely inducement for increased production, according to Lincoln.

Lincoln indicated that the modifications in the oil quota system were temporary, and said Nixon had under "final review" plans for major changes in the system which included a proposal to auction oil import allocations to the highest bidder.

Interior Secretary Rogers C. B. Morton had told the Senate Interior Committee Jan. 10 that "it is no longer feasible to set [oil] quota levels in a fixed relationship to domestic production or demand."

Morton revealed that the nation's current "energy trade deficit" of $4 billion a year would reach $20 billion by 1985 when imported oil would total 50% of use. "We're in a bind until we get in hand the economic and technologic techniques to fully exploit our own coal and oil resources," Morton said.

In testimony before the committee Jan. 10, Lincoln attributed the fuel oil shortage to federal price restrictions and manufacturing practices. "We froze the price of gasoline too high and of heating oil too low. The result is the refineries have been on a gasoline binge since last summer."

Energy group seeks quota change. A petrochemical energy group representing 19 corporate members petitioned the White House Jan. 23 to create a new category of crude oil imports quotas as an incentive for firms to produce low sulphur residual oil and naphtha and to construct new refineries.

The group opposed the Administration's plan to auction future increases in oil import quotas with the proceeds going to the government.

The group's proposal was aimed at providing the industry with extra supplies of natural gas, a "feedstock" or raw material for the energy industry. It was hoped that industrial users of natural gas would switch to the more readily available residual oil produced under the new quota system.

To induce this changeover, natural gas prices would be allowed to rise. "Tax incentives . . . to recover fuel oil premiums and/or conversion costs" would finance the fuel switch and the Environmental Protection Agency would be encouraged to relax air quality standards to permit coal production, which would then be used as another natural gas substitute, according to the industry proposal.

Legislation introduced in the House and Senate Jan. 18 by a coalition of Northeastern and Midwestern lawmakers called for a 90-day suspension of oil import quotas.

Sen. Thomas J. McIntyre (D, N.H.) and 18 co-sponsors then introduced a bill Feb. 26 to abolish the oil import quota system. The measure had 100 co-sponsors in the House.

The Exxon Corp. reversed a long-standing position April 13 and urged the Administration to suspend quotas on imported crude oil and products.

'Hardship' imports aided. Nixon March 23 removed the ceiling on the volume of oil imports available as "hardship" allocations by the Oil Import Appeals Board to small, independent distributors. Previously the board had limited imports to no more than 60,000 barrels a day.

Distributors had filed 220 applications with the board by March 23 seeking rights to more than 1 million barrels daily of crude and refined foreign oil products.

Large refiners, allowed to import 2.7 million barrels a day, were not affected by the executive order.

The Administration also appealed to domestic refiners to increase gasoline production. In the five weeks ending March 16, gasoline production "averaged only 42.5 million barrels a week," which was only 1 million barrels greater than during 1972, according to Darrell M. Trent, acting director of the Office of Emergency Preparedness.

"Consumption of gasoline is up 7% over last year and gasoline stocks of 215 million barrels are down 25 million barrels from last year," Trent said. (There are 42 gallons in a barrel.)

Oil reserve proposed. Sen. Henry M. Jackson (D, Wash.), chairman of the Senate Interior Committee, introduced legislation April 16 which would create a 90-day national oil reserve, made up in part from domestic oil derived from wells on federal lands.

Jackson claimed that the U.S. currently had only a five-day pipeline reserve.

Budget. President Nixon submitted his fiscal 1974 budget to Congress Jan. 29. The proposals called for increased research and exploration to increase the energy supply.

Some 3 million acres of Gulf of Mexico offshore oil and gas land would be leased in three regions ranging from Texas to Florida, although bonus payments from the sales were estimated at only $1.8 billion, $300 million less than in the current fiscal year, when more promising parcels were leased.

The Interior Department would receive $129 million for research on exploitation of fossil fuel, up from $107 million. A special $25 million research fund would be set up, to study potentially more abundant or less polluting energy sources such as liquified coal, solar heat and geothermal steam.

The Atomic Energy Commission (AEC) planned to increase research on nuclear power by $171 million to $974 million, including work on a fast breeder demonstration power plant. The AEC would increase permit and license charges to utilities from $18.4 million to $32.1 million.

Administration energy aides named. President Nixon Feb. 23 selected Charles J. DiBona, 41, as his special consultant on energy matters.

DiBona, president of the Center for Naval Analyses, a private firm performing research under government contract in association with the University of Rochester (N.Y.), joined other presidential assistants John D. Ehrlichman, Henry A. Kissinger, and Treasury Secretary George P. Shultz on a special White House energy committee.

William E. Simon, deputy secretary of the Treasury, was named chairman of the interdepartmental Oil Policy Committee Feb. 7.

Fuel research developments. A new electrolytic process that could reduce by 30% the amount of electricity used in aluminum smelting and eliminate the process' hazardous pollutants was announced Jan. 11 by the Aluminum Co. of America (Alcoa).

A spokesman said that although the new process had required 15 years of research at an expense of $25 million, future costs for aluminum smelting would be reduced because of labor savings as well as energy demand cutbacks. The aluminum industry used about 10% of the nation's industrial electricity, according to the New York Times Jan. 12.

Alcoa expected to begin construction of a 15,000 ton a year unit utilizing the new method.

The Interior Department Jan. 15 awarded the Westinghouse Electric Corp. an $8.2 million contract to study the conversion of coal to a low heat content, nonpolluting gas for electric power generation.

The project was aimed at providing utilities with a coal-based fuel free of sulphur, which was the major source of the industry's air pollution and present in most eastern coal deposits.

Mine leases opposed. The Tribal Council of the Northern Cheyenne Reservation in Montana unanimously instructed the Bureau of Indian Affairs (BIA) to cancel billions of dollars in strip mining leases granted by the agency on reservation land, charging that BIA had not obtained fair prices and had not prepared adequate environmental impact statements, it was reported March 31.

The action came a few days after the Peabody Coal Co. announced it would commit $700 million to exploit its 16,000-acre lease for a massive coal gassification project. Cheyenne leaders said the Consolidated Coal Co., which had only acquired exploration permits from BIA, had offered much more generous terms in 1972 for a project on a similar scale.

BIA officials said the tribe had already received $2.5 million in payments on leases and permits it now wished to declare "null and void."

United Mine Workers President Arnold R. Miller asked Congress April 6 to adopt a "crash program" for extracting clean gas from the nation's 1.3 trillion tons of coal reserves.

"While coal miners lose their jobs because air pollution standards prohibit burning high sulphur coal, the Administration cuts by $10 million funds allotted to the Environmental Protection Agency in the 1974 budget for the development of technology to reduce sulphur emmissions," Miller declared.

Trust cases. The Supreme Court ruled Feb. 22 that private power firms were subject to antitrust prosecution if they engaged in monopolistic or anti-competitive practices. The Otter Tail Power Co. had claimed exemption from the Sherman Antitrust Act because it was regulated by the Federal Power Commission under the Federal Power Act.

Otter Tail, which supplied power to parts of Minnesota, North Dakota, and South Dakota, was sued when it refused to sell power generated at government facilities to municipal power companies.

The High Court ended a 15-year legal battle March 5 when it ordered El Paso Natural Gas Co. to divest itself of its subsidiary Pacific Northwest Pipeline Co. Without delivering an opinion, the court affirmed 6–0 a federal district court plan that would turn over $290 million worth of energy resources, pipelines, and property to a group headed by Apco Oil Corp. of Oklahoma City.

It was the fifth time the El Paso case, which began in 1957 when the Justice Department filed an antitrust suit, had appeared before the court. After ordering the divestiture in 1964, the court rejected three succeeding plans for doing it. The plan finally agreed to by the court ordered division of bonded indebtedness and provided for a means of transfer of assets as well as reimbursement of El Paso.

Oil industry controls reimposed. The Administration acted March 6 to reimpose mandatory price controls on the nation's 23 largest manufacturers of crude oil, gasoline, heating oil and other refinery products. The Cost of Living Council (CLC) set a 1% limit for the first year of Phase 3 on average price increases which could be undertaken without government approval. Price hikes of 1.5% on a weighted annual average basis would be permitted if justified by higher costs.

Natural gas producers were not covered by the action but 95% of the petroleum industry, or those companies with sales in excess of $250 million annually, were included in the controls program.

The CLC decision resulted from its hearings held Feb. 7–9 to investigate price increases in home heating oil. According to the Wall Street Journal March 7, those increases had been as large as 9% in January during the recent energy crisis.

The CLC did not ask for a rollback on the January increases because the group said higher costs justified the increases.

Mandatory controls had not been reimposed on any segment of the economy since Phase 3 of the economic stabilization program had been announced Jan. 11.

Gasoline prices were expected to rise because of the fuel's short supply and a loophole built into the new price controls structure by the CLC when it failed to limit single product price increases within the 1%–1.5% ceiling.

Oil company developments—Mobil Oil Co. and Atlantic Richfield Co. raised bulk prices for gasoline sold to distributors by as much as 10%, the Wall Street Journal reported March 28. Cities Service Co. also boosted prices, effective May 1.

Gas rationing programs were announced by: the Exxon Co., U.S.A. May 29; Gulf Oil Co. May 29; Sun Oil Co. April 24; Standard Oil Co. of California May 7; Shell Oil Co. May 30; Standard Oil Co. of Ohio May 24; Texaco, Inc., May 30; and Mobil March 30.

Exxon May 15 announced a 30% expansion in its refinery capacity, at a cost of $400 million. An additional 350,000 barrels a day would be produced, bringing total capacity to more than 1.5 million barrels. Standard Oil of California had announced a $140 million refinery expansion project May 14.

Gulf announced May 17 it would reduce the octane ratings for two of its three grades of gasoline in order to increase the fuel yield from a barrel of crude oil and raise supplies.

Connally's mission. President Nixon revealed during a press conference held March 2 that former Treasury Secretary John B. Connally Jr., during a private world tour as an attorney, had undertaken at the President's request "some informal discussions with leaders in various parts of the world." Nixon said Connally was "very knowledgeable in the field of energy, and without getting involved in anything involving his client-attorney relationship he is studying the situation with regard to energy from the private sector and is making recommendations to me and to our energy group."

U.S., Antarctic oil, gas finds. Two major exploration ventures revealed evidence of sedimentary deposits with oil-bearing potential extending from New Jersey to Maine 30–50 miles offshore and deposits of natural gas beneath the Ross Sea near Antarctica.

The U.S. Geological Survey reported on the Atlantic coast explorations April 10. It said there were indications that the oil-bearing potential of portions of the Atlantic Outer Continental Shelf approached the density of Gulf Coast deposits and covered a wider area than California oil fields.

But, an Administration spokesman said that despite widespread commercial interest, leases would not be issued for the

North Atlantic littoral until resolution of a court case brought by 13 Eastern states seeking return on their offshore mineral rights. The Supreme Court was considering the states' petition.

Dennis Hayes, a Columbia University geologist, announced the Antarctic discovery March 22. He said the test drilling, the first ever undertaken of the Antarctic sea floor, revealed that "minor amounts" of natural gas components existed at a depth of 450–600 feet.

The discovery could prove important, Hayes said, because the deposits were located in unusually shallow water. In addition, researchers had concluded on the basis of examination of cores taken from sea floor drillings that Australia had broken away from Antarctica 50 million years ago. Major natural gas fields existed off the coasts of New Zealand and Tasmania, in areas which had been near the Ross Sea shelf before the continents separated.

Commercial exploitation of Antarctic fields could prove difficult, however, because a 1959 treaty declaring the area international territory, had no provisions for commercial development.

Presidential Message

Nixon offers energy policy. President Nixon outlined the Administration's energy policy in a message to Congress April 18. Among the executive actions undertaken to deal with the "possibility of occasional energy shortages and some increase in energy prices," Nixon emphasized measures designed to stimulate development of domestic energy supplies.

Action included:

■ An end to the 14-year-old mandatory quota on oil imports, effective May 1. The system would be replaced by a federal licensing program aimed at reducing crude oil shortages for refineries and providing incentives for expansion of domestic refining capacity.

The new "license fee quota" system, actually protective tariffs, would end tariffs of 10½¢ on a 42-gallon barrel of crude oil and 1¼¢ a gallon on gasoline. Quotas for 1973 could be filled without payment of duties, but over a seven-year period, the quota restriction would be reduced to zero.

Oil imports in excess of the authorized 1973 level and above the phased out future limit would be subject to license fees, starting at 10½¢ a barrel for crude oil and 52¢ a barrel for gasoline. The schedule would rise semiannually until November 1975 when rates would be 21¢ for crude oil, 63¢ for gasoline and 63¢ for other finished petroleum products. (The presidential action was authorized under national security provisions of the Trade Expansion Act.)

■ A planned threefold increase by 1979 in the sale of offshore leases for oil and gas exploration by the Interior Department. (The Santa Barbara channel off California was specifically excluded.) Nixon claimed that "by 1985 this accelerated leasing rate could increase annual energy production by an estimated 1.5 billion barrels of oil" (representing 16% of the nation's estimated oil needs, and 20% of natural gas requirements). Exploration of the continental shelf below a depth of 200 meters in the Gulf of Mexico would begin in 1974.

■ Establishment of an Office of Energy Conservation in the Interior Department "to educate consumers on ways to get the greatest return on their energy dollar" because, Nixon said, "we as a nation must develop a national energy conservation ethic." The President said the federal government would devise a "voluntary system of energy efficient labels for major home applicances" to encourage the manufacture of more efficient products.

Emphasizing the Administration's primary focus on increased energy supplies, Nixon argued, "Energy conservation is a national necessity, but I believe that it can be undertaken most effectively on a voluntary basis."

The President asked Congress to act on several legislative proposals. The White House sought legislation to end federal regulation of the wellhead price of natural gas in order to stimulate exploration.

New gas reserves and existing supplies newly committed to interstate markets would be free of the Federal Power Commission's artificially low price ceilings immediately, Nixon said. Wells already under contract for interstate delivery would become exempt from federal regulations when contracts expired. Gradually rising prices resulting from a lifting of restrictions would provide incentives for "increasing supplies and reducing inefficient usage," Nixon said.

The Administration asked Congress to extend to the oil industry another tax subsidy in the form of a tax credit for exploratory oil and gas drilling. (Treasury Secretary George P. Shultz conceded that tax revenues lost because of the credit could total $60 million a year.) The White House sought a 7% credit for "dry holes" and a 12% credit for "wet holes." No change in the 22% oil depletion allowance was mentioned.

The White House also asked authorization for the Interior Department to license offshore terminals for 250,000 ton supertankers, a proposal which Nixon claimed would reduce pollution by utilizing "fewer but larger tankers and deep water facilities."

The President reaffirmed support for construction of an Alaskan oil pipeline, adding that a Canadian transport route would take longer to build.

Nixon offered no major programs to speed development of nuclear power plants, although he asked Congress to shorten licensing procedures which had delayed plant operations.

The Administration would study the environmental effects of tapping oil shale reserves and harnessing geothermal energy before offering specific policy recommendations, Nixon said.

Nixon termed coal "our most abundant and least costly domestic source of energy" but offered no policy directives for expanding coal development. In encouraging broader use of the fuel, Nixon suggested that states delay implementation of secondary air pollution controls, assuring them that the Environmental Protection Agency would postpone that air safety deadline.

Text of the President's message:

CONCERNING ENERGY RESOURCES

To the Congress of the United States:

At home and abroad, America is in a time of transition. Old problems are yielding to new initiatives, but in their place new problems are arising which once again challenge our ingenuity and require vigorous action. Nowhere is this more clearly true than in the field of energy.

As America has become more prosperous and more heavily industrialized, our demands for energy have soared. Today, with 6 percent of the world's population, we consume almost a third of all the energy used in the world. Our energy demands have grown so rapidly that they now outstrip our available supplies, and at our present rate of growth, our energy needs a dozen years from now will be nearly double what they were in 1970.

In the years immediately ahead, we must face up to the possibility of occasional energy shortages and some increase in energy prices.

Clearly, we are facing a vitally important energy challenge. If present trends continue unchecked, we could face a genuine energy crisis. But that crisis can and should be averted, for we have the capacity and the resources to meet our energy needs if only we take the proper steps—and take them now.

More than half the world's total reserves of coal are located within the United States. This resource alone would be enough to provide for our energy needs for well over a century. We have potential resources of billions of barrels of recoverable oil, similar quantities of shale oil and more than 2,000 trillion cubic feet of natural gas. Properly managed, and with more attention on the part of consumers to the conservation of energy, these supplies can last for as long as our economy depends on conventional fuels.

In addition to natural fuels, we can draw upon hydroelectric plants and increasing numbers of nuclear powered facilities. Moreover, long before our present energy sources are exhausted, America's vast capabilities in research and development can provide us with new, clean and virtually unlimited sources of power.

Thus we should not be misled into pessimistic predictions of an energy disaster. But neither should we be lulled into a false sense of security. We must examine our circumstances realistically, carefully weigh the alternatives—and then move forward decisively.

WEIGHING THE ALTERNATIVES

Over 90 percent of the energy we consume today in the United States comes from three sources: natural gas, coal and petroleum. Each source presents us with a different set of problems.

Natural gas is our cleanest fuel and is most preferred in order to protect our environment, but ill-considered regulations of natural gas prices by the Federal Government have produced a serious and increasing scarcity of this fuel.

We have vast quantities of coal, but the extraction and use of coal have presented such persistent environmental problems that, today, less than 20 percent of our energy needs are met by coal and the health of the entire coal industry is seriously threatened.

Our third conventional resource is oil, but domestic production of available oil is no longer able to keep pace with demands.

In determining how we should expand and develop these resources, along with others such as nuclear power, we must take into account not only our economic goals, but also our environmental goals and our national security goals. Each of these areas is profoundly affected by our decisions concerning energy.

If we are to maintain the vigor of our economy, the health of our environment, and the security of our energy resources, it is essential that we strike the right balance among these priorities.

The choices are difficult, but we cannot refuse to act because of this. We cannot stand still simply because it is difficult to go forward. That is the one choice Americans must never make.

The energy challenge is one of the great opportunities of our time. We have already begun to meet that challenge, and realize its opportunities.

NATIONAL ENERGY POLICY

In 1971, I sent to the Congress the first message on energy policies ever submitted by an American President. In that message I proposed a number of specific steps to meet our projected needs by increasing our supply of clean energy in America.

Those steps included expanded research and development to obtain more clean energy, increased availability of energy resources located on Federal lands, increased efforts in the development of nuclear power, and a new Federal organization to plan and manage our energy programs.

In the twenty-two months since I submitted that message, America's energy research and development efforts have been expanded by 50 percent.

In order to increase domestic production of conventional fuels, sales of oil and gas leases on the Outer Continental Shelf have been increased. Federal and State standards to protect the marine environment in which these leases are located are being tightened. We have developed a more rigorous surveillance capability and an improved ability to prevent and clean up oil spills.

We are planning to proceed with the development of oil shale and geothermal energy sources on Federal lands, so long as an evaluation now underway shows that our environment can be adequately protected.

We have also taken new steps to expand our uranium enrichment capacity for the production of fuels for nuclear power plants, to standardize nuclear power plant designs, and to ensure the continuation of an already enviable safety record.

We have issued new standards and guidelines, and have taken other actions to increase and encourage better conservation of energy.

In short, we have made a strong beginning in our effort to ensure that America will always have the power needed to fuel its prosperity. But what we have accomplished is only a beginning.

Now me must build on our increased knowledge, and on the accomplishments of the past twenty-two months, to develop a more comprehensive, integrated national energy policy. To carry out this policy we must:

—increase domestic production of all forms of energy;
—act to conserve energy more effectively;
—strive to meet our energy needs at the lowest cost consistent with the protection of both our national security and our natural environment;
—reduce excessive regulatory and administrative impediments which have delayed or prevented construction of energy-producing facilities;
—act in concert with other nations to conduct research in the energy field and to find ways to prevent serious shortages; and
—apply our vast scientific and technological capacities—both public and private—so we can utilize our current energy resources more wisely and develop new sources and new forms of energy.

The actions I am announcing today and the proposals I am submitting to the Congress are designed to achieve these objectives. They reflect the fact that we are in a period of transition, in which we must work to avoid or at least minimize short-term supply shortages, while we act to expand and develop our domestic supplies in order to meet long-term energy needs.

We should not suppose this transition period will be easy. The task ahead will require the concerted and cooperative efforts of consumers, industry, and government.

DEVELOPING OUR DOMESTIC ENERGY RESOURCES

The effort to increase domestic energy production in a manner consistent with our economic, environmental and security interests should focus on the following areas:

Natural Gas

Natural gas is America's premium fuel. It is clean-burning and thus has the least detrimental effect on our environment.

Since 1966, our consumption of natural gas has increased by over one-third, so that today natural gas comprises 32 percent of the total energy we consume from all sources. During this same period, our proven and available reserves of natural gas have decreased by a fifth. Unless we act responsibly, we will soon encounter increasing shortages of this vital fuel.

Yet the problem of shortages results less from inadequate resources than from ill-conceived regulation. Natural gas is the fuel most heavily regulated by the Federal Government—through the Federal Power Commission. Not only are the operations of interstate natural gas pipelines regulated, as was originally and properly intended by the Congress, but the price of the natural gas supplied to these pipelines by thousands of independent producers has also been regulated.

For more than a decade the prices of natural gas supplied to pipelines under this extended regulation have been kept artificially low. As a result, demand has been artificially stimulated, but the exploration and development required to provide new supplies to satisfy this increasing demand have been allowed to wither. This form of government regulation has contributed heavily to the shortages we have experienced, and to the greater scarcity we now anticipate.

As a result of its low regulated price, more than 50 percent of our natural gas is consumed by industrial users and utilities, many of which might otherwise be using coal or oil. While homeowners are being forced to turn away from natural gas and toward more expensive fuels, unnecessarily large quantities of natural gas are being used by industry.

Furthermore, because prices within producing States are often higher than the interstate prices established by the Federal Power Commission, most newly discovered and newly produced natural gas does not enter interstate pipelines. Potential consumers in non-producing States thus suffer the worst shortages. While the Federal Power Commission has tried to alleviate these problems, the regulatory framework and attendant judicial constraints inhibit the ability of the Commission to respond adequately.

It is clear that the price paid to producers for natural gas in interstate trade must increase if there is to be the needed incentive for increasing supply and reducing inefficient usage. Some have suggested additional regulation to provide new incentives, but we have already seen the pitfalls in this approach. We must regulate less, not more. At the same time, we cannot remove all natural gas regulations without greatly inflating the price of gas currently in production and generating windfall profits.

To resolve this issue, I am proposing that gas from new wells, gas newly-dedicated to interstate markets, and the continuing production of natural gas from expired contracts should no longer be subject to price regulation at the wellhead. Enactment of this legislation should stimulate new exploration and development. At the same time, because increased prices on new unregulated gas would be averaged in with the prices for gas that is still regulated, the consumer should be protected against precipitous cost increases.

To add further consumer protection against unjustified price increases, I propose that the Secretary of the Interior be given authority to impose a ceiling on the price of new natural gas when circumstances warrant. Before exercising this power, the Secretary would consider the cost of alternative domestic fuels, taking into account the superiority of natural gas from an environmental standpoint. He would also consider the importance of encouraging production and more efficient use of natural gas.

Outer Continental Shelf

Approximately half of the oil and gas resources in this country are located on public lands, primarily on the Outer Continental Shelf (OCS). The speed at which we can increase our domestic energy production will depend in large measure on how rapidly these resources can be developed.

Since 1954, the Department of the Interior has leased to private developers almost 8 million acres on the Outer Continental Shelf. But this is only a small percentage of these potentially productive areas. At a time when we are being forced to obtain almost 30 percent of our oil from foreign sources, this level of development is not adequate.

I am therefore directing the Secretary of the Interior to take steps which would triple the annual acreage leased on the Outer Continental Shelf by 1979, beginning with expanded sales in 1974 in the Gulf of Mexico and including areas beyond 200 meters in depth under conditions consistent with my oceans policy statement of May, 1970. By 1985, this accelerated leasing rate could increase annual energy production by an estimated 1.5 billion barrels of oil (approximately 16 percent of our projected oil requirements in that year), and 5 trillion cubic feet of natural gas (approximately 20 percent of expected demand for natural gas that year).

In the past, a central concern in bringing these particular resources into production has been the threat of environmental damage. Today, new techniques, new regulations and standards, and new surveillance capabilities enable us to reduce and control environmental dangers substantially. We should now take advantage of this progress. The resources under the Shelf, and on all our public lands, belong to all Americans, and the critical needs of all Americans for new energy supplies require that we develop them.

If at any time it is determined that exploration and development of a specific shelf area can only proceed with inadequate protection of the environment, we will not commence or continue operations. This policy was reflected in the suspension of 35 leases in the Santa Barbara Channel in 1971. We are continuing the Santa Barbara suspensions, and I again request that the Congress pass legislation that would provide for appropriate settlement for those who are forced to relinquish their leases in the area.

At the same time, I am directing the Secretary of the Interior to proceed with leasing the Outer Continental Shelf beyond the Channel Islands of California if the reviews now underway show that the environmental risks are acceptable.

I am also asking the Chairman of the Council on Environmental Quality to work with the Environmental Protection Agency, in consultation with the National Academy of Sciences and appropriate Federal agencies, to study the environmental impact of oil and gas production on the Atlantic Outer Continental Shelf and in the Gulf of Alaska. No drilling will be undertaken in these areas until its environmental impact is determined. Governors, legislators and citizens of these areas will be consulted in this process.

Finally, I am asking the Secretary of the Interior to develop a long-term leasing program for *all* energy resources on public lands, based on a thorough analysis of the Nation's energy, environmental, and economic objectives.

Alaskan Pipeline

Another important source of domestic oil exists on the North Slope of Alaska. Although private industry stands ready to develop these reserves and the Federal Government has spent large sums on environmental analyses, this project is still being delayed. This delay is not related to any adverse judicial findings concerning environmental impact, but rather to an outmoded legal restriction regarding the width of the right of way for the proposed pipeline.

At a time when we are importing growing quantities of oil at great detriment to our balance of payments, and at a time when we are also experiencing significant oil shortages, we clearly need the two million barrels a day which the North Slope could provide—a supply equal to fully one-third of our present import levels.

In recent weeks I have proposed legislation to the Congress which would remove the present restriction on the pipeline. I appeal to the Congress to act swiftly on this matter so that we can begin construction of the pipeline with all possible speed.

I oppose any further delay in order to restudy the advisability of building the pipeline through Canada. Our interest in rapidly increasing our supply of oil is best served by an Alaskan pipeline. It could be completed much more quickly than a Canadian pipeline; its entire capacity would be used to carry domestically owned oil to American markets where it is needed; and construction of an Alaskan pipeline would create a significant number of American jobs both in Alaska and in the maritime industry.

Shale Oil

Recoverable deposits of shale oil in the continental United States are estimated at some 600 billion barrels, 80 billion of which are considered easily accessible.

At the time of my Energy Message of 1971, I requested the Secretary of the Interior to develop an oil shale leasing program on a pilot basis and to provide me with a thorough evaluation of the environmental impact of such a program. The Secretary has prepared this pilot project and expects to have a final environmental impact statement soon. If the environmental risks are acceptable, we will proceed with the program.

To date there has been no commercial production of shale oil in the United States. Our pilot program will provide us with valuable experience in using various operational techniques and acting under various environmental conditions. Under the proposed program, the costs both of development and environmental protection would be borne by the private lessee.

Geothermal Leases

At the time of my earlier Energy Message, I also directed the Department of the Interior to prepare a leasing program for the development of geothermal energy on Federal lands. The regulations and final environmental analysis for such a program should be completed by late spring of this year.

If the analysis indicates that we can proceed in an environmentally acceptable manner, I expect leasing of geothermal fields on Federal lands to begin soon thereafter.

The use of geothermal energy could be of significant importance to many of our western areas, and by supplying a part of the western energy demand, could release other energy resources that would otherwise have to be used. Today, for instance, power from the Geysers geothermal field in California furnishes about one-third of the electric power of the city of San Francisco.

New technologies in locating and producing geothermal energy are now under development. During the coming fiscal year, the National Science Foundation and the Geological Survey will intensify their research and development efforts in this field.

Coal

Coal is our most abundant and least costly domestic source of energy. Nevertheless, at a time when energy shortages loom on the horizon, coal provides less than 20 percent of our energy demands, and there is serious danger that its use will be reduced even further. If this reduction occurs, we would have to increase our oil imports rapidly, with all the trade and security problems this would entail.

Production of coal has been limited not only by competition from natural gas—a competition which has been artificially induced by Federal price regulation—but also by emerging environmental concerns and mine health and safety requirements. In order to meet environmental standards, utilities have shifted to natural gas and imported low-sulphur fuel oil. The problem is compounded by the fact that some low-sulphur coal resources are not being developed because of uncertainty about Federal and State mining regulations.

I urge that highest national priority be given to expanded development and utilization of our coal resources. Present and potential users who are able to choose among energy sources should consider the national interest as they make their choice. Each decision against coal increases petroleum or gas consumption, compromising our national self-sufficiency and raising the cost of meeting our energy needs.

In my State of the Union Message on Natural Resources and the Environment earlier this year, I called for strong legislation to protect the environment from abuse caused by mining. I now repeat that call. Until the coal industry knows the mining rules under which it will have to operate, our vast reserves of low-sulphur coal will not be developed as rapidly as they should be and the under-utilization of such coal will persist.

The Clean Air Act of 1970, as amended, requires that primary air quality standards—those related to health—must be met by 1975, while more stringent secondary standards—those related to the "general welfare"—must be met within a reasonable period. The States are moving very effectively to meet primary standards established by the Clean Air Act, and I am encouraged by their efforts.

At the same time, our concern for the "general welfare" or national interest should take into account considerations of national security and economic prosperity, as well as our environment.

If we insisted upon meeting both primary and secondary clean air standards by 1975, we could prevent the use of up to 155 million tons of coal per year. This would force an increase in demand for oil of 1.6 million barrels per day. This oil would have to be imported, with an adverse effect on our balance of payments of some $1.5 billion or more a year. Such a development would also threaten the loss of an estimated 26,000 coal mining jobs.

If, on the other hand, we carry out the provisions of the Clean Air Act in a judicious manner, carefully meeting the primary, health-related standards, but not moving in a precipitous way toward meeting the secondary standards, then we should be able to use virtually all of that coal which would otherwise go unused.

The Environmental Protection Agency has indicated that the reasonable time allowed by the Clean Air Act for meeting secondary standards could extend beyond 1975. Last year, the Administrator of the Environmental Protection Agency sent to all State governors a letter explaining that during the current period of shortages in low-sulphur fuel, the States should not require the burning of such fuels except where necessary to meet the primary standards for the protection of health. This action by the States should permit the desirable substitution of coal for low-sulphur fuel in many instances. I strongly support this policy.

Many State regulatory commissions permit their State utilities to pass on increased fuel costs to the consumer in the form of higher rates, but there are sometimes lags in allowing the costs of environmental control equipment to be passed on in a similar way. Such lags discourage the use of environmental control technology and encourage the use of low-sulphur fuels, most of which are imported.

To increase the incentive for using new environmental technology, I urge all State utility commissions to ensure that utilities receive a rapid and fair return on pollution control equipment, including stack gas cleaning devices and coal gasification processes.

As an additional measure to increase the production and use of coal, I am directing that a new reporting system on national coal production be instituted within the Department of the Interior, and I am asking the Federal Power Commission for regular reports on the use of coal by utilities.

I am also stepping up our spending for research and development in coal, with special emphasis on technology for sulphur removal and the development of low-cost, clean-burning forms of coal.

Nuclear Energy

Although our greatest dependence for energy until now has been on fossil fuels such as coal and oil, we must not and we need not continue this heavy reliance in the future. The major alternative to fossil fuel energy for the remainder of this century is nuclear energy.

Our well-established nuclear technology already represents an indispensable source of energy for meeting present needs. At present there are 30 nuclear power plants in operation in the United States; of the new electrical generator capacity contracted for during 1972, 70 percent will be nuclear powered. By 1980, the amount of electricity generated by nuclear reactors will be equivalent to 1.25 billion barrels of oil, or 8 trillion cubic feet of gas. It is estimated that nuclear power will provide more than one-quarter of this country's electrical production by 1985, and over half by the year 2000.

Most nuclear power plants now in operation utilize light water reactors. In the near future, some will use high temperature gascooled reactors. These techniques will be supplemented during the next decade by the fast breeder reactor, which will bring about a 30-fold increase in the efficiency with which we utilize our domestic uranium resources. At present, development of the liquid metal fast breeder reactor is our highest priority target for nuclear research and development.

Nuclear power generation has an extraordinary safety record. There has never been a nuclear-related fatality in our civilian atomic energy program. We intend to maintain that record by increasing research and development in reactor safety.

The process of determining the safety and environmental acceptability of nuclear power plants is more vigorous and more open to public participation than for any comparable industrial enterprise. Every effort must be made by the Government and industry to protect public health and safety and to provide satisfactory answers to those with honest concerns about this source of power.

At the same time, we must seek to avoid unreasonable delays in developing nuclear power. They serve only to impose unnecessary costs and aggravate our energy shortages. It is discouraging to know that nuclear facilities capable of generating 27,000 megawatts of electric power which were expected to be operational by 1972 were not completed. To replace that generating capaity we would have to use the equivalent of one-third of the natural gas the country used for generating electricity in 1972. This situation must not continue.

In my first Energy Special Message in 1971, I proposed that utilities prepare and publish long-range plans for the siting of nuclear power plants and transmission lines. This legislation would provide a Federal-State framework for licensing individual plants on the basis of a full and balanced consideration of both environmental and energy needs. The Congress has not acted on that proposal. I am resubmitting that legislation this year with a number of new provisions to simplify licensing, including one to require that the Government act on all completed license applicaitons within 18 months after they are received.

I would also emphasize that the private sector's role in future nuclear development must continue to grow. The Atomic Energy Commission is presently taking steps to provide greater amounts of enriched uranium fuel for the Nation's nuclear power plants. However, this expansion will not fully meet our needs in the 1980's; the Government now looks to private industry to provide the additional capacity that will be required.

Our nuclear technology is a national asset of inestimable value. It is essential that we press forward with its development.

The increasing occurrence of unnecessary delays in the development of energy facilities must be ended if we are to meet our energy needs. To be sure, reasonable safeguards must be vigorously maintained for protection of the public and of our environment. Full public participation and questioning must also be allowed as we decide where new energy facilities are to be built. We need to streamline our governmental procedures for licensing and inspections, reduce overlapping jurisdictions and eliminate confusion generated by the government.

To achieve these ends I am taking several steps. During the coming year we will examine various possibilities to assure that all public and private interests are impartially and expeditiously weighed in all government proceedings for permits, licensing and inspections.

I am again proposing siting legislation to the Congress for electric facilities and for the first time, for deepwater ports. All of my new siting legislation includes provision for simplified licensing at both Federal and State levels. It is vital that the Congress take prompt and favorable action on these proposals.

Encouraging Domestic Exploration

Our tax system now provides needed incentives for mineral exploration in the form of percentage depletion allowances and deductions for certain drilling expenses. These provisions do not, however, distinguish between exploration for new reserves and development of existing reserves.

In order to encourage increased exploration, I ask the Congress to extend the investment credit provisions of our present tax law so that a credit will be provided for all exploratory drilling for new oil and gas fields. Under this proposal, a somewhat higher credit would apply for successful exploratory wells than for unsuccessful ones, in order to put an additional premium on results.

The investment credit has proven itself a powerful stimulus to industrial activity. I expect it to be equally effective in the search for new reserves.

IMPORTING TO MEET OUR ENERGY NEEDS
Oil Imports

In order to avert a short-term fuel shortage and to keep fuel costs as low as possible, it will be necessary for us to increase fuel imports.

At the same time, in order to reduce our long-term reliance on imports, we must encourage the exploration and development of our domestic oil and the construction of refineries to process it.

The present quota system for oil imports—the Mandatory Oil Import Program—was established at a time when we could produce more oil at home than we were using. By imposing quantitative restrictions on imports, the quota system restricted imports of foreign oil. It also encouraged the development of our domestic petroleum industry in the interest of national security.

Today, however, we are not producing as much oil as we are using, and we must import ever larger amounts to meet our needs.

As a result, the current Mandatory Oil Import Program is of virtually no benefit any longer. Instead, it has the very real potential of aggravating our supply problems, and it denies us the flexibility we need to deal quickly and efficiently with our import requirements. General dissatisfaction with the program and the apparent need for change has led to uncertainty. Under these conditions, there can be little long-range investment planning for new drilling and refinery construction.

Effective today, I am removing by proclamation all existing tariffs on imported crude oil and products. Holders of import licenses will be able to import petroleum duty free. This action will help hold down the cost of energy to the American consumer.

Effective today, I am also suspending direct control over the quantity of crude oil and refined products which can be imported. In place of these controls, I am substituting a license-fee quota system.

Under the new system, present holders of import licenses may import petroleum exempt from fees up to the level of their 1973 quota allocations. For imports in excess of the 1973 level, a fee must be paid by the importer.

This system should achieve several objectives.

First, it should help to meet our immediate energy needs by encouraging importation of foreign oil at the lowest cost to consumers, while also providing incentives for exploration and development of our domestic resources to meet our long-term needs. There will be little paid in fees this year, although all exemptions from fees will be phased out over several years. By gradually increasing fees over the next two and one-half years to a maximum level of one-half cent per gallon for crude oil and one and one-half cents per gallon for all refined products, we should continue to meet our energy needs while encouraging industry to increase its domestic production.

Second, this system should encourage refinery construction in the United States, because the fees are higher for refined products than for crude oil. As an added incentive, crude oil in amounts up to three-fourths of new refining capacity may be imported without being subject to any fees. This special allowance will be available to an oil company during the first five years after it builds or expands its refining capacity.

Third, this system should provide the flexibility we must have to meet short and long-term needs efficiently. We will review the fee level periodically to ensure that we are imposing the lowest fees consistent with our intention to increase domestic production while keeping costs to the consumer at the lowest possible level. We will also make full use of the Oil Import Appeals Board to ensure that the needs of all elements of the petroleum industry are met, particularly those of independent operators who help to maintain market competition.

Fourth, the new system should contribute to our national security. Increased domestic production will leave us less dependent on foreign supplies. At the same time, we will adjust the fees in a manner designed to encourage, to the extent possible, the security of our foreign supplies. Finally, I am directing the Oil Policy Committee to examine incentives aimed at increasing our domestic storage capacity or shut-in production. In this way we will provide buffer stocks to insulate ourselves against a temporary loss of foreign supplies.

Deepwater Ports

It is clear that in the foreseeable future, we will have to import oil in large quantities. We should do this as cheaply as we can with minimal damage to the environment. Unfortunately, our present capabilities are inadequate for these purposes.

The answer to this problem lies in deepwater ports which can accommodate those larger ships, providing important economic advantages while reducing the risks of collision and grounding. Recent studies by the Council on Environmental Quality demonstrate that we can expect considerably less pollution if we use fewer but larger tankers and deepwater facilities, as opposed to the many small tankers and conventional facilities which we would otherwise need.

If we do not enlarge our deepwater port capacity it is clear that both American and foreign companies will expand oil transshipment terminals in the Bahamas and the Canadian Maritime Provinces. From these terminals, oil will be brought to our conventional ports by growing numbers of small and medium size transshipment vessels, thereby increasing the risks of pollution from shipping operations and accidents. At the same time, the United States will lose the jobs and capital that those foreign facilities provide.

Given these considerations, I believe we must move forward with an ambitious program to create new deepwater ports for receiving petroleum imports.

The development of ports has usually been a responsibility of State and local governments and the private sector. However, States cannot issue licenses beyond the three-mile limit. I am therefore proposing legislation to permit the Department of the Interior to issue such licenses. Licensing would be contingent upon full and proper evaluation of environmental impact, and would provide for strict navigation and safety, as well as proper land use requirements. The proposed legislation specifically provides for Federal cooperation with State and local authorities.

CONSERVING ENERGY

The abundance of America's natural resources has been one of our greatest advantages in the past. But if this abundance encourages us

to take our resources for granted, then it may well be a detriment
to our future.

Commonsense clearly dictates that as we expand the types and
sources of energy available to us for the future, we must direct equal
attention to conserving the energy available to us today, and we must
explore means to limit future growth in energy demand.

We as a nation must develop a national energy conservation ethic.
Industry can help by designing products which conserve energy and
by using energy more efficiently. All workers and consumers can help
by continually saving energy in their day-to-day activities: by turning out lights, tuning up automobiles, reducing the use of air conditioning and heating, and purchasing products which use energy
efficiently.

Government at all levels also has an important role to play, both
by conserving energy directly, and by providing leadership in energy
conservation efforts.

I am directing today that an Office of Energy Conservation be
established in the Department of the Interior to coordinate the energy
conservation programs which are presently scattered throughout the
Federal establishment. This office will conduct research and work
with consumer and environmental groups in their efforts to educate
consumers on ways to get the greatest return on their energy dollar.

To provide consumers with further information, I am directing the
Department of Commerce, working with the Council on Environmental Quality and the Environmental Protection Agency, to develop
a voluntary system of energy efficiency labels for major home appliances. These labels should provide data on energy use as well as a
rating comparing the product's efficiency to other similar products. In
addition, the Environmental Protection Agency will soon release the
results of its tests of fuel efficiency in automobiles.

There are other ways, too, in which government can exercise leadership in this field. I urge again, for example, that we allow local officials to use money from Highway Trust Fund for mass transit purposes. Greater reliance on mass transit can do a great deal to help us
conserve gasoline.

The Federal Government can also lead by example. The General
Services Administration, for instance, is constructing a new Federal
office building using advanced energy conservation techniques, with a
goal of reducing energy use by 20 percent over typical buildings of the
same size. At the same time, the National Bureau of Standards is
evaluating energy use in a full-size house within its laboratories. When
this evaluation is complete, analytical techniques will be available to
help predict energy use for new dwellings. This information, together
with the experience gained in the construction and operation of the
demonstration Federal building, will assist architects and contractors
to design and construct energy-efficient buildings.

Significant steps to upgrade insulation standards on single and
multi-family dwellings were taken at my direction in 1971 and 1972,
helping to reduce heat loss and otherwise conserve energy in the residential sector. As soon as the results of these important demonstration
projects are available, I will direct the Federal Housing Administra-

tion to update its insulation standards in light of what we have learned and to consider their possible extension to mobile homes.

Finally, we should recognize that the single most effective means of encouraging energy conservation is to ensure that energy prices reflect their true costs. By eliminating regulations such as the current ceiling on natural gas prices and by ensuring that the costs of adequate environmental controls are equitably allocated, we can move toward more efficient distribution of our resources.

Energy conservation is a national necessity, but I believe that it can be undertaken most effectively on a voluntary basis. If the challenge is ignored, the result will be a danger of increased shortages, increased prices, damage to the environment and the increased possibility that conservation will have to be undertaken by compulsory means in the future. There should be no need for a nation which has always been rich in energy to have to turn to energy rationing. This is a part of the energy challenge which every American can help to meet, and I call upon every American to do his or her part.

RESEARCH AND DEVELOPMENT

If we are to be certain that the forward thrust of our economy will not be hampered by insufficient energy supplies or by energy supplies that are prohibitively expensive, then we must not continue to be dependent on conventional forms of energy. We must instead make every useful effort through research and development to provide both alternative sources of energy and new technologies for producing and utilizing this energy.

For the short-term future, our research and development strategy will provide technologies to extract and utilize our existing fossil fuels in a manner most compatible with a healthy environment.

In the longer run, from 1985 to the beginning of the next century, we will have more sophisticated development of our fossil fuel resources and on the full development of the Liquid Metal Fast Breeder Reactor. Our efforts for the distant future center on the development of technologies—such as nuclear fusion and solar power—that can provide us with a virtually limitless supply of clean energy.

In my 1971 Energy Special Message to the Congress I outlined a broadly based research and development program. I proposed the expansion of cooperative Government-industry efforts to develop the Liquid Metal Fast Breeder Reactor, coal gasification, and stack gas cleaning systems at the demonstration level. These programs are all progressing well.

My budget for fiscal year 1974 provides for an increase in energy research and development funding of 20 percent over the level of 1973.

My 1974 budget provides for creation of a new central energy fund in the Interior Department to provide additional money for nonnuclear research and development, with the greatest part designated for coal research. This central fund is designed to give us the flexibility we need for rapid exploitation of new, especially promising energy technologies with near-term payoffs.

One of the most promising programs that will be receiving increased funding in fiscal year 1974 is the solvent refined coal process which will produce low-ash, low-sulphur fuels from coal. Altogether, coal research and development and proposed funding is increased by 27 percent.

In addition to increased funding for the Liquid Metal Fast Breeder Reactor, I am asking for greater research and development on reactor safety and radioactive waste disposal, and the production of nuclear fuel.

The waters of the world contain potential fuel—in the form of a special isotope of hydrogen—sufficient to power fusion reactors for thousands of years. Scientists at the Atomic Energy Commission now predict with increasing confidence that we can demonstrate laboratory feasibility of controlled thermonuclear fusion by magnetic confinement in the near future. We have also advanced to the point where some scientists believe the feasibility of laser fusion could be demonstrated within the next several years. I have proposed in my 1974 budget a 35 percent increase in funding for our total fusion research and development effort to accelerate experimental programs and to initiate preliminary reactor design studies.

While we look to breeder reactors to meet our mid-term energy needs, today's commercial power reactors will continue to provide most of our nuclear generating capacity for the balance of this century. Although nuclear reactors have had a remarkable safety record, my 1974 budget provides additional funds to assure that our rapidly growing reliance on nuclear power will not compromise public health and safety. This includes work on systems for safe storage of the radioactive waste which nuclear reactors produce. The Atomic Energy Commission is working on additional improvements in surface storage and will continue to explore the possibility of underground burial for long-term containment of these wastes.

Solar energy holds great promise as a potentially limitless source of clean energy. My new budget triples our solar energy research and development effort to a level of $12 million. A major portion of these funds would be devoted to accelerating the development of commercial systems for heating and cooling buildings.

Research and development funds relating to environmental control technologies would be increased 24 percent in my 1974 budget. This research includes a variety of projects related to stack gas cleaning and includes the construction of a demonstration sulphur dioxide removal plant. In addition, the Atomic Energy Commission and the Environmental Protection Agency will continue to conduct research on the thermal effects of power plants.

While the Federal Government is significantly increasing its commitment to energy research and development, a large share of such research is and should be conducted by the private sector.

I am especially pleased that the electric utilities have recognized the importance of research in meeting the rapidly escalating demand for electrical energy. The recent establishment of the Electric Power Research Institute, which will have a budget in 1974 in excess of $100 million, can help develop technology to meet both load demands and environmental regulations currently challenging the industry.

Historically the electric power industry has allocated a smaller portion of its revenues to research than have most other technology-dependent industries. This pattern has been partly attributable to the reluctance of some State utility commissions to include increased research and development expenditures in utility rate bases. Recently the Federal Power Commission instituted a national rule to allow the recovery of research and development expenditures in rates. State regulatory agencies have followed the FPC's lead and are liberalizing their treatment of research and development expenditures consistent with our changing national energy demands.

I am hopeful that this trend will continue and I urge all State utility commissions to review their regulations regarding research and development expenditures to ensure that the electric utility industry can fully cooperate in a national energy research and development effort.

It is foolish and self-defeating to allocate funds more rapidly than they can be effectively spent. At the same time, we must carefully monitor our progress and our needs to ensure that our funding is adequate. When additional funds are found to be essential, I shall do everything I can to see that they are provided.

INTERNATIONAL COOPERATION

The energy challenge confronts every nation. Where there is such a community of interest, there is both a cause and a basis for cooperative action.

Today, the United States is involved in a number of cooperative, international efforts. We have joined with the other 22 member-nations of the Organization for Economic Cooperation and Development to produce a comprehensive report on long-term problems and to develop an agreement for sharing oil in times of acute shortages. The European Economic Community has already discussed the need for cooperative efforts and is preparing recommendations for a Community energy policy. We have expressed a desire to work together with them in this effort.

We have also agreed with the Soviet Union to pursue joint research in magnetohydrodynamics (MHD), a highly efficient process for generating electricity, and to exchange information on fusion, fission, the generation of electricity, transmission and pollution control technology. These efforts should be a model for joint research efforts with other countries. Additionally, American companies are looking into the possibility of joint projects with the Soviet Union to develop natural resources for the benefit of both nations.

I have also instructed the Department of State, in coordination with the Atomic Energy Commission, other appropriate Government agencies, and the Congress to move rapidly in developing a program of international cooperation in research and development on new forms of energy and in developing international mechanisms for dealing with energy questions in times of critical shortages.

I believe the energy challenge provides an important opportunity for nations to pursue vital objectives through peaceful cooperation. No chance should be lost to strengthen the structure of peace we are seeking to build in the world, and few issues provide us with as good an opportunity to demonstrate that there is more to be gained in pursuing our national interests through mutual cooperation than through destructive competition or dangerous confrontation.

Federal Energy Organization

If we are to meet the energy challenge, the current fragmented organization of energy-related activities in the executive branch of the Government must be overhauled.

In 1971, I proposed legislation to consolidate Federal energy-related activities within a new Department of Natural Resources. The 92nd Congress did not act on this proposal. In the interim I have created a new post of Counsellor to the President on Natural Resources to assist in the policy coordination in the natural resources field.

Today I am taking executive action specifically to improve the Federal organization of energy activities.

I have directed the Secretary of the Interior to strengthen his Department's organization of energy activities in several ways.

—The responsibilities of the new Assistant Secretary for Energy and Minerals will be expanded to incorporate all departmental energy activities;

—The Department is to develop a capacity for gathering an analysis of energy data;

—An Office of Energy Conservation is being created to seek means for reducing demands for energy;

—The Department of the Interior has also strengthened its capabilities for overseeing and coordinating a broader range of energy research and development.

By Executive order, I have placed authority in the Department of the Treasury for directing the Oil Policy Committee. That Committee coordinates the oil import program and makes recommendations to me for changes in that program. The Deputy Secretary of the Treasury has been designated Chairman of that Committee.

Through a second Executive order, effective today, I am strengthening the capabilities of the Executive Office of the President to deal with top level energy policy matters by establishing a special energy committee composed of three of my principal advisors. The order also reaffirms the appointment of a Special Consultant, who heads an energy staff in the Office of the President.

Additionally, a new division of Energy and Science is being established within the Office of Management and Budget.

While these executive actions will help, more fundamental reorganization is needed. To meet this need. I shall propose legislation to establish a Department of Energy and Natural Resources (DENR) building on the legislation I submitted in 1971, with heightened emphasis on energy programs.

This new Department would provide leadership across the entire range of national energy. It would, in short, be responsible for administering the national energy policy detailed in this message.

CONCLUSION

Nations succeed only as they are able to respond to challenge, and to change when circumstances and opportunities require change.

When the first settlers came to America, they found a land of untold natural wealth, and this became the cornerstone of the most prosperous nation in the world. As we have grown in population, in prosperity, in industrial capacity, in all those indices that reflect the constant upward thrust in the American standard of living, the demands on our natural resources have also grown.

Today, the energy resources which have fueled so much of our national growth are not sufficiently developed to meet the constantly increasing demands which have been placed upon them. The time has come to change the way we meet these demands. The challenge facing us represents one of the great opportunities of our time—an opportunity to create an even stronger domestic economy, a cleaner environment, and a better life for all our people.

The proposals I am submitting and the actions I will take can give us the tools to do this important job.

The need for action is urgent. I hope the Congress will act with dispatch on the proposals I am submitting. But in the final analysis, the ultimate responsibility does not rest merely with the Congress or with this Administration. It rests with all of us—with government, with industry and with the individual citizen.

Whenever we have been confronted with great national challenges in the past, the American people have done their duty. I am confident we shall do so now.

RICHARD NIXON.

THE WHITE HOUSE, *April 18, 1973.*

Nixon Administration Policy

International considerations. A new relationship between the U.S. and its Western European allies, Canada and Japan was proposed in a major policy address delivered in New York April 23 by Henry A. Kissinger, President Nixon's national security adviser.

Kissinger spoke at the Associated Press luncheon of the annual meeting of the American Newspaper Publishers Association. In his address called "The Year of Europe," he said that America "proposes to its Atlantic partners that, by the time the President travels to Europe toward the end of the year, we will have worked out a new Atlantic Charter setting the goals for the future." The "blueprint" for this program, Kissinger asserted, "builds on the past without becoming its prisoner; deals with the problems our success has created; creates for the Atlantic nations a new relationship whose progress Japan can share." Since the Atlantic community cannot be "an exclusive club," Japan must be "a principal partner" in "our common enterprise," he said.

The need for a new approach in America's relations with its allies stemmed from "new realities" produced by the successes of the past, Kissinger said. He cited these new conditions as follows:

"The revival of Western Europe is an established fact as is the historic success of its movement toward economic unification.

"The East-West strategic military balance has shifted from American preponderance to near equality, bringing with it the necessity for a new understanding of the requirements of our common security.

"Other areas of the world have grown in importance. Japan has emerged as a major power center. In many fields 'Atlantic' solutions to be viable must include Japan.

"We are in a period of relaxation of tensions. But as the rigid divisions of the past two decades diminish, new assertions of national identity and national rivalry emerge.

"Problems have arisen, unforeseen a generation ago, which require new types of cooperative action. Insuring the supply of energy for industrialized nations is an example."

In his conclusion, Kissinger again touched on the energy issue:

"We are prepared to work cooperatively on new common problems we face. Energy, for example, raises the challenging issues of assurance of supply, impact of oil revenues on international currency stability, the nature of common political and strategic interests and long-range relations of oil-consuming to oil-producing countries. This could be an area of competition; it should be an area of collaboration."

Rural electric loan bill enacted. President Nixon signed a bill May 11 that revived the low-cost loan program of the Rural Electrification Administration (REA) for rural electric and telephone systems. The direct federal loan program for the REA had been terminated by the Nixon Administration effective Jan. 1. The bill was cleared by a 93–3 Senate vote May 9 and a 363–25 House vote May 10.

The bill converted the program from direct loans to insured loans from a revolving fund utilizing the more than $4 billion in REA assets. The 2% rate of the previous direct REA loans was retained for sparsely populated areas, estimated to be 20% of prospective borrowers. Other borrowers would pay 5%.

The revision of the program originated in the House, which was advised May 8 by Agriculture Secretary Earl L. Butz that the Administration would accept the change if a requirement also under consideration would be dropped. The requirement was that REA make loans at a level set by Congress. The conference committee of House and Senate members agreed to remove the requirement in return for a pledge by the Administration for REA loans during fiscal 1974–76 at levels at least as high as those budgeted for fiscal 1974. This would be a loan level of $758 million—$618 million for electric loans and $140 million for telephone loans. At least $105 million was scheduled for the 2% loans for sparsely populated areas.

Bill authorizes oil rations. President Nixon April 30 signed a compromise bill extending the Economic Stabilization Act until April 30, 1974.

One of the provisions authorized the President to ration crude oil and petroleum products in the event of regional shortages.

Gasoline shortage. Spot shortages of gasoline developed across the nation during the Memorial Day weekend, highlighting a worsening energy crisis involving crude oil, gasoline and natural gas which affected airlines, farmers, builders, motorists, utility users and other consumer groups. According to the Wall Street Journal, U.S. demand for gas was rising 6%–7% a year.

A voluntary "share the [gas] shortage" plan was announced May 10 by President Nixon, who rejected fuel rationing as a solution to the crude oil and gasoline supply problem.

The plan required suppliers to provide independent gas stations and other petroleum products users with the same percentages of refinery output as was sold in the base period—Oct. 1, 1971–Sept. 30, 1972.

The Interior Department's Office of Oil and Gas was empowered to examine complaints, hold public hearings and impose mandatory allocation requirements on suppliers found to be in violation.

In addition to the new prorated system, the office was authorized to "assign" suppliers with allocations of up to 10% for priority customers—such as municipal transport systems, fire departments, hospitals, police, farmers, commercial transportation services and "residents in states or parts of states not well served by major oil companies and unable to obtain sufficient crude oil or products."

Prices charged by refiners to "unaffiliated customers" could not "exceed normal refinery rack prices charged by major companies to new contract customers." Independent refiners would also be protected by price guidelines.

Independent gasoline stations, which had been threatened with a total cutoff in supplies, received immediate relief from the Administration plan. (The Office of Emergency Preparedness had reported May 9 that 562 gas stations across the country were closed by shortages, while another 1,376 faced imminent closings. The Midwest was the area hardest hit by the gas shortage.)

The Oil Import Appeals Board May 3 issued import tickets to 17 oil firms in 10 states which would allow them to bring into the country an additional 128 million gallons of gas. The tickets permitted imports free of license fees upon demonstration of hardship.

■ The Cost of Living Council (CLC) May 11 denied the first application by a large oil company for a price increase. Ashland Oil, Inc.'s request for a 2.4% price hike was rejected.

The Wall Street Journal reported May 14 that the CLC had eased restrictions on the oil industry in an effort to provide an incentive for major companies to resell crude oil and gasoline to independent refiners and jobbers. According to the new rule, price increases paid by the major companies for crude oil could be passed on in the resale and would not be included in their 1.5% ceiling for price increases set by the CLC.

■ A Federal Trade Commission (FTC) official, Alan S. Ward, told Congress May 11 that an FTC investigation "strongly suggests that major [oil] company control of refinery capacity and pipelines has contributed in a major way" to the current shortage, prompting the commission to study the "possibility for antitrust enforcement action."

■ Darrell M. Trent, acting director of the Office of Emergency Preparedness, said March 26 that a reduction in highway speed limits to 50 miles per hour would increase engine efficiency and curb gasoline use by an estimated 7%.

■ Treasury Secretary George P. Shultz disclosed May 29 that the Administration was considering an increase in the federal excise tax on gasoline in an effort to curb demand and slow a booming economy by increasing federal revenues while reducing excessive consumption.

Antitrust investigation begins. Prompted by charges that the current energy crisis was a "deliberate, conscious contrivance" of the major petroleum and natural gas producing companies, the antitrust and monopoly subcommittee of the Senate Judiciary Committee conducted hearings June 8–27 to investigate the circumstances of the alleged gasoline, oil and natural gas shortages.

The attorneys general of Connecticut, New York, Michigan and North Carolina told the subcommittee June 8 that the gasoline shortage was an artificial one, aimed at driving independent refiners and retailers out of the petroleum market.

"There is no energy crisis," a Florida official declared. "There is a competition crisis." Other state officials urged the breakup of major oil corporations into independent production and marketing companies.

The subcommittee heard testimony June 11 from representatives of gasoline station owners, who also urged government antitrust action against the major oil companies.

The subcommittee extended its investigations June 11 into the Federal Power Commission's (FPC) attempted destruction of natural gas statistics (reported in the Washington Post June 10), compiled by the FPC and based on industry reports, which indicated that natural gas reserves were 9% less than originally predicted by gas producing companies. The FPC report, released May 17, showed gas reserves in 1970 to be at 261.6 trillion cubic feet, compared with the American Gas Association's (AGA) estimate of 286.7 trillion cubic feet. Discrepancies with statistics released March 18 by the AGA were related to a difference in dates and the FPC's decision not to compute reserves in underground storage.

The data, together with the association's announcement of rapidly declining gas reserves, were principal factors in the FPC's May 30 decision to approve a 73% increase in the wellhead price of gas. According to the FPC, and stated as well in the President's energy message, unduly low ceiling prices for gas had discouraged exploration and expansion of current reserves. (The price increase was delayed when President Nixon June 13 announced a 90-day price freeze.)

Other officials in the FPC and also at the Federal Trade Commission (FTC) who were accumulating evidence that the natural gas industry was noncompetitive and could not be trusted to set prices according to market values, opposed the officially adopted FPC statistics and approval of the price increase.

Sen. Philip A. Hart (D, Mich.), chairman of the antitrust subcommittee, in a letter to the FPC May 18, had asked to see the documents which indicated the nation faced a diminishing supply of natural gas; however, Lawrence R. Mangen, an aide to Thomas R. Joyce, chief of the FPC's Bureau of Natural Gas, subsequently attempted to burn the documents. The effort failed only when government incinerators were unavailable or inoperative.

The FPC agreed June 22 "under compulsion and under protest" to turn the papers over to Hart's subcommittee after the Congressional group had issued a subpoena June 21 for the documents.

(One of the controversial documents was described in the Wall Street Journal June 20. According to the FPC, the nation's four largest gas producing companies controlled nearly 50% of all gas reserves by mid-1972, excluding Alaska. The eight largest companies controlled nearly 68% of all reserves. The FPC did not name the companies.)

FPC chairman John N. Nassikas testified June 26, insisting that the gas shortage was real and that "workable competition" existed among gas producers; however, Nassikas admitted that the deregulation of natural gas wellhead prices would not necessarily lead to increased supplies of gas by stimulating exploration efforts.

Rep. George E. Brown Jr. (D, Calif.) told the subcommittee June 26 that FPC commissioner Rush Moody Jr., who had voted with the majority to increase wellhead gas prices, should have disqualified himself from the decision because the commissioner's law firm in Texas had represented Texaco, Inc.

At hearings June 27 an FTC official charged that natural gas producers regularly under-reported actual gas reserves in order to mislead the public about gas supplies and cause government agencies to approve price increases. Sen. Henry M. Jackson (D, Wash.) had asked the FTC June 1 to accelerate its investigation, ongoing since 1971, of alleged anticompetitive activities among petroleum and natural gas producing companies.

Nine major oil companies refused to comply with the FTC request for information documenting their claims of an energy supply crisis, thereby forcing the Justice Department June 5 to seek compliance under a court order.

Antitrust suit against Texaco. The Justice Department filed a civil antitrust suit June 12 against the second largest oil company and the leading gasoline marketer in the U.S., Texaco, Inc., and Coastal States Gas Producing Co. of Texas, the nation's largest source of gasoline for independent gasoline operators. The companies were accused of the illegal restraint of trade and anticompetitive practices in the sale of gasoline and fuel oil to independent marketers.

The government cited a March 1971 contract between the two companies which made Texaco's sale of crude oil to Coastal conditional on Coastal's agreement to sell "substantial amounts" of refined products to Texaco.

The output purchase contract deprived Coastal of the right to sell refined products to others and deprived competitors of Texaco from buying substantial amounts of refined products from Coastal, the suit alleged.

The government cited another portion of the Texaco-Coastal agreement, in which Texaco had the option to purchase Coastal's refinery in Texas, as evidence of anticompetitive practices. It charged that Texaco's acquisition of the refinery operations would foreclose independent marketers from a substantial supply source, and would end competition between Texaco and Coastal in the manufacture and sale of No. 2 fuel oil used in home heating. No. 2 fuel had been in short supply during the winter of 1972–73.

Florida files oil antitrust suit. Florida filed a federal antitrust class action suit July 9 against 15 major oil companies, alleging they had conspired to create a gasoline shortage in the state in order to create a monopoly and increase gasoline prices. The suit asked divestiture of crude oil operations and restrictions limiting the oil companies to refining, distributing and selling petroleum products.

Florida also asked triple damages as compensation for alleged losses from reduced gasoline and sales taxes due to a decline in tourist travel.

Named as defendants were the Exxon Corp., Texaco Inc., Gulf Oil Corp., Mobil Oil Corp., Standard Oil Co. of California, Standard Oil Co. (Indiana), Shell Oil Co.,

Atlantic Richfield Co., Phillips Petroleum Co., Continental Oil Co., Sun Oil Co., Union Oil Co. of California, Cities Service Co., Standard Oil Co. (Ohio) and Marathon Oil Co.

SEC judge blocks utility merger. A Securities and Exchange Commission (SEC) administrative law judge July 20 rejected a proposal by the American Electric Power Co. (AEP), the nation's largest utility holding company and through its subsidiary (Ohio Power) the largest electric system in Ohio, to merge with Columbus and Southern Ohio Electric Co., another of the state's largest.

In rejecting AEP's claim that the acquisition would "contribute significantly to the development and growth of AEP as an integrated electric utility system," Judge Irving Schiller declared, "it was the Congressional desire to curb the developing concentration of power in the utility industry." If the takeover were permitted, he continued, "such a precedent would permit every large system to acquire a smaller system at will by simply contending that the acquisition would contribute to its growth or is ideally suited for affiliation." AEP's plan was "detrimental to the public interest," he concluded.

The ruling on AEP's five-year attempt to acquire Southern-Ohio through an exchange of stock was regarded as a landmark decision. It was the largest case which the SEC had been asked to consider under the Public Utilities Holding Company Act of 1935.

Nixon revises policy units. President Nixon June 29 appointed Colorado Gov. John A. Love (R) director of a new White House office—the Energy Policy Office— responsible for formulating and coordinating the Administration's energy policies. The appointment was announced in a presidential statement outlining the Administration's plans and policies for dealing with what the President termed the nation's "serious energy problem."

In his statement, Nixon asked Congress to create a new Cabinet level Department of Energy and Natural Resources and an independent Energy Research and Development Administration to lead long-range efforts to deal with energy matters.

In action designed to provide immediate action on the energy problem, Nixon announced the launching of a five-year program for energy research and development, costing $10 billion, to begin in fiscal 1975. Meanwhile, he said, $100 million would be added to the $770-million energy research budget for fiscal 1974. He urged a nationwide, voluntary conservation effort to reduce personal energy consumption by 5% over a 12 month period.

"The federal government will take the lead in this effort by reducing its anticipated consumption by 7% during this same period," Nixon said.

In directives issued to federal departments and agencies, Nixon ordered reduction in air conditioning levels, the use of smaller and more efficient cars and a cutback in official travel. The Transportation Department was asked to seek a reduction in "flight speeds, and, where possible, the frequency of commercial airline flights."

The Defense Department, the largest single energy consumer in the executive branch, had cutback its energy demands by 10% from 1972, the President reported.

Nixon urged governors and state legislatures to reduce highway speed limits and he reiterated his support for federal legislation giving states the options to make capital investments in mass transit systems rather than in highway construction. Nixon also appealed to motorists to utilize "energy-efficient cars" in an effort to overcome the gasoline shortage, which he said could reach up to 5% of total demand by the end of the summer.

Nixon emphasized that the "conservation of existing energy resources is not a proposal. It is a requirement that will remain with us indefinitely, and it is for this reason that I believe that the American people must develop an energy conservation ethic." The President said, "America faces a serious energy problem. While we have only 6% of the world's population, we consume one-third of the world's energy output. The supply of domestic energy resources available to us is not keeping pace with our ever-growing demand, and unless we act swiftly and effectively, we could face a genuine energy crisis in the foreseeable future."

Love, whose appointment did not require Senate approval, would serve as a presidential assistant as well as director of the new White House Energy Policy Office.

The proposed Energy Department would assume most of the present activities of the Department of Interior, as well as energy and natural resource programs in the Agriculture Department, the Army Corps of Engineers, the Commerce Department, the Transportation Department and related agencies.

The present Atomic Energy Commission (AEC) would form the basis for the proposed Energy Administration, which would also assume the functions of the Interior Department's Office of Coal Research. Nixon asked that the five member AEC be reconstituted in a separate Nuclear Energy Commission to carry out atomic licensing and regulatory duties.

Controls to replace freeze. The Administration's Phase 4 economic control program was outlined by President Nixon in a statement issued July 18. The Phase 4 plan, which would end the current price freeze, was based on mandatory compliance with a system of price controls similar to the controls in effect under Phase 2 and abandoned in January.

The price freeze, announced June 13 as a stopgap measure to deal with Phase 3's largely unsuccessful program of voluntary economic guidelines, would remain in effect until Aug. 12 for the industrial and service sectors;

Nixon also reiterated his belief that the system of economic controls should be "selective." The Phase 2 rule exempting firms with fewer than 60 employes from wage restrictions was reinstituted. Public utilities, the lumber industry and the price of coal sold under long-term contract were also decontrolled.

At a briefing for reporters following the President's announcement, Treasury Secretary George P. Shultz provided details for the special regulations imposed on the petroleum industry. New guidelines, to be issued July 19, would provide separate price ceilings for two categories: crude oil; and gasoline, heating oil and diesel fuel. Exemptions in increased crude oil production, it was hoped, would provide the industry with incentives to increase petroleum supplies.

A Cost of Living Council fact sheet detailing the changes listed these rules for the "petroleum sector":

Petroleum Sector

Prices remain frozen until Aug. 12 at which time Phase 4 regulations become effective.

Regulations to be issued July 19 for public comment. These regulations will provide:

Price ceilings for gasoline heating oil and diesel fuel. The ceiling price is computed as the seller's actual cost of the product plus the dollar and cents markup applied to a retail sale of the same product on Jan. 10, 1973.

A ceiling price for crude oil.

Increased crude production and equivalent amount of old oil to be exempted.

Ceiling prices and octane ratings to be posted on each gasoline pump.

Petroleum industry gets Phase 4 rules. Final Phase 4 regulations for the petroleum industry were released Aug. 17. The Cost of Living Council's rules, which took effect Aug. 20, replaced a price freeze which had been extended one week, although the freeze on gasoline and diesel fuel prices was not scheduled to be lifted until Aug. 31.

Price ceilings were ordered for all petroleum products but increased costs of imported home heating oil could be passed along to consumers and wholesalers could pass through higher costs on imported crude oil. Despite this flexibility, CLC officials predicted that price rollbacks for crude and heating oil could result from the price ceilings.

In an effort to mitigate the impact of the recent oil shortage, retailers' markups were restricted to their minimum on Jan. 10, before the shortage became widespread. (Ceiling prices were based on retailers' costs of buying gasoline on Aug. 1 plus their markup.)

Gasoline and diesel fuel retailers also were required to post the ceiling prices as well as octane ratings for each grade of gasoline.

Gasoline price freeze extended. The Cost of Living Council (CLC) Aug. 30 announced a third extension in three weeks of a price freeze on retail gasoline, thereby postponing the start of controversial Phase 4 regulations governing gasoline sales until Sept. 7. The new rules, which were contested by gasoline dealers, would force some price rollbacks.

The CLC's delaying action was intended to forestall threats of widespread gasoline station shutdowns during the Labor Day weekend as well as to reduce confusion about Phase 4 ceiling and octane rating requirements arising out of several court decisions on the Phase 4 issue.

The gasoline price freeze had first been extended beyond Aug. 12 when an across-the-board freeze on most of the nation's businesses was lifted. Another extension had been announced Aug. 16, to terminate Aug. 31, but a federal district court in Washington had issued a temporary restraining order Aug. 24.

The court had ruled that Phase 4 rules for gasoline dealers were "arbitrary and capricious" and could not take effect as scheduled. The class action suit against the CLC had been brought on behalf of 165,000 brand name gasoline dealers, who claimed to sell 75% of the nation's gasoline.

The dealers had argued that other small businesses were exempt from price freezes and that Phase 4 rules would permit gasoline wholesalers to raise prices.

A special appeals court, established in 1972 to deal with economic control cases, was convened in Washington Aug. 29 and it upheld the government's position. The court issued a stay in the lower court decision and allowed the CLC to implement Phase 4 rules for gasoline dealers.

CLC Chairman John T. Dunlop charged Aug. 30 that gasoline costs were "one of the biggest contributors to inflation" because retail prices had increased at an annual rate of 17.5% during the first half of 1973. The CLC had claimed in June that more than 1,000 gasoline stations were overcharging customers by an average 2¢ a gallon.

Wellhead gas price hike OKd. The Federal Power Commission (FPC) Aug. 7 approved an increase in the wellhead price of natural gas from the Permian Basin in Texas and New Mexico. The rate hike would nearly double the cost of gas for consumers. Ceiling prices on "new" gas contracts, dated since October 1968, would increase from 16.5¢ per thousand cubic feet to 35¢; "old" gas wells (pre-1968) would be subject to a rate of 23¢ ceiling per thousand cubic feet, compared with a previous high of 14.5¢.

Producers' annual revenues were expected to increase by as much as $50 million as a result of the agency's 3-0 decision. The rate increases, overturning levels set in 1965, were intended to stimulate "further exploration and development" of natural gas, which was in short supply, according to the FPC.

Nixon urges Congressional action. President Nixon urged legislators to enact seven energy bills before Congress in order to avert a winter heating oil shortage and reduce the nation's dependence on Middle East oil producers, who threatened to limit the availability of petroleum supplies to U.S. consumers as political retaliation for U.S. support of Israel.

Nixon's remarks were included in an opening statement at a press conference in Washington Sept. 5.

"If the Congress does not act upon these proposals, it means that we will have an energy crisis. Not perhaps just this winter, but perhaps, certainly, later on as well," Nixon said.

"And if the Congress does not act upon these proposals which, in effect, have as their purpose increasing the domestic capacity of the United States to create its energy, it means that we will be at the mercy of the producers of oil in the Mideast," he added.

Despite the President's attempt to shift responsibility for the energy crisis to Congress, none of the bills named would directly increase short-term supplies of oil in the U.S. The legislation would end federal regulation of wellhead natural gas prices; authorize deep water offshore ports for oil supertankers; set standards for determining sites of nuclear power plants and end delays over the locations of proposed power stations; set standards for strip mining and encourage coal production; authorize the Alaska pipeline; authorize production from a Naval Reserve in California (to finance the 1971 suspension of offshore leases in the Santa Barbara, Calif. channel) and allow exploration of a Naval Reserve in Alaska; and strengthen the ability of the Interior Department's Bureau of Land Management to authorize mineral development on 450 million acres of federal land. Nixon reportedly also sought Congressional action on his proposal to create a Cabinet level energy department and an independent agency for energy research.

Mandatory fuel allocation systems proposed. Administration proposals for a mandatory allocation system for propane fuel was announced Aug. 31 by the White House Office of Energy Policy in anticipation of a winter shortage.

Customers using propane supplies in agricultural production, food processing, residential cooking and heating, mass transit vehicles and buildings housing medical and nursing patients, would be assured priority assignments. Other customers would receive supplies in proportion to their purchases in a base period from September 1970 through April 1973.

The Administration also urged the Cost of Living Council to permit price increases in propane fuel.

The action represented the second use of executive authority to issue mandatory oil allocation rulings since Congress passed the measure April 30.

The energy office had proposed another mandatory allocation system Aug. 27 aimed at averting a winter heating oil shortage by increasing the use of alternative fuels which were in greater supply. Increased air pollution could result from the decision, however.

In areas already meeting federal air pollution standards, electrical utilities and factories would be unable to reduce the sulphur content in their fuel supplies by: converting from coal to cleaner burning oils; changing from residual oils to the less polluting home heating oils; and increasing the quality of home heating oils and other distillate products in a blend with residual fuel oils.

Administration spokesmen argued that the rulings constituted only a "temporary halt" in the trend toward cleaner air. Rollbacks in the "gains already made in reducing the sulphur content of fuels under the Clean Air Act" would not result from the allocation proposal, spokesmen said.

An "automatic exception" to the rules barring conversion was granted users of natural gas who switched to fuel oil, "provided that alternative fuels, such as coal, cannot presently be utilized."

Rationing of heating oil considered—
John A. Love, director of the President's energy office, revealed Sept. 6 that the Administration was drafting a rationing program for heating oil and other distillate fuels such as diesel fuel. Winter shortages of 10%–15% were feared. (Supplies were 4% below demand during the 1972 winter.)

Love said "the situation this winter is very tight, although it is difficult to forecast because of variables" such as weather conditions and the availability of oil imports.

Analysts attributed the worsening fuel shortage to increased consumption by industries operating near capacity, utilities converting to low sulphur distillates, and natural gas shortages prompting shifts to fuel oil.

Supplies of fuel oil were low in the Caribbean and could prove to be tight in Europe, resulting in export restrictions from those areas. In any case, oil refined in Europe contained generally higher quantities of sulphur than strict East Coast air pollution standards permitted. Imports of 700,000 barrels of distillate a day would be required to fill the gap between domestic consumption and supplies, and complex price control regulations could deter distributors from importing higher-cost supplies.

(Texaco, Inc. notified Massachusetts officials Aug. 13 that it would be unable to meet industrial and residential demands for winter fuel. Shortages of 11%–15% for the state were predicted. The company asked that state air pollution standards be eased to allow the increased use of high sulphur content fuel.)

Love had said Aug. 9 that the Administration would retain a system of voluntary fuel oil allocation because the energy situation was not critical enough to warrant imposition of mandatory controls.

Mandatory plans failed to "dampen demand, they don't increase supply, . . . they simply [switch] any actual shortage from one user to another," Love declared.

Despite warnings that "horrendous" actions requiring mandatory allocation should be considered only in the event of a catastrophe, Love said the energy office would propose a standby mandatory plan assuring that refineries would receive 90% of their crude oil capacity. Wholesalers would receive supplies on a prorated basis using calendar 1972.

Coal gasification regulation limited.

The FPC ruled Sept. 4 that it had no regulatory authority over production and transportation of synthetic gas derived from coal, unless the artificial gas was mixed with natural gas and also was transported or sold in interstate commerce.

The FPC earlier had ruled that the regulation of synthetic gas made from naphtha was not within its jurisdiction.

The decision resulted from applications filed with the government by three companies—El Paso Natural Gas Co., Transwestern Pipeline Co. and Pacific Lighting Corp.—to build and operate coal gasification plants in New Mexico.

The gasification process involved mining coal, its purification to remove ash and pyritic sulphur, and its conversion to a gas in a process utilizing a cylindrical vessel called a gasifier. Under controlled temperatures and pressure, coal in the gasifer combined with steam and oxygen to form methane. When impurities were removed, a synthetic gas remained.

Exxon Corp. announced Sept. 5 that it would spend $10 million to finance further research in the coal gasification and liquefaction process. (Liquefaction involved the conversion of pulverized coal into a low sulphur fuel oil and a synthetic crude oil.)

The company said that since 1966 it had spent an estimated $20 million on coal research and anticipated that total research costs, including construction of large scale pilot plants, would reach $145 million.

As the price raises continued, Standard Oil Co. (Ohio) Oct. 9 announced a 1¢ a gallon increase in the price of gasoline sold to company-owned stations and a 1.2¢ a gallon increase for gasoline purchased by independent dealers. It was Sohio's second major increase in three weeks—wholesale and retail gasoline prices had been raised 1¢ a gallon Sept. 21.

Exxon also announced price increases Oct. 9. Heating oil and kerosene prices would be .3¢ a gallon higher, diesel fuel was up 1.15¢ a gallon and propane was raised 2¢ a gallon. Exxon had announced previous increases Sept. 6.

John T. Dunlop, director of the Cost of Living Council, said Oct. 3 that if gasoline dealers passed on every price increase from their suppliers, the system of price controls would be unenforceable.

Despite the warning, gasoline dealers across the country continued to curtail service in sporadic, but effective, protests against the Administration's policy of price controls.

Phase 4 rules upheld—

The Temporary Emergency Court of Appeals upheld the Administration's Phase 4 gasoline rules Oct. 10 in a case that was brought by independent gasoline dealers to protest CLC guidelines.

Retailers had objected to government rules requiring them to absorb higher costs while permitting refiners to raise wholesale gasoline prices because of higher importing costs.

Under new regulations proposed Oct. 15, retailers would be permitted automatic pass-throughs for prices that were increased by their wholesale distributors. In an effort to slow the accelerating rise in gasoline prices, the council said refiners would be allowed to raise wholesale prices only in full penny increments.

New rise—

The CLC permitted an immediate 1.5¢ per gallon increase in the price of retail gasoline Oct. 15. (Home heating oil and diesel fuel prices also were allowed to increase.)

Mandatory fuel allocation ordered.

John A. Love, director of the White House Energy Policy Office, announced Oct. 2 that President Nixon had ordered mandatory allocation of propane and distillate fuel supplies, including home heating oil, diesel fuel, jet fuel and kerosene.

The program, last in use during World War II, took effect immediately for propane, but delays were reported in drawing up rules for allocation of distillate fuels. Allocation was scheduled to terminate at the end of April 1974.

Love emphasized that the system would not increase supplies of fuel, but would reduce regional shortages by insuring an equitable distribution of supplies.

Added to the priority list of propane users already announced were natural gas utilities which used propane air mixtures to supplement pipeline gas supplies during peak periods of demand.

Officials who were drawing up regulations for distillates said no attempt would be made to set priorities at the federal level.

Love then told the House Select Committee on Small Businesses Oct. 9 that some form of fuel rationing would be required for the "next three-five" years until domestic fuel supplies increased to meet demand.

Love met with President Nixon later in the day to discuss means of conserving fuel supplies. Following the meeting Nixon released a "citizen action guide to energy conservation."

The Administration's opposition to compulsory fuel rationing, weight taxes for cars that consume large amounts of gasoline and the mandatory labeling of appliances to publicize their energy efficiency was well known; but, consumers were asked to voluntarily reduce thermostats by 4 degrees. Total savings from that action could reduce the expected shortage by 400,000 barrels a day, according to government experts.

Distillate fuel allocation set. The Administration's mandatory allocation plan for distillate fuels—home heating oil, jet fuel, diesel fuel, kerosene, range oil, gas oil and stove oil—would take effect Nov. 1, it was announced Oct. 12.

Unlike Administration allocation rules for propane fuel, the government did not issue priority lists for distribution of distillate fuels to customers. Supplies would be apportioned to consumers on the basis of month-to-month sales during 1972. The Interior Department's Office of Oil and Gas would administer the program.

The federal government would be authorized to set aside 10% of available supplies for distribution to each state for hardship cases. Any unused state reserves would be returned to the supply pool for proportional allocation to prevent states from accumulating or deferring their hardship reserves. Oil companies also would be permitted to make shifts of up to 5% in distribution of supplies from one region to another.

The nation's airlines were most immediately affected by the allocation program. Texaco Inc., a major supplier of jet fuel, notified at least seven airlines that it would reduce deliveries by an average of 15% during the last three months of 1973, it was reported Oct. 5. The Administration's allocation plans would redistribute the shortage to some extent, but not entirely. Under allocation rules, supplies would be based on 1972 purchases. During 1973, the airlines had expanded flight schedules and were requiring more jet fuel.

Trans World Airlines announced Oct. 12 that it would cut back flights by 5%, effective immediately. The reduction, which officials termed "temporary," would cause 40 flights to be canceled or re-routed.

(In an announcement Oct. 18, TWA said it had agreed with United Air Lines and American Airlines to eliminate 44 daily flights starting Nov. 1 in an effort to conserve 6.6 million gallons of fuel a month.)

Following announcement of the Administration's allocation plans, the Civil Aeronautics Board urged the nation's airlines to take joint action in implementing a fuel conservation plan. In 1972, jet fuel had comprised 5% of the total 6 billion barrels of petroleum used by U.S. consumers.

House OKs broad fuel allocation plan. The House voted 337-72 Oct. 17 for a far more extensive fuel allocation plan than the Administration's limited · redistribution program.

The bill, which had been introduced by Rep. Torbert H. Macdonald (D, Mass.), would require allocation of crude oil and all refined petroleum products, including gasoline, heavy residual oils, butane and refined lubricating oils, all of which were not covered by the Administration plan. Distribution would be on a pro rata basis of 1972 sales, according to the House plan.

The bill would authorize the President to set aside priority supplies for essential health and civic services under a plan that would take effect 25 days after the bill's enactment. The bill also provided for dollar for dollar pass-throughs to all retail customers of costs incurred at the wholesale level by dealers in crude oil and all refined petroleum products.

An amendment that would have exempted crude oil at the wellhead from mandatory allocation was overwhelmingly defeated.

The bill was sent to conference with the Senate, which had passed two similar measures. President Nixon had threatened to veto the Macdonald bill. John A. Love, director of the White House Energy Policy Office, said in a message to the House Oct. 16 that the legislation would "have a serious negative impact on the availability of supply."

Gas price deregulation extended. The Federal Power Commission (FPC) ruled Sept. 15 that natural gas pipelines could purchase gas on an emergency basis for 180 days without submitting the purchase price to the FPC for approval. The action, which observers interpreted as another step toward deregulation of interstate natural gas prices, was intended to alleviate temporary winter shortages, FPC spokesmen said.

Gasoline, heating oil prices raised. Gasoline prices were permitted to rise 1¢–2.5¢ a gallon Oct. 1 as a result of a Cost of Living Council (CLC) ruling Sept. 28. President Nixon Sept. 24 had asked the council to expedite processing of price and cost information received from gasoline dealers.

At the same time, the CLC permitted increases of 1¢–2¢ a gallon in heating oil under new rules that would allow distributors to pass through higher costs for domestic fuel.

Gasoline price increases (as well as higher prices for diesel fuel) took effect as part of new, streamlined rules for setting ceiling prices. According to the new markup formula, retailers based their ceiling prices on a May 15 selling price plus any cost increases since then. Prices were higher in May than in January, which was the former markup date.

Gasoline dealers across the country had launched widespread shutdowns to protest ceiling price restrictions, which, they said, required them to absorb price increases when refiners were permitted to pass through higher costs for imported fuel. CLC Director John T. Dunlop warned retailers to use the larger markup allowance as a "cushion" against expected cost increases between Oct. 1 and the next periodic review of gasoline prices by the CLC. According to the council, every penny per gallon increase in gasoline brought $1 billion in revenues to gasoline dealers.

Refiners lift wholesale price—Three major oil companies Oct. 1 announced increases in the price of gasoline sold to retailers. Shell Oil Co. planned .2¢ a gallon hikes; Atlantic Richfield Co. sought 1.5¢ a gallon increases; Phillips Petroleum Co. would raise prices .5¢ a gallon.

Gulf Oil Corp. announced its second increase in eight days Sept. 27 and warned that other price hikes "must be anticipated." Another .5¢ a gallon increase would be added to the .9¢ increase already planned.

John A. Love, director of the White House Energy Policy Office, predicted Sept. 30 that gasoline prices would rise 10%–15% during the next year.

Alaskan Oil Problem

Courts bar pipeline permits. The U.S. Court of Appeals for the District of Columbia Feb. 9 ordered that a lower court prevent Interior Secretary Rogers C. B. Morton from issuing permits for construction of the trans-Alaska oil pipeline, since the right-of-way requested by the Alyeska Pipeline Service Co. exceeded limits set by Congress in the Mineral Leasing Act of 1920.

The act limited the right-of-way to 25 feet on each side of the pipe, plus the diameter of the pipe, for a total in this case of 54 feet. Alyeska had requested a path as wide as 146 feet in some places, necessitated by topographic and climatic factors.

In the majority opinion, supported on the right-of-way issue by six of the seven judges, Judge J. Skelly Wright wrote that the court had "no more power" to authorize an exception to the law than it had "power to increase Congressional appropriations to needy recipients."

The judges refused to rule on the environmental questions raised by the opponents of the pipeline. A U.S. district court judge had lifted a two-year injunction against the project in August 1972, ruling that the environmental issues had been adequately considered. In the latest ruling, the majority said it had avoided the problem because of "a desire to expedite our decision," and because the environmental situation might change by the time Congress decided to amend the Mineral Leasing Act, if it did so at all.

The court added that "developments with respect to the Canadian alternative," a prospective pipe route from the Alaska North Slope across Canada to the U.S. Midwest, "may well moot anything we may now say." Rep. Les Aspin (D, Wis.) said the court's decision would allow Congressional debate on the Canada route, which he said had grown more attractive in light of the Midwest winter fuel shortage, it was reported Feb. 12.

The Supreme Court April 2 refused to review the decision.

The effect was to force proponents of the pipeline to petition Congress for amendments to the 1920 Mineral Leasing Act.

After a 45-minute meeting with President Nixon April 5, Interior Secretary Rogers C. B. Morton said the President had ordered him to go "all-out" to obtain Congressional approval of the amendments, since Nixon considered the pipeline "vital to the national interest" in view of the energy shortage and increasing oil imports.

Morton said Nixon opposed an alternate pipeline route through Canada, which Morton said would be more ecologically damaging, more expensive and take longer to build, since it would be four times as long as the proposed Alaska pipeline.

Senate OKs Alaska pipeline. By a vote of 77–20 July 17, the Senate passed a bill to authorize licensing of construction of a 789-mile, $3.5 billion Alaskan oil pipeline from the North Slope fields to the port of Valdez.

The key votes before final passage came on an amendment by Alaska's Senators Mike Gravel (D) and Ted F. Stevens (R) to "immunize" the project from court challenges by environmentalists. The amendment, declaring the pipeline to be in conformity with the National Environmental Policy Act, was tentatively passed, 49–48. On a motion to reconsider, a 49–49 tie vote was broken by Vice President

Spiro T. Agnew, making passage of the amendment final.

The main portion of the bill overturned the right-of-way size restrictions of the 1920 Mineral Leasing Act.

In preliminary action July 13, the Senate had defeated 61–29 an amendment by Birch Bayh (D, Ind.) and Walter F. Mondale (D, Minn.) to delay licensing of the project for eight months while the U.S. negotiated with Canada on a possible alternative pipeline route through Canada to the Midwest.

In debate July 17, Mondale said the State Department had "misrepresented the position of the Canadian government" by withholding until July 16 a Canadian statement that Canada would not insist on majority control of an alternate route.

The bill was sent to the House, where an Interior subcommittee approved a similar measure July 17 by a vote of 13–3.

B.C. plan for U.S. oil. British Columbia Premier David Barrett advanced a plan March 13 by which Alaskan oil and gas could be transported through his province by rail to the U.S.

Barrett's proposal, whose cost was estimated at $4 billion, would involve construction of a railroad from Alaska's North Slope oil fields through the Yukon to the British Columbia border, where rail connections to the U.S. could be made.

Energy Minister Donald S. MacDonald said March 14 that Barrett's idea was "interesting" and added that the government "will examine it fully before making any recommendation to Washington." He indicated that the proposal's "most glaring deficiency" was that the railroad could not be used for taking oil and gas from the Canadian north.

MacDonald's evaluation was confirmed March 14 in the House of Commons by Prime Minister Pierre Elliott Trudeau, who said "since it would serve only the U.S. oil deposits" in Alaska, Barrett's plan might be "less preferable" than a pipeline up the Mackenzie Valley.

Environmental Issue

Sulfur fuel ban upheld. The Supreme Court Jan. 17 refused to review an action by the Environmental Protection Agency ordering the Delmarva Power & Light Co. to cease burning high-sulfur fuel in order to conform with Delaware air pollution rules. The Getty Oil Co., Delmarva's fuel supplier, had challenged the order. Getty had since agreed to supply low-sulfur fuel im-mediately and to switch to 1% sulfur content fuel by Jan. 1, 1974, it was re-ported Jan. 16.

Pollution waiver sought. Oil distributors announced Jan. 26 that they had asked Pennsylvania, New Jersey, New York, Connecticut, Massachusetts and Rhode Island to reduce air pollution standards in order to permit the burning of home heating oil with a sulfur content higher than currently allowable.

Low sulfur fuel had been in short supply during the recent energy crisis.

New York City acted Jan. 24 to permit Texaco, Inc. to sell fuel oil with a sulfur level eight to ten times higher than the law allowed. The variance was effective for 45 days but it included a pollution tax on every barrel of high sulfur oil sold.

Texaco lost its waiver bid in New Jersey, where it supplied 12% of the home fuel, according to the Washington Post Jan. 26.

Texas plant in water accord. The Hous-ton Lighting and Power Co. and the Jus-tice Department submitted an agreement to a U.S. district court judge in Victoria, Tex. Jan. 26 that would prevent construc-tion of three new generating units at the company's Cedar Bayou generating com-plex, to protect the ecology of Trinity Bay, a major wildlife breeding ground.

The company agreed to continue to op-erate a 2,600 acre cooling pond to reduce the temperature of water used in cooling its condensers in the three units already operating or under construction, before discharging the water into the bay. The company would also be required to oper-ate protective screens at its water intake point to protect marine life, and to con-duct a sampling and monitoring program at more than 20 points in the bay and bayou to determine the effect on marine life.

Under the agreement, the Environmen-tal Protection Agency's regional adminis-trator in Dallas would be empowered to order the company to take necessary re-medial action if he determined that the company's operations were contributing to significant environmental damage in the bay. Remedial action could include a reduction or temporary cessation of oper-ation at one or more of the generating units.

W.Va. grants delay. The West Virginia Air Pollution Control Commission bowed to a demand by Gov. Arch A. Moore Jr. and granted a two-year delay Feb. 1 on 1975 fly ash and sulfur dioxide emission limits for the American Electric Power System.

The utility had claimed that it could not obtain enough low-sulfur coal, and threatened to import coal from Western states in competition with the local product. Environmentalists replied that the state had enough reserves of adequate coal, and said high-moisture Western coal would produce excessive fly ash in power plants built for dryer coal.

Easing of sulfur limit proposed. The EPA proposed May 7 to rescind one of its "secondary" (protection of vegetation) standards on industrial emissions of sulfur oxides, mostly by copper smelters and utilities.

The standard to be revoked specified an average annual maximum of 60 micro-grams of sulfur oxides per cubic meter of air. Left unchanged was a short-term secondary ceiling of 1,300 micrograms per cubic meter per hour, averaged over three hours, not to be exceeded more than once a year. "Primary" standards, designed to protect human health, were not affected by the proposal.

A deadline of July 1, 1975, had been set for states to comply with primary stan-dards, but there was no firm deadline for the secondary standards.

The EPA said the proposal was made after "extensive re-evaluation" of sci-entific data, which led to the conclusion that vegetation damage was caused by short-term peak concentrations of sulfur oxides, rather than continuous exposure to lower levels during the growing season. The agency said it was also acting in response to a court appeal filed by the Kennecott Copper Corp. shortly after the standards were set in April 1971.

EPA Assistant Administrator Robert Sansom said the change had nothing to do with Administration efforts to allow the burning of plentiful, high-sulfur coal to avoid aggravating oil shortages.

Storm King plant OKd. The New York State Court of Appeals, the state's highest court, March 14 approved the state's decision that Consolidated Edison's planned Storm King water power plant would not endanger water quality standards. But the 10-year fight over the plant continued, as environmentalists promised to petition the Federal Power Commission (FPC) to consider alternate plans they said would be less environmentally damaging.

The state had ruled that Hudson River water quality would not be affected because the Storm King plant itself would not burn fuel or raise water temperatures. The plant would be designed to use electricity generated at other sites in off-peak hours to pump Hudson water up to a storage reservoir. During peak demand, the water would be used to generate two million kilowatts of power within two minutes.

Henry L. Diamond, New York's environment commissioner, said he had approved the plant on the grounds of water quality, but that he believed the plant might cause irreparable environmental damage by defacing the Hudson River gorge. He said he hoped the conservationists would win, since Storm King was "the symbol of the first legal victory for the environment," a 1965 U.S. Court of Appeals ruling that the FPC must consider scenic and historic resources before approving a new facility.

EPA asks LA gas rationing. Environmental Protection Agency (EPA) Administrator William D. Ruckelshaus Jan. 15 announced a proposed air pollution control plan for the Los Angeles basin, which included a gasoline rationing system designed to reduce automobile use by up to 82% during summer months by 1977. But Ruckelshaus said he had "grave

doubts" that such a plan was feasible, and said it had been issued only to meet a court order and to stimulate further debate on how to meet federal pollution guidelines or on whether to adjust the standards to the special social and meteorological conditions of the area.

The EPA had been ordered by a U.S. district court in November 1972 to come up with a plan for the region, the only area in the country for which a state had not submitted a plan for reaching the national standard of .08 parts per million (ppm) of photochemical oxidants.

Under the plan, rationing would be undertaken beginning in 1975, with rations changing each month and enforced either by consumer coupons or limited distribution to retailers. The plan included other measures, but without rationing, Ruckelshaus said, pollution would still exceed the limit by one-half. The other measures, in addition to controls on stationary pollution sources and on air-craft, included conversion of commercial fleet vehicles to low-pollution fuels, including natural gas or liquid petroleum gas, costing $500–$800 for each vehicle, installation of 1975 model emission control devices on all pre-1975 cars, at a cost of $200–$300 each, and annual inspections.

Ruckelshaus said the proposal would have severe effects on the region, which he said was uniquely dependent on automobiles for transportation and for its "lifestyle," with nearly 6 million motor vehicles for its 10 million population.

Excess auto gas use charged. Environmental Protection Agency (EPA) Administrator William D. Ruckelshaus said Jan. 24 that luxury "power options" and excess weight in U.S. cars commonly caused losses of up to 20% in gas mileage, in contrast to the 7% reduction that might be caused by installation of emission control devices.

Ruckelshaus listed average mileage reductions of 9% for air conditioners, 6% for automatic transmission, and 10% for omitting steel-belted radial tires. A weight reduction of 500 pounds could reduce gas mileage by 14%, and a drop in average U.S. auto weight to 2,500 pounds would reduce crude oil imports by 2.1 million barrels a day over expected 1985 needs.

Impact statements on tankers. The federal government agreed to file environmental impact statements for future contracts under the oil tanker subsidy program, it was reported Jan. 11.

In addition to providing design modifications to reduce the dangers of spills, the government agreed to review 10 of the 18 contracts already approved.

The settlement was reached in a suit filed by environmentalists who had charged the government with failing to comply with the National Environmental Policy Act.

Tough oil spill law upheld. The Supreme Court April 18 unanimously reversed a lower court decision barring states from fixing stiffer penalties for maritime oil spills than those ordered by the federal government.

Justice William O. Douglas wrote for the court that a state could exercise police power over maritime activities as long as the state law did not present "a clear conflict with federal law."

"To rule as the district court has done is to allow federal admiralty jurisdiction to swallow most of the police power of the states over oil spillage—an insidious form of pollution of vast concern to every coastal city or port," Douglas reasoned.

The 1970 Florida law in question had made shippers and dockers liable for oil spillage damages without reference to who was negligent, and it forced them to have insurance or to post a surety bond to meet this responsibility. Federal law required proof of negligence or misconduct.

The ruling did not question the validity of an 1851 law limiting liability of a vessel owner to "the value of such vessels and freight pending."

GAO asks action on offshore oil. In a report released July 7, the General Accounting Office (GAO) urged stronger inspection and regulation of oil and gas operations on offshore federal leases. Rep. Henry S. Reuss (D, Wis.), who had requested the report, said the GAO had found "serious deficiencies" in federal procedures against water pollution from oil wells in both Gulf Coast and Pacific regions.

The GAO said the Interior Department's Geological Survey, responsible for policing the operations, had failed to follow prescribed enforcement actions in the Gulf Coast region and had been ineffective in correcting improper equipment in the Pacific. The report suggested that the Interior Department consider shutting down deficient Pacific wells.

Impact of shale development reported. An Interior Department report Aug. 30 said the fuel potential of oil in shale deposits would offset the environmental damage caused by exploitation of the deposits. The report was the final environmental impact statement on the department's proposal to lease six tracts of federal land for a prototype extraction project. The tracts totaled 30,720 acres in Colorado, Utah and Wyoming.

(When shale is broken up and heated, organic material emerges as a form of crude oil that can be refined into most petroleum products.)

The proposed lease contracts would require compliance with environmental standards and efforts by commercial leaseholders to restore the land damaged by shale mining and processing.

The report said, however, that there would be "unavoidable" environmental damage and changes in the social and economic conditions in the production areas. Full-scale development of the deposits would "probably exert a cumulative influence" on the environment "for more than a hundred years."

Land would be disrupted by processing the shale, the report said: surface disposal of processed shale from a single tract could fill six canyons of about 700 acres to a depth of 250 feet. Vegetation would be destroyed and some animal species would be depleted.

Water requirements of the industry would reduce underground water supplies, and the quality of surface water would be lowered. The report estimated that full-scale development could increase the salt content of the Colorado River 1.5%.

The report said urbanization of rural areas would be an inevitable result of development. The prototype project alone would bring to the region an estimated 34,-000 people.

The department said the initial program could produce up to 250,000 barrels of oil a day. If the harmful effects were not excessive, production could be expanded to a million barrels a day.

Power plant, smelter rules eased. The Environmental Protection Agency (EPA) issued revised air pollution rules Sept. 6 for electric utilities and nonferrous-metal smelters which would retain the agency's "primary" emission standards (protection of human health) but allow an alternative to the previously required installation of expensive "scrubbing" equipment on plant smokestacks to meet the standards.

According to old rules, companies had to decide under state deadlines whether to install the scrubbing mechanism to curb pollution or to burn costly low-sulfur fuels. The new alternative would allow many plants to continue emitting pollutants at current levels until company-operated monitors determined that maximum allowable concentrations might be reached. Plants would then be required to cut back operations or close temporarily.

Critics of the rules had predicted blackouts in urban areas if power plants were allowed to emit pollutants until the stagnant air conditions common in summer months forced cutbacks in operations when power demand was highest. Acting EPA Administrator John R. Quarles Jr. said Sept. 6, however, that utilities in large urban areas would not be allowed to take advantage of the alternative.

Some critics said the rules would stifle the incentive to develop better plant equipment to clean emissions. Laurence I. Moss, president of the conservationist Sierra Club, also noted that the rules would put antipollution enforcement "in the hands of industry."

Nixon asks relaxation of clean air rules. President Nixon held a two-hour conference with 15 Administration officials Sept. 8 to discuss problems of energy supplies, development of alternative sources of fuel and increased production from U.S. oil reserves. Following the meeting, the President made an appearance at a White House press conference on the energy situation conducted by John A. Love, director of the Office of Energy Policy.

The President urged states to relax antipollution laws in an effort to overcome the expected shortage of fuel oil during the winter. Although the federal government had no power to limit state laws, Nixon said Love would meet with governors and local officials from the Northeast and Midwest during the next two weeks to lobby for modifications in environmental standards. States would be asked to permit the burning of higher sulfur grade oil by utilities and other large users which had been diverting short supplies of low sulfur content heating oil from residential consumers. The Administration also planned to allow increased imports of higher sulfur fuel oil.

Eased sulfur standards could increase heating oil supplies by 200,000–400,000 barrels a day, Administration spokesmen said.

Nixon also said he was determined to give "new impetus" to development of nuclear power and licensing of new plants. He repeated his request that Congress give approval to seven energy bills "before the end of the year." Nixon emphasized the importance of one of those measures—authorization for development of the Naval Reserve in Elk Hills, Calif., which was believed to contain one billion barrels or 2% of domestic crude oil needs.

Nixon also noted that the U.S. possessed an estimated half of the known world reserves of coal, but lacked the technology to convert the fuel into a clean energy supply.

Construction of deep water ports for oil supertankers also was essential, Nixon said, but he warned of heavy U.S. dependence on fuel imports.

Train differs with White House. Russell E. Train, who was confirmed by an 85–0 Senate vote Sept. 10 as administrator of the Environmental Protection Agency, said Sept. 11 that he would oppose relaxation of clean air standards unless the Administration implemented a mandatory allocation program for fuel oil at the same time that antipollution rules were eased.

A mandatory program, Train said, would insure that low sulfur oil was allocated for areas where air pollution levels were high.

Another EPA official, Assistant Administrator Robert L. Sansom, had objected to Administration efforts to link environmental restrictions to the worsening energy shortage. "The heating oil problem is going to be blamed on the environmentalists instead of on the restrictions on oil imports," Sansom said Aug. 20. He urged the White House to deal with the immediate shortage by lifting oil import restrictions.

Quality council reports. The Council on Environmental Quality, in its fourth annual report to President Nixon and Congress, said Sept. 17 that land use control was "probably the most important single factor in improving the quality of the environment in the United States."

Supporting the deregulation of natural gas wellhead prices, the council said that although consumer gas prices would rise, any increases would be offset by development of new supplies.

The council reported mixed results on air quality: while emissions of sulfur dioxide and hydrocarbons decreased about 2.5% from 1970 to 1971, particulate emissions increased almost 6%. There was little change in carbon monoxide and nitrogen oxide emissions.

The report expressed concern over the increasing worldwide concentration of carbon dioxide in the air, which it attributed to the burning of fossil fuels. Continued increases, the report noted, could create a "greenhouse effect," trapping solar heat and altering climate.

Regarding radiation hazards, the council said Atomic Energy Commission regulations provided sufficient safeguards for people living near nuclear power plants.

After meeting with Nixon to deliver the report, council members said Nixon had told them that even with the current concern with the energy crisis, he regarded environmental protection as a major consideration. Nixon had asked Sept. 8 for relaxation of air pollution rules to help deal with fuel shortages.

States warn of oil retaliation. Governors of three major oil and natural gas producing states—Louisiana, Texas and Oklahoma—warned that shipments to other states could be cut off unless restrictions on refining and exploration drilling were relaxed.

Gov. Edwin W. Edwards spoke at the Southern Governors Conference Sept. 25. (An estimated 75% of the fuel produced in the U.S. originated in Southern states.) "We have the police power to regulate where our fuel goes," he declared, adding it was unfair that other states with oil drilling or oil refining capabilities limited their efforts during the current energy crisis because of fears that the environment would be contaminated.

Senate OKs strip mine curbs. By a vote of 82-8 Oct. 9, the Senate approved and sent to the House a bill establishing federal regulation of strip mining of coal while leaving primary enforcement responsibilities with states.

A major provision would require that all surface-mined land be restored to "approximate original contour." Before final passage, the Senate defeated, 62-29, an attempt by Sen. James B. Allen (D, Ala.) to weaken the provision by allowing restoration of stripped land to "another configuration" compatible with surrounding terrain.

The Interior Department would have six months to adopt federal standards for control, which would serve as minimum guidelines for states and would be imposed if state plans were unacceptable.

Strip mining would be prohibited in areas where land reclamation was unfeasible. Within 22 months after the bill's enactment, mining on other land would be allowed only under permit, before which miners would be required to submit land repair plans and post performance bonds.

The bill authorized $100 million for use by the Interior Department in obtaining and repairing land already stripped, and $5 million a year in grants to states to aid in developing regulatory plans.

Court backs TVA dam. U.S. District Court Judge Robert L. Taylor ruled in Knoxville Oct. 25 that the Tennessee Valley Authority (TVA) could resume construction of the controversial $69 million Tellico Dam project in eastern Tennessee.

Rejecting arguments by environmentalist groups and property owners, Taylor ruled that the environmental impact statement filed by the TVA met the requirements of the National Environmental Policy Act. The court, Taylor said, could not "substitute its judgment for that of the TVA" as to the wisdom of a project on which $35 million had already been spent.

Canada

U.S. oil talks 'useful.' Recent talks with U.S. officials on prospects for a bilateral sharing of oil supplies had gone "in a way that we think is useful to us," Deputy Energy Minister Jack Austin said Jan. 17.

Austin emphasized that no final decision had been reached on oil sharing. Energy Minister Donald S. Macdonald had said the previous week that his government would continue a policy of reserving supplies against the present and future needs of Canada, and then selling any surplus . . . into American . . . or other markets."

Oil exports to U.S. curbed. Energy Minister Macdonald announced Feb. 15 that beginning as of March 1, oil exports to the U.S. would have to be licensed by the National Energy Board.

Warning of a possible oil shortage, Macdonald declared that "recent levels of export demand for our oil have been such as to strain the capacity of our oil production and transportation systems and to threaten the continuity of supply of Canadian oil to domestic refiners." Canada supplied about 25% of America's crude oil imports.

The New York Times cited U.S. oil experts Feb. 16 as expressing confidence that the new measure would not lead to a reduction of the current daily quota of 675,000 barrels of crude oil that Canada currently sent to refiners east of the Rockies.

In keeping with the policy that Macdonald had announced, the National Energy Board Feb. 27 reduced by 3.7%

the amount of crude oil to be delivered to U.S. customers in March.

The daily amount authorized was 1.235 million barrels, with the cutback affecting the Inter-Provincial Pipeline Co., which was to deliver 824,770 barrels a day to the Midwest market.

Energy exports up, control pledged. Statistics Canada said Canadian energy exports to the U.S. rose 32% in the first quarter over year earlier levels to $521.4 million, it was reported May 30.

The increase was paced by a 37% rise in the value of crude petroleum exports, partly the result of a price increase from $2.92 to $3.15 a barrel.

Energy Minister Macdonald said May 28 that he would impose controls if exports began to cause gasoline shortages in Canada. Earlier that day, Capital City Gas and Fuels said it would have to close 16 of its 24 retail gasoline stations in Ontario because major refiners, all owned by U.S. firms, had exported surplus gas rather than selling it to competitive independent Canadian chains.

Gas, fuel oil exports curbed. Energy Minister Donald Macdonald announced June 14 that exports of gasoline and heating oil would be cut "basically right down to zero" as of midnight June 15, to avoid a possible energy shortage.

Macdonald said the mandatory controls were necessary because of the failure of voluntary restraints. Under the plan, no exports would be allowed without a special permit from the National Energy Board.

Gasoline exports had increased during the first four months of 1973 to 872,000 barrels from 17,000 barrels during the comparable period in 1972, while heating oil exports had increased from one million barrels to 1.8 million barrels in the same period. High gasoline prices in the U.S., combined with a worldwide shortage of crude petroleum, raised the prospect that Canadian refiners would continue producing gasoline beyond their usual shift in August to winter heating oil production, Macdonald said.

All three opposition parties welcomed the new measures.

Propane, fuel oil export curbed. The government announced Oct. 11 that it was placing controls on the export of propane, butane and heavy fuel oils effective Oct. 15.

The National Energy Board said the controls reflected unusually heavy U.S. purchases of the products, and Canada's fear that U.S. shortages and price increases could spread to Canada. The board said the controls would allow established export levels to continue.

Energy report issued. The federal government issued a 300-page energy report June 28 that predicted adequate domestic fuel supplies though at higher costs to Canadians in the future, but warned that Canada could not be relied upon to solve the U.S. energy crisis.

The report was tabled in the House of Commons by Energy, Mines and Resources Minister Donald Macdonald, who said he hoped it would begin a nationwide debate that would lead to a national energy policy.

The study predicted that natural gas prices would double by 1980, and that gasoline and fuel oil prices would double by 1990. The study recommended consideration of a plan to link gas prices to oil prices, which could lead to immediate increases for Canadian consumers.

Increased energy exports to the U.S. were foreseen, that could raise the Canadian contribution from 6% to at most 12% of U.S. oil and gas needs. However, the report called for an increase in export prices, claiming that the price of Canadian gas was well below world market levels. Foreign ownership and control of Canadian energy industries, which stood at 35% overall and 91% for oil and gas, could be reduced by creation of a national petroleum company. A greater Canadian role in services, research and processing would mitigate the effects of foreign participation, the study suggested.

The report said a Mackenzie Valley natural gas pipeline could be built without "a major upheaval," but construction of both gas and oil pipelines along the route would entail "stringent monetary and fiscal restraints and would involve a reordering of economic and social priorities."

The pipeline supplying Alberta crude oil to Ontario should not be extended to Eastern Canada, the report said. Macdonald told a news conference June 28 that the East could continue to rely on cheaper imports.

Environmental costs—In a section on ecology, the energy study estimated the cost of protecting the environment from damage through energy production and consumption at $7 billion–$10 billion over the next 10 years.

The largest amount, $4 billion–$7 billion, would be spent to limit pollution from transportation, while about $1.8 billion would be spent by the electric power industry. The report said costs would be borne by "those who damage the environment or who produce or use polluting products," rather than "the public purse."

The study did not cover the effects of Arctic energy exploration and exploitation, which it said would "be undertaken from the beginning under strict environmental controls," making it impossible to identify the additional costs of environmental protection.

Ontario urges national energy policy. Ontario Premier William Davis called April 3 for a national conference, under chairmanship of the federal government, to formulate an energy policy for all of Canada.

Davis, who was addressing members of the Canadian Petroleum Association in Calgary, Alberta, was making his second trip to that province since January to discuss Alberta's plans to increase the price of natural gas by 10 cents per thousand cubic feet. Proposed rebates for local users would then have the effect of creating a two-tier price system—one for Alberta and a higher one for the rest of the country.

Davis told the oilmen in Calgary that he was not so much worried about the proposed raise as about "the much larger increases that are frequently suggested." He said that as a result of a meeting earlier in the day at Edmonton Airport with Alberta Premier Peter Lougheed he was "satisfied that men of good will can sort these things out." Although the two provinces might not agree completely, he added, "we do have a mutual interest in seeing and aiding in formation of a far-seeing national energy policy that considers all regional interests and that relates in very effective terms to a national industrial strategy."

Lougheed declared April 3 that Alberta was open to "any different views that may be expressed either by Ontario, the federal government or by other provincial governments. But we are pretty firm that we don't intend to be put in a position over an extended period of time that we sell our depleting resources below value."

Premiers call for energy meeting. Premiers of the provinces met in Charlottetown, P.E.I. Aug. 9–10, and called for a national conference to develop an energy policy for Canada.

The premiers Aug. 9 asked the federal government to convene the energy conference early in 1974 to plan federal-provincial structures to supervise conservation and rationalization of energy supplies.

B. C. ups royalties. The British Columbian government announced a new rate structure for oil and gas royalties May 11 that would yield about 30% more to the provincial treasury than the old rates.

Oil royalty rates would increase to 10%–40%, depending on the size of the well, compared with 5%–16% under the old rates. Natural gas production would be taxed at a straight 15%, but not less than three quarters of a cent for each 1,-000 cubic feet of gas produced.

Alberta gas prices up. Alberta Mines and Minerals Minister William Dickie said Aug. 3 prices for natural gas obtained by producing companies had increased an average of over 50% since the province announced its gas pricing policy in 1972.

Dickie said only TransCanada Pipe-Lines Ltd. of Toronto had not adequately followed the new policy, with prices ranging from substantially below to marginally below the province's recommended levels. Dickie said prices had risen over 50%, from 16¢ to 24¢ and 25¢ a thousand cubic feet in over half of all natural gas contracts.

Government OKs gas price hike. The National Energy Board had approved a natural gas export price increase under a contract between Westcoast Transmission, Ltd. and El Paso Natural Gas Co., a board official said Aug. 23. The increase was 5¢ per 1,000 cubic feet, or about 10%.

Gas pipeline delayed. A court in Yellow-knife, Northwest Territories adjourned May 16 after having heard arguments on land claims brought that week by the Indian Brotherhood of the Northwest Territories, involving 400,000 square miles. The court continued a land freeze imposed in April. The move was expected to delay further natural gas exploration, and could delay construction of a Mackenzie Valley natural gas pipeline.

A 25-firm consortium, Canadian Gas Study Ltd., had planned to file for a construction permit by the end of 1973, to build a 48-inch pipeline to bring natural gas from the Mackenzie Delta and from the North Slope of Alaska to meet an expected natural gas shortage in Canada by 1980. The project also hinged on approval in the U. S. of a plan to exploit Alaska oil, since gas would not be shipped from the North Slope until oil from that region moved to market, and since inclusion of Alaska gas as well as Mackenzie Delta gas was considered necessary to the pipeline's economic success.

Pipeline's value questioned—The Economic Impact Committee of the Cabinet level Task Force on Northern Oil Development, in a confidential report revealed May 22, found that the proposed Mackenzie Valley gas pipeline "will not make a major long-term contribution to the Canadian economy in terms of employment or personal incomes," and might worsen inflationary pressures and harm Canada's international trade position.

The report estimated that total employment generated directly or indirectly by the project would never exceed 1%–1½% of the Canadian labor force during the years of construction. Furthermore, since the project would be a one-time effort, construction "could have a destabilizing effect on employment trends."

Federal revenues, the committee said, would be "minimal," and might even decrease, because of tax deferral provisions relating to the pipe's operation, and because gas royalties would not, under current law, be charged on Alaska gas moving through the line.

Construction would cause a net increase in demand on the Canadian dollar of up to $600 million over three years, the report said, causing inflationary pressures that could increase the costs of more labor-intensive Canadian products in foreign markets.

The committee said financing of the $4.5 billion Canadian portion of the pipeline would "put some strain on Canadian and world financial markets," and government insistence on majority Canadian ownership could push up interest rates. Even if Canadians held majority ownership, the committee warned, foreign investors, "especially the American-controlled members of the pipeline consortium," might be able to secure effective financial control through more active participation.

Most of the natural gas would be marketed in the U.S., the report said, and "for Canada to accept anything less than the maximum possible return would be to subsidize the U.S. user."

Arguments against the plan were criticized in a Department of Energy, Mines and Resources study reported June 29. The study said the project would reduce unemployment to below 4% by 1976, and would not necessarily place unmanageable strains on the Canadian dollar.

Pipeline expansion set. The federal Cabinet approved two programs July 6 for a $66 million capacity expansion in the crude oil pipeline from Alberta to Ontario and the U.S., and a $35 million improvement in the British Columbia natural gas pipeline system.

The Cabinet denied that the Alberta project would lead to increased exports to the U.S., and said the expansion was needed to assure that existing export obligations would be met and to assure a spare capacity for shipments to Ontario.

Further pipeline plans were announced by Prime Minister Pierre Elliott Trudeau as he presented his anti-inflation program in Parliament Sept. 4.

To control energy cost inflation, Trudeau announced a major shift in oil policy that would bring Alberta crude oil to refineries in Quebec through a new pipeline, to supply gasoline and heating oil needs in the eastern provinces. Trudeau also asked the oil industry to refrain from price increases to consumers until Jan. 30, 1974, and said the government would propose legislation to insure that U.S. oil price increases were not automatically passed on to Canada.

Quebec and the maritime provinces had in the past been supplied by oil imported from Venezuela and, to a lesser extent, the Middle East, which had been less expensive than Alberta crude. But increases in international oil prices "would put Canadian oil into competition with" imported oil, Trudeau said, and extension of the Alberta-Ontario oil pipeline system (which passed through U.S. territory) to the East "would give additional security against international disruption of supply."

To prevent U.S. price increases from affecting Canada, Trudeau said the government might impose an oil export tax or set up a national oil marketing board. Energy Minister Donald Macdonald said in Parliament Sept. 5 that the fuel price freeze was, in effect, voluntary, and only applied west of the Ottawa Valley, the dividing line between Canadian and imported oil use. He said the government had no authority to restrain retail gasoline dealers and was not planning to submit legislation asking for that authority.

Mackenzie gas find reported. The Toronto Globe and Mail reported Sept. 7 that four Imperial Oil, Ltd. discovery wells in the Taglu field of the Mackenzie delta were found to contain about 6.5 trillion cubic feet of gas, the largest field so far found in Canada.

However, the Glove and Mail reported, total proven reserves were still substantially below the 15–25 trillion cubic feet thought needed to justify a solely Canadian gas pipeline from the Mackenzie delta without a spur from Alaska North Slope gas fields.

In a related development, the Wall Street Journal reported April 6 that Shell Canada Ltd. regarded as "a significant discovery" a natural gas find made in January in the Canadian Arctic. The gas was found at Shell's Niglintgak 30 well site in the MacKenzie Delta area.

Oil exports to U.S. to be taxed. The government announced Sept. 13 that it would institute a two-price system for crude oil starting Oct. 1, that would include a 40¢ a barrel export tax on oil destined for the U.S. market.

The government decision was announced a few hours after the National Energy Board denied applications to export oil to the U.S. in October at the pre-tax Alberta wellhead price of about $4 a barrel, since the price was not "just and reasonable." The 40¢ tax, applied to current exports of 1.2 million barrels a day, mostly to the U.S. Midwest, would add about $15 million a month to the government treasury.

The move was supported by eastern Canadians, who had been paying 30% higher prices for crude oil since Jan. 1, largely because of pressures from the U.S. market. But Alberta Premier Peter Lougheed, whose province provided 75% of the oil exports, called the action Sept. 14 "the most discriminatory action taken by a federal government against a particular province in the entire history of confederation."

The U.S. State Department said Sept. 14 that the tax "comes as a surprise and most certainly is not welcome information." It criticized Canada for not first consulting U.S. refiners.

Alberta moves to control fuels. Alberta Premier Peter Lougheed appointed his government as the pricing agency for Alberta oil Oct. 4 and cancelled all existing royalty arrangements, in response to federal imposition of an oil export tax.

Lougheed charged that Alberta could lose up to $300 million in revenue because of the tax, whose proceeds would go to the federal treasury, and said exploration activity would suffer.

Federal Energy Minister Donald Macdonald said any royalty increase could be absorbed by the oil companies, whom he said had enjoyed "a very substantial increase in returns in the last year."

Macdonald agreed Oct. 2 to convene a national conference on energy matters, to resolve such problems as the Alberta oil controversy and a dispute between Alberta and Ontario over natural gas deliveries and prices.

Oil sands plans set. Shell Canada, Ltd. and Shell Exploration, Ltd. reported Sept. 1 they planned to build a $700 million oil extraction plant to produce 100,000 barrels a day from the Athabasca oil sands in northern Alberta.

The plant, which would have to be approved by the Energy Resources Conservation Board, would double current production from the sands, estimated to contain over half the oil reserves in North America.

Middle East & North Africa

Kuwait signs participation pact. Kuwait signed a participation agreement Jan. 8 with two Western oil firms under which it would receive an initial 25% interest in existing concessions on its territory.

The pact, involving the Gulf Oil Corp. and British Petroleum Co., Ltd., would eventually increase the interest to 51% in 10 years. Kuwait became the third Persian Gulf state to sign the agreement negotiated Oct. 5, 1972. Saudi Arabia and Abu Dhabi had signed similar agreements late in 1972.

Iraq-Lebanon oil line reopens. Iraq and Lebanon agreed Jan. 13 to reopen the 20-mile section of the Iraq Petroleum Co.'s pipeline running through Lebanese territory. The pipeline had been shut when the Baghdad government had seized the assets of the Western oil firm June 1, 1972. The announcement gave no details of the accord, which would go into affect when ratified by the two governments.

Syrian-Iraqi transit fee agreement. Under an agreement signed in Damascus Jan. 18, Syria was to recieve a 50% increase in the transit fees on the shipment of Iraqi crude oil through Syria to Mediterranean ports. Syria would be paid

$150 million annually, compared with the $100 million paid by the Iraq Petroleum Co. prior to nationalization of its assets by Iraq. The agreement on the trans-Syrian pipeline use, retroactive to June 1972, was for 15 years. Financial arrangements were to be renegotiated after Dec. 31, 1975.

Iraq oil seizure pact. Iraq and the Iraq Petroleum Co. reached agreement on the June 1, 1972 nationalization of the western oil firm's oil fields and assets, President Ahmed Hassan al-Bakr announced in Baghdad Feb. 28.

Under terms of the pact, the company was to pay Iraq about $350 million as settlement of debts. The company accepted 15 million tons of Iraqi crude oil as compensation for the nationalization. Seven million tons were to be delivered in 1973, the remainder the following year. IPC also agreed to relinquish concession of its affiliate, the Mosul Petroleum Co., without compensation. IPC was to increase production of its Basra field to 80 million tons by 1976.

The company also agreed to hand over to Iraq its refinery and a pipeline section in Lebanon, with compensation to be paid in future oil deliveries. Lebanon, however, refused to permit Iraq to take over the installations and entered into negotiations with Baghdad. A Beirut announcement said at the conclusion of the talks March 5 that Lebanon had decided to take control of the refinery and pipeline. Pumping operations at the refinery in Tripoli began March 6 for the first time since its shutdown following Iraq's nationalization of IPC in 1972.

Japanese firm in pact. Japan Line, a tanker operator, reported Feb. 7 that it had concluded an eight-year contract with Abu Dhabi for direct imports of crude oil from the Persian Gulf sheikdom. It was the first time a Mideast oil-producing country had contracted for a direct sale of crude oil to Japan. Previously most Mideast oil had been supplied to Japan through U.S. and European oil firms.

The terms of the Abu Dhabi agreement called for shipments of 83.9 million barrels of oil in the first three years, increasing to 730 million barrels in 1980. The oil was to cost $2.38 per 8.38 barrels.

Iran nationalizes all oil. The nationalization of Iran's foreign-operated oil industry was announced March 20 by Shah Mohammed Reza Pahlevi. The order placed under government control all facilities of a consortium of British, U.S., French and Dutch companies. The action in effect nullified Iran's current 25-year oil contract, which had been scheduled to remain in effect until 1979.

The shah had warned the consortium Jan. 23 of his refusal to renew the contract after 1979 unless it agreed to increase its current production of 4.5 million barrels a day to 8.3 million barrels a day by 1977. He said the companies should turn over their operations to Iran now and sign a 20-year or 25-year contract under which they would purchase oil from Iran.

The shah disclosed March 16 that the Western firms had "handed over to us total and real operations of the oil industry of Iran with the ownership of all installations." He said major foreign oil companies would be only "buyers of our oil and we will not be paid less than what other Persian Gulf oil producers are getting for their oil."

Iran had first nationalized its oil fields in 1951, but the Western consortium regained virtual control of the country's oil industry under an agreement signed in 1954.

Iran gained "full and real control" of the oil industry within its borders under a 20-year agreement signed in Teheran May 24 with the Western consortium.

The National Iranian Oil Co. took control of all operations and facilities of the Western firms, with the consortium acting as technical adviser. The 30,000 square miles in which the consortium had rights would be reduced by 30%, but the remaining area contained all the current producing fields. Oil production was to be increased by eight million barrels a day by 1976. The National Iranian Oil Co. would receive enough oil to meet Iran's internal needs and certain agreed quantities for export.

The agreement would officially come into force when approved by the Iranian Parliament.

Members of the consortium (with interest in parentheses): British Petroleum Co. (40%), Royal Dutch Shell (14%), Gulf Oil Corp., Exxon Corp., Mobil Oil Corp., Texaco Inc., and Standard Oil of Calif.

(7% each), Cie. Francaise des Petroles (6%), and the Iricon group (5%).

U.S.-owned oil tanks bombed in Lebanon. An American-owned oil installation near Saida, on the Lebanese coast south of Beirut, was bombed and set ablaze April 14. Four oil tanks were blasted by explosive charges set by a group of 20 armed men who overpowered guards at the Zahrani oil terminal. One of the tanks was destroyed, two others were badly damaged, and the fourth was slightly damaged.

A previously unknown group calling itself the Lebanese Revolutionary Guard took credit for the attack, describing it as "a blow against American support for Israel." The statement accused the Lebanese army of "failing to protect the country against Israeli raids."

Palestinian guerrillas denied a Lebanese government statement intimating that they were responsible for the raid on the oil terminal. The Palestinians said Israelis had carried out the attack, but Israel denied the charge.

The blasted tanks were owned by the Trans-Arabian Pipeline Co. (Tapline) and the Mediterranean Refinery Co. Pumping of crude oil at the refinery resumed April 16.

Another Tapline installation four miles south of the Zahrani terminal was the target of an unsuccessful saboteur attack April 16. Explosive charges went off under a pipeline, but caused only dents and did not affect the flow of oil.

OPEC-Western oil agreement. A group of Western oil companies agreed to increase posted prices of crude oil by 11.9% to eight nations of the 11-member Organization of Petroleum Exporting Countries under an accord signed in Geneva June 2.

The pact, which would expire at the end of 1975, settled a dispute involving the organization's demand for a price increase to compensate for the devaluation of the dollar in February. The OPEC was to receive an immediate rise of 6.1%, making a total increase since the February devaluation of 11.9%. The agreement also provided for a new formula under which posted prices would reflect more rapidly and fully any future changes in the value of the dollar.

The price increases applied to Iran, Iraq, Abu Dhabi, Qatar, Kuwait, Saudi Arabia, Libya and Nigeria. The three other OPEC members—Algeria, Venezuela and Indonesia—were not part of the accord because they had their own individual pricing agreements.

Representatives of the oil companies and OPEC had held unsuccessful price talks in Tripoli, Libya May 7–9.

Saudis link oil flow to U.S. policy. Saudi Arabian Petroleum Minister Sheik Ahmad Zaki Yamani was reported to have warned Secretary of State William P. Rogers and other U.S. officials April 16 that his country would find it difficult to increase its oil production for export to the U.S. if Washington did not help in settling the Middle East problem to the satisfaction of the Arab states.

Yamani also was reported to have said that the oil wells were vulnerable to attacks by terrorists, and that such activities could only be curbed by a political settlement. Rogers replied that the U.S. opposed terrorism and that its halt was necessary if progress was to be made in achieving a peaceful resolution of the conflict.

Yamani reiterated in a Washington Post interview April 18 that there was a good chance that Saudi Arabia could increase its current production quota of 7.2 million barrels of oil a day to 20 million barrels by 1980 if the U.S. created "the right political atmosphere." The minister said he was specifically referring to U.S. policy toward Israel.

Arabs briefly halt oil flow. Libya, Iraq, Kuwait and Algeria temporarily halted the flow of oil to the West May 15 as a symbolic protest against the continued existence of Israel as a state. Although the stoppage was to have lasted only an hour, Libya shut its pumps for 24 hours. The action was taken in response to an appeal issued by a Pan-Arab Trade Union Conference in Cairo earlier in May.

Egyptian President Anwar Sadat had called on Arab states May 14 to use their oil to pressure the U.S. to drop its support of Israel. Speaking to Parliament, Sadat said: "The case is one of protracted struggle and not only on the Suez Canal battle.

There is the battle of America's interests, the battle of energy, the battle of Arabs."

Libyan leader Muammar el-Qaddafi had predicted May 13 that oil would be used as "a weapon of Arab self-defense." Speaking at a news conference in Tripoli, Qaddafi charged that an American oil firm operating in Libya, the Oasis Oil Co., employed seven Israeli agents on its staff. Oil industry sources in Libya and an Israeli government official denied the charges May 14.

Libya seizes oil firms. A series of nationalizations of U.S.-owned (and other foreign-owned) oil companies was started by Libya June 11 in retaliation for Washington's support of Israel.

The seizure of the Bunker Hunt Oil Co. of Dallas, Tex. was announced in a speech by Libya's leader, Col. Muammar el-Qaddafi, at a rally in Tripoli commemorating the third anniversary of the ouster of U.S. forces from nearby Wheelus Air Base. Describing U.S. oil firms as an extension of the U.S.' "policy of domination" in the Mid East, he said the U.S. "deserved a strong slap in the face." He warned "The time might come where there will be a real confrontation with oil companies and the entire American imperialism."

Egyptian President Anwar Sadat and Ugandan president Idi Amin also spoke.

Sadat was in Tripoli to discuss the forthcoming union between Egypt and Libya. He said June 12 that the nationalization of Bunker Hunt was "the beginning of a battle against American interests in the whole Arab region. America must realize that it cannot protect its interests if it continues defying the Arab nation and supporting Israel without limitations."

The Libyan government had first asked Bunker Hunt in the fall of 1972 for more than 50% control. The company refused and negotiations between the two sides had continued inconclusively since. As a prelude to expropriation, the Libyans had stopped Bunker Hunt from producing oil May 24.

A Bunker Hunt spokesman in Dallas said June 11 that the firm would "pursue all legal remedies."

The U.S. State Department disclosed June 12 that Libya had been informed of the U.S. government's reaction to the nationalization of Bunker Hunt. Department spokesman John King said the U.S. "has a right under international law to expect" owners of nationalized property to "receive prompt, adequate and effective compensation from the nationalizing government."

Libya took 51% control of the American-owned Oasis Oil Co. under an agreement announced Aug. 16 by Petroleum Minister Ezzedin Mabrouk. The pact provided for Libyan compensation to Oasis of $135 million in four installments without interest.

One-third of Oasis was owned by Marathon Oil Co., another third was jointly owned by Shell Oil and Amerada Hess Oil Co., and the remaining third was owned by Continental Oil Co. Oasis signed the agreement on behalf of all the companies, except Shell, which said it would take the matter under further consideration.

Libya Sept. 1 announced the nationalization of 51% of all foreign oil firms operating in the country. The decree was followed by another announcement Sept. 2 that Libya would increase the price of its oil from the current $4.90 a barrel to $6 and would refuse to accept U.S. dollars in payment for the oil.

The nationalization decree affected the Libyan subsidiaries of Exxon, Mobil Oil, Texaco, Standard Oil of California, Atlantic Richfield, W. R. Grace, Royal Dutch Shell and Gelsenberg. The action followed a breakdown in negotiations between the companies and the government, which had set Aug. 25 as the deadline for the takeover. In the talks the firms had rejected Libya's demand for 51% participation for fear that such an agreement would precipitate similar demands in the Persian Gulf states, which had a 25% participation pact with the companies.

The Libyan decree gave the firms 30 days to decide on their course of action. Tripoli radio said the government would compensate the companies, with the amount being fixed by a committee of Libyan officials. Each of the companies would be operated by a three-man board comprising two Libyans and one company executive. One of the Libyans would be named president of the board, and majority rule would prevail.

The decision on the price increase was announced Sept. 2 by Premier Abdel Salam Jalloud. He said Libya would "no longer accept payment in U.S. dollars" for its oil because "the dollar has lost its value and we want currency that is convertible into gold."

The U.S. State Department asserted Sept. 4 that the oil company takeover "does not comport" with Libya's obligations to comply with the agreements it had made with the U.S. firms.

The seriousness of the Libyan action and prospects for counteraction was expressed by two American oil experts Sept. 4. John Lichtblau of the Petroleum Industry Foundation warned, "If there is a cutback in Libyan oil there will be a worldwide shortage of supply not only because of a tight crude situation but because of transportation difficulties." With "tanker rates at record levels and ships in short supply," the existing difficulty in heating homes on the U.S.' East Coast would be further aggravated, Lichtblau said.

Oil economist Walter J. Levy said "consuming interests are disorganized and demoralized, and a destructive situation exists in the international oil trade." Levy noted that because of failure of oil companies and oil-consuming nations to cooperate in the past two years, "now nothing exists that will insure supplies for the consumers."

Other U.S. government and oil industry sources expressed the belief that Libya's oil seizure would lead to higher world oil prices generally and heighten pressure on the U.S. to alter its support of Israel.

Six of the nationalized companies submitted a letter of protest to Libya Sept. 7. The statement was released in New York and London by the U.S.-based Atlantic Richfield, Exxon, Mobil, Standard Oil of California and Texaco, and Royal Dutch Shell of London. The firms said they were not "willing to accept terms imposed unilaterally in contravention of valid agreements." They said "each company individually intends to take" action "to protect its rights."

The statement questioned Libya's intentions to provide compensation, saying that past experience showed that the Tripoli government had no "respect for the rights of the companies."

The Washington Post reported Sept. 12 that the firms affected by the Libyan

action were putting pressure on the U.S. government to take countermeasures, including the possibility of a boycott against Libya. The companies' efforts were said to be directed by John J. McCloy, whose New York law firm represented the oil group to which the firms belonged.

McCloy met Aug. 9 with State Department officials and Presidential adviser Henry A. Kissinger. McCloy later disclosed that "We expressed our hope to the State Department that the adventure of the Libyans would not succeed." The department indicated, he said, that it was discussing the situation with other governments.

Libyan Petroleum Minister Ezzedin Mabrouk had warned Sept. 5 that his country would fight if necessary to keep control of its nationalized oil industry. In an interview in Kuwait, where he had attended a conference of Arab oil ministers, Mabrouk said Libya would support any country at the conference "taking steps in protection of . . . [its oil] interests."

Saudis warn U.S. of oil cutoff. King Faisal of Saudi Arabia warned the U.S. government in a statement Aug. 30 that it would be "extremely difficult" for his country to continue supplying oil to the U.S. because of the U.S.' "complete support of Zionism against the Arabs." In another statement made public the same day Faisal, cautioned Arabs against the use of oil as a political weapon.

In response to these apparent conflicting remarks, U.S. Administration oil experts said Saudi Arabia was releasing "contradictory signals" because it still had not worked out a definitive oil policy.

In his first statement, Faisal said in a television interview that his country was "deeply concerned that if the United States does not change its policy in the Middle East and continues to side with Zionism, then, . . . such course of action will affect our relations with our American friends because it will place us in an untenable position in the Arab world."

The Beirut weekly Al Hawadess carried statements Aug. 30 by Faisal and his son Prince Saud al Faisal, undersecretary of the Petroleum Ministry, warning against slogans "which deliberately intend to push the Arabs to gamble with their strongest weapon [oil]." "No one is asking where we would get the money we need if we cut off the oil, not only for supporting our country, but also for providing assistance to our brothers on the front lines" with Israel, the king said.

Saud said the U.S. would not be affected by an immediate halt in Arab oil exports, because total U.S. dependence on Arab oil would not occur before the end of the 1970s. Only Western Europe and Japan, with whom the Arabs had no quarrel, would be hurt by an oil cutoff, the prince said.

Another warning by Faisal that he would use his oil in retaliation against U.S. support of Israel appeared in the Sept. 3 issue of Newsweek magazine. Faisal said the U.S. must refrain "from giving unlimited aid to Israel ... If no American response is forthcoming, one of our conditions for increasing our [oil] production will not have been satisfied."

Saudis want price increase. Saudi Arabian Petroleum Minister Ahmed Zaki Yamani called Sept. 7 for a revision of the 1971 Teheran Agreement to permit higher prices for the oil-producing countries.

The Teheran accord, aimed at regulating oil prices, expired in 1975. Yamani said the pact was "dead" and proposed that oil producers and major Western companies hold new talks in October to consider an increase in posted prices for oil exports, which determined the governments' revenue. Yamani also demanded an increase in the annual 2.5% price rise allowed to compensate for inflation.

The minister warned that unless the oil companies cooperated in amending the Teheran agreement, "we would have to exercise our rights on our own."

Juan de Onis, in an April 4 dispatch to the New York Times from Riyadh, had quoted Yamani as saying: "We are in a position to dictate prices, and we are going to be very rich."

Arab oil ministers meet. Oil ministers of 10 Arab states met in Kuwait Sept. 4 to formulate a common policy on using oil resources as a diplomatic weapon against Israel.

The conferees reportedly had failed to reach agreement on joint action as radical and conservative states posed sharply differing views. It was decided instead to have the matter "left to the heads of state of countries concerned," who were to participate in a meeting in Algiers Sept. 5 of nonaligned nations.

Iraq, among the radical states, proposed a cut in oil production for 10 years during which a minimum amount would be exported to the West to finance development plans. Conservative states such as Saudi Arabia and Kuwait cautioned against use of oil as a political weapon. They recommended suspension of any action pending an American response to Saudi King Faisal's warning Aug. 30 to the U.S.

Other states participating in the conference were Libya, Qatar, Syria, Bahrein, Abu Dhabi, Algeria and Egypt.

OECD meets on possible crisis. The oil committee of the 24-nation Organization for Economic Cooperation and Development held a secret meeting in Paris Sept. 11 to study means of pooling their petroleum resources in the event of an Arab cutback of oil exports.

Participating in the talks were representatives of the U.S., Japan, France, the Netherlands, Britain, Canada, West Germany, Italy and Norway. A conference source said the meeting was not contemplating a "united consumer front" but merely discussing "coordination of mutual assistance in the event of an emergency."

Nixon on oil & Mid East. President Nixon discussed various issues Sept. 5 during a press conference. He was asked what the Administration was "doing to meet these threats from the Arab countries to use oil as a club to force a change in our Middle East policy?"

"We are having discussions with some of the companies involved," Nixon said. As for the Arab countries involved, the problem was that "it's tied up with the Arab-Israeli dispute." That was why, in talking to Henry A. Kissinger, both before he nominated him as secretary of state and since, "that we have put at the highest priority moving toward making some progress toward the settlement of that dispute."

Another problem was "the radical elements that presently seem to be on the ascendancy in various countries in the Mideast, like Libya. Those elements, of course, we are not in a position to control, although we may be in a position to influence. Influence them for this reason: oil without a market . . . doesn't do a country much good. We and Europe are the market. And I think that the responsible Arab leaders will see to it that if they continue to up the price, if they continue to expropriate, if they do expropriate without fair compensation, the inevitable result is that they will lose their markets and other sources will be developed."

Nixon was asked if it was possible "that the threat of limiting the supply of oil would cause a moderation in U.S. support of Israel?"

He said "to suggest that we are going to relate our policy toward Israel, which has to do with the independence of that country to which we are dedicated, to what happens on Arab oil, I think would be highly inappropriate." Both sides were at fault, he said. "Both sides need to start negotiating. That is our position. We're not pro-Israel; and we're not pro-Arab. And we're not any more pro-Arab because they have oil and Israel hasn't. We are pro-peace. And it's the interest of the whole area for us to get those negotiations off dead center. That is why we will use our influence with Israel; and we will use our influence—what influence we have—with the various Arab states."

Oil imports soar—The U.S. was importing a record 1.2 billion barrels of oil daily from the Middle East, according to Deputy Treasury Secretary William E. Simon, who also was chairman of the Oil Policy Committee, Sept. 21.

Overall imports of oil had risen 32% in one year to six million barrels a day, Simon said. In 1972, the Middle East had exported 800,000 barrels a day to the U.S. Price increases by Middle East suppliers, combined with increased demand, would result in record 1973 costs of $7.5 billion for oil imports. Higher tanker rates and a shortage of supertankers also contributed to the higher costs.

Israel presses oil search. A report submitted Sept. 5 to the Israel Petroleum Institute proposed a five-year oil exploration program in Israel and the occupied Arab areas. The recommendations submitted by a committee of geologists called for immediate implementation of the plan, saying there was reasonable promise of finding oil in commercial quantities.

Israel had been operating the Egyptian-Italian oil field at Abu Rhodeis in the Sinai since 1967, according to the London Times Sept. 6. That field had produced 5,-300,000 tons in 1972, about 80% of the country's annual domestic consumption.

U.S., Iran in natural gas project. Units of the Transco Companies Inc. and the Iranian National Oil Co. signed a long term contract Sept. 27 in New York in an arrangement that would provide natural gas products to the East Coast for 22 years.

The $650 million venture included construction in Iran of a collection and processing plant for natural gas products, which would be shipped to the U.S. for conversion to pipeline quality gas. The Iranian and U.S. firms would be equal partners in the project, which expected initial deliveries by 1976 and full production two years later.

It was the second major natural gas contract signed by the state-owned Iranian company and a U.S. firm. In October 1972, International Systems & Controls Corp. had agreed to build a $700 million liquified natural gas project in Iran.

Germans to build refinery—Economy Minister Hans Friderichs announced Oct. 9, on a visit to Teheran, that West Germany would build an oil refinery in Iran with a capacity of 25 million tons a year, primarily aimed for West German consumption.

The refinery would be built at Bushire, a port on the Persian Gulf, at an initial cost of $500 million.

The two countries had also reached agreement on construction of a steelworks in southern Iran fed by natural gas.

U.S. loan to Algeria. The U.S. Export-Import Bank approved March 26 a $402 million package of loans and financial guarantees for the import of one billion cubic feet of Algerian liquefied natural gas daily to the U.S. for 25 years.

A guarantee of $52 million in commercial bank credits was to cover U.S. technical assistance for the project. The remaining $350 million would pay for construction by a U.S. firm of a liquefication plant at Arzew, near Oran. Sonatrach, the state-owned petroleum company, would furnish 10% of the costs, with $315 million to be divided equally between a direct Export-Import Bank loan and a loan from a commercial banking group headed by First National City Bank of New York.

In a related development, the U.S. Federal Power Commission March 30 removed what was considered the last legal obstacle to the huge gas deal by ruling in favor of a petition from the Sierra Club, a U.S. group concerned with environmental issues, which had asked that the El Paso Natural Gas Co. be required to use a tunnel instead of a pipeline trestle to bring the gas ashore at Cove Point, Md. The tunnel was expected to cost El Paso $23 million.

Algeria-Italy gas pipe set—The first underwater pipeline between North Africa and Europe would begin delivering natural gas from Algeria to Italy in 1978, it was reported Oct. 22.

The pipeline, to be built at a cost of $700 million, would pass through Tunisia and run for over 90 miles beneath the Mediterranean to Sicily. Deliveries would reach a volume of up to 120 billion cubic feet by 1980.

Other gas sales—A Spanish company and Sonatrach, the Algerian state oil company, reached agreement in principle on annual sales of about 50 billion cubic feet of Algerian natural gas a year for 20 years, it was reported Oct. 11.

Sonatrach and Ruhrgas of West Germany, the largest natural gas distribution company in Europe, signed a protocol Sept. 24 for the annual sale of 130 billion cubic feet of Algerian natural gas beginning in 1979 and continuing for twenty years.

French oil accord—An accord had been signed May 30 between Sonatrach, the Algerian national oil company, and Total-Algerie, a French concern, under which

$33.5 million would be invested in oil research and prospecting in the Sahara.

Other Oil & Energy Producers

Geothermal energy studied. World energy experts met at U.N. headquarters in New York Jan. 8–10 and generally agreed that one of the most promising new sources of relatively nonpolluting power was geothermal energy, the natural heat of the earth's core.

About 250 persons participated in the seminar, sponsored by the U.N. Department of Economic and Social Affairs with the assistance of the Center for Energy Information, a nonprofit foundation supporting the development of nonconventional energy sources.

According to one optimistic estimate, geothermal energy might become within 50 years a resource even more significant than petroleum. At least 80 nations were thought to have geological conditions indicating a substantial reservoir of the energy.

However, experts noted that there were a number of obstacles to full development, including the continuing reluctance of governments and industry to take geothermal energy seriously, the lack of systematic exploration of its potential, and the failure of most nations to exchange information on the subject.

World energy bank formed. A major consortium bank, called the International Energy Bank (IEB), had been formed in London to help finance development of new and expansion of existing energy supplies, the Journal of Commerce reported July 27.

Initial capital for the bank was $50 million. Participating in IBE were European, U.S. and Japanese banks. IEB would concentrate initially on oil and gas development, particularly in the North Sea.

Latin American unit. Ministers of energy and hydrocarbons from 21 Latin American and Caribbean nations including Cuba met in Quito, Ecuador April 2–5 to coordinate the region's power policies. They agreed to establish a Latin American Energy Organization, to be headquartered in Quito, and asked Venezuela to prepare a report on the formation of a Latin American body to finance power development projects.

Argentina. *Salto Grande loan*—The Inter-American Development Bank (IDB) had authorized an $80 million loan to help finance construction by Argentina and Uruguay of a 1,620,000 kilowatt hydroelectric power station at Salto Grande on the Uruguay River, it was reported Jan. 17.

The station would provide power for a 110,000 square mile market area in Uruguay and northeast Argentina with a population of 5.2 million. The total cost of its first stage was estimated at $432 million, of which the IDB loan would cover 18.5%, suppliers' credits 23.6%, and the Argentine and Uruguayan governments 57.9%.

The loan for the project, the largest ever granted by the IDB, was set in December 1972. The bank awarded a record $807 million in credits in 1972, of which $213 million went to Brazil, according to an IDB report Jan. 15.

Australia. *Export controls set*—Minerals and Energy Minister Reginald F. X. Connor announced Jan. 31 that the federal government had imposed export controls on all Australian minerals.

Connor said the move was designed to insure that mineral export prices were set at a reasonable level in relation to those of other countries.

The low price of Queensland coal exports was the main factor behind the federal action, according to official sources cited in news reports Jan. 31. Japan, in particular, paid less for Queensland coal than for coal from New South Wales or elsewhere. As a result, coal exports from Queensland had increased by 51.3% to 11.8 million tons in 1972, while New South Wales exports fell by 9.9% to 11.5 million tons, according to a Jan. 31 report.

The government already controlled oil, natural-gas and uranium exports.

National gas pipeline to be built—Deputy Prime Minister Lance Barnard announced

Feb. 20 that the federal government would build a nationwide natural gas pipeline linking the South Australian and northwest shelf gas fields with industrial and urban areas in Western Australia, the Northern Territory and the east coast.

The project, envisaged as one of Australia's biggest development undertakings, would be administered by a National Pipelines Authority. A total of $A190 million would be spent on the first stage, linking Sydney with the Gidgealpa gas fields in South Australia. The scheme would be financed by sales of national development bonds and an upgraded Australian Industry Development Corp.

Unofficial estimates on the total cost of the project ranged from $A850 million–$A1.5 billion.

The government would take over the Australian Gas Light Company's plan to construct an 840-mile natural gas pipeline from South Australia to Sydney, Minister for Minerals and Energy Reginald Connor confirmed Feb. 21. The line would correspond to the first stage of the national grid.

Gas & oil discoveries—Radio Australia reported July 10 a major find of natural gas on Barrow Island, off the northwest coast. The island was not far from the areas where the Woodside-Burmah consortium had found large natural gas reserves underseas. Barrow Island had been established as a major oil field in 1966 with reserves of about 200 million barrels of oil.

Woodside-Burmah Oil was reported Feb. 16 to have made the biggest oil strike thus far on the northwest continental shelf. A well registered a flow of light-grade crude oil at a rate of 2,730 barrels a day. Gas flowed there at a rate of 2.7 million cubic feet a day.

Oil authority proposed—Minister for Minerals and Energy Reginald Connor announced April 12 plans to introduce legislation creating a national petroleum authority that would be involved in all aspects of the oil and gas industry.

While the proposed authority would be concerned with more efficient use of resources, particularly in refining, Connor emphasized the government's first priority would be the exploitation of offshore reserves. He insisted the federal government had undoubted sovereignty over offshore exploration.

Under the plan, the authority would use its own personnel and equipment for exploration, grant contracts, operate in partnership with private companies and, in some cases, buy company shares. Connor added that since 50% of exploration already came from public funds, the government should have equity in the industry.

Country party leader John Douglas Anthony attacked the plan as a step towards nationalization.

Offshore bill approved—The House of Representatives Sept. 19 approved the government's controversial offshore legislation asserting federal sovereignty over all offshore areas. The opposition Country party opposed the principle of federal sovereignty; the Liberal party supported the principle. But Liberal party national development spokesman David Fairbairn maintained that the states should administer the legislation and that the federal and state governments should jointly allocate oil exploration permits and production licenses, with either government having veto rights. It was the second time in 1973 that the House had considered the legislation.

Government to buy offshore output—Woodside-Burmah Oil NL, the major partner in the consortium developing the vast gas deposits off the coast of West Australia, disclosed Oct. 2 that the government planned to control the production and marketing of all gas and oil found in the fields. The government had told the company it would buy all the oil and gas at the wellhead.

Minerals and Energy Minister Reginald Connor said Oct. 3 that the government's main aim in controlling gas and oil production was to insure that the reserves were fully integrated into an overall national program. He said the action would be taken in accordance with the Pipeline Authority Act.

Connor stressed Oct. 16 that Woodside-Burmah considered its primary responsibility to its nearly 85% overseas interests, while the government's "primary responsibility is to the Australian people. These resources are their birthright." He

had charged Oct. 10 that Woodside-Burmah wanted to "rip out and rape" the northwest shelf resources.

Burmah Oil, a British company, owned 54% of Woodside-Burmah, which in turn held 50% of the shares of the exploration consortium. The other shares were held 70% by the Australian public and 30% by other overseas interests.

Connor also announced Oct. 16 government plans to build a major refinery and petrochemical plant at Dampier, Western Australia. He said two Australian companies would handle the project. He also estimated that the admitted reserves of the northwest shelf oil and gas fields were worth nearly $A6 billion, although he added that the government had not yet learned the full extent of Woodside-Burmah's reserves. He assured Woodside-Burmah's investors that they would get a "handsome return on their investment," although not the astronomical returns that would come from unrestricted exports.

Groups buy into U.S.-owned rigs—An Australian public company, Ampol Petroleum Ltd., and the government-sponsored Australian Industries Development Corp. were reported Oct. 2 to have bought a 50% interest in offshore oil drilling rigs owned by the Ocean Drilling and Exploration Co. of the U.S. One of the rigs was currently operating on the northwest shelf; the other was being built at Kwinana, near Perth.

Bolivia. President Hugo Banzer announced April 11 that oil had been discovered in the southern province of Gran Chaco. He said initial production at each well drilled there was expected to be 600 barrels daily. In other economic developments:

The government had asked the Inter-American Development Bank (IDB) for credits totaling $200 million for a three-year development program, mainly for oil and mining. The IDB Sept. 21 approved a $46.5 million loan to the state oil firm YPFB to increase its refining capacity from 21,000 barrels a day to 41,500 barrels a day by 1978.

Brazil. Gen. Ernesto Geisel, president of the state oil monopoly Petrobras, said in his annual report to stockholders

March 23 that the company's profits had risen by 42.9% in 1972. He added, however, that expenditures for oil imports had risen, and warned that the rise in world oil prices would make Petrobras vulnerable to foreign suppliers.

Petrobras, which made a $1.8 billion profit in 1972, would step up prospecting in Acre, on the Peruvian frontier; increase its range and volume of lubricants with a view to replacing all imports; and bring offshore wells in the northeast into production in 1973, it was reported Jan. 12.

Braspetro, the international subsidiary of the state oil firm Petrobras, had signed a $10 million exploration agreement with the Egyptian General Petroleum Corp., Latin America reported Jan. 26. The pact provided for Braspetro to prospect in a 12,500 square kilometer area east of the Nile valley, with a right to 50% of the profits on oil finds. The favorable terms were ascribed in part to a recent attempt by Brazil to mediate in the Middle East conflict.

Electric power expansion—The World Bank April 12 approved a $20 million loan to Servicos de Electricidade to help finance expansion of electric power subtransmission and distribution facilities in Rio de Janeiro, Sao Paulo and surrounding areas. The project, costing the equivalent of $382.6 million, would help meet rising demand and provide service for more than 700,000 new residential customers.

Centrais Electricas de Sao Paulo S.A., the utility supplying electricity for Sao Paulo state, had been granted private joint financing totaling $248 million from French, British, Japanese and other banks for construction of the Gua Vermelha power station and extension of the Ilha Solteira station, both part of the Urusupunga electric complex in northeastern Sao Paulo, it was announced March 22.

Brazil Aug. 1 signed an agreement for a $125 million World Bank credit to help build the Itumbiara hydroelectric project.

The Inter-American Development Bank approved a $54.2 million loan to Brazil Sept. 13 and a $40 million credit Oct. 4. The first would help build the second stage of the electrical transmission system to connect the Ilha Solteira hydroelectric plant with the city of Sao Paulo. The second would help finance

broad-scale development in the petrochemical and chemical industries.

Brazil-Paraguay project—Brazil and Paraguay signed a partnership agreement April 26 to build what was expected to be the world's largest hydroelectric project, at Itaipu on the Parana River between the two countries.

The pact was signed in Brasilia by Foreign Minister Mario Gibson Barboza and his Paraguayan counterpart, Raul Sapena Pastor. President Emilio G. Medici and Paraguayan President Alfredo Stroessner attended the ceremony.

Argentina, which had opposed the project for two years, recalled its ambassador from Brasilia to protest the agreement. Argentine officials had demanded prior assurance that projects such as Itaipu would not harm navigation downstream in Argentina, or jeopardize other Parana hydroelectric schemes planned at Yacyreta-Apipe and Corpus between Argentina and Paraguay.

The Itaipu project, with a planned capacity of 10 million kilowatts, reportedly would cost $2 billion. It was planned by the International Engineering Co. of the U.S. and Electroconsult SpA of Italy.

A dam 720 feet high and 4,600 feet long would create a 515-square-mile reservoir, half to cover Brazilian territory and half Paraguayan territory.

Brazil and Paraguay agreed to set up a joint company to build and run the project, and to divide the power production equally. Paraguay was expected to have a surplus, which it would sell to Brazil for use in the developing industrial areas of Sao Paulo and other states. Brazil's power consumption had been growing by 10%-12% annually.

Before the agreement was signed, Brazil and Argentina had made efforts to resolve their conflict, agreeing to exchange advice and information about their hydroelectric projects and plans. However, Argentina had subsequently protested the way in which Brazil warned that it was temporarily restricting the flow of Parana waters to fill the reservoir of another big hydroelectric project, at Ilha Solteira in Sao Paulo state.

Colombia. Technicians as well as industrialists and political leaders meeting in Medellin concluded March 4 that

Colombia would have to invest $480 million in the next 10 years to find new sources of petroleum to overcome the current energy crisis. The government moved to increase investment in oil exploration by saying it would allow foreign firms to increase the price of hydrocarbons extracted from sources found in the next five years.

Ecuador. *Concessions void*—The government Jan. 1 took back 1.27 million acres of oil concessions granted to the U.S.-based Texaco-Gulf consortium during the administration of ex-President Jose Maria Velasco Ibarra.

The action conformed with a June 1972 decree ordering oil companies to return 60% of their concessions to the state upon completion of exploration. A Texaco-Gulf request for "adequate compensation" was rejected Jan. 10 on grounds that the consortium, as stipulated in the decree, had selected the acreage to be returned.

Natural Resources Minister Gustavo Jarrin Ampudia said Jan. 4 that the returned concessions contained proved reserves of 35 million barrels of crude, representing at least $70 million in revenues for the government.

(Ecuador's ambassador to the U.S. had said his government's November 1972 cancellation of another U.S. consortium's oil concessions in the Gulf of Guayaquil was an "isolated" action designed to facilitate investigation and determination of the circumstances under which the concessions were granted, it was reported Jan. 10. He said there had been no discussion of compensation to the consortium because its concessions had only been "revoked," not "expropriated" or "confiscated.")

Ecuador was disclosed March 1 to have canceled the Minas and Yasuni oil concessions in Oriente province, in which three U.S. firms held an interest. Amerada Hess Corp. held 57% working interest in the concessions and was operator for itself and Hamilton Bros. Oil Co., with 6%, Kirby Industries Inc., with 5%, and others.

Neither Amerada Hess nor Kirby explained the cancellation. However, both said their efforts to negotiate oil and gas concession agreements with the Ecuadorean government had not been successful.

President Guillermo Rodriguez Lara and Venezuelan President Rafael Caldera agreed Feb. 6, during Caldera's visit to Ecuador, to a technical cooperation program between their two petroleum industries. Ecuador was second only to Venezuela among Latin American oil producers.

Production income tax set—The government May 31 decreed an 86% tax on the oil production income of firms which had obtained exploration rights to Ecuadorean concessions and then transferred the concessions to other companies for exploitation.

Natural Resources Minister Capt. Gustavo Jarrin Ampudia said the measure would prevent companies from making large profits on concessions after they stopped work on them.

Among the firms affected were Phoenix Canada Oil Co. of Canada and Norsul Oil & Mining Ltd. of the U.S.

Each held a 1% production participation in the Coca concession in Oriente Province, one of two which supplied the more than 200,000 barrels of oil produced daily by the Texaco-Gulf consortium.

Ecuador exported more than 30 million barrels of petroleum during the first five months of exports by the Texaco-Gulf consortium in Oriente, the Miami Herald reported Feb. 10.

The government would receive 80% of income from oil exports as of Jan. 1, the Times of the Americas reported Jan. 10.

The government raised the reference price for crude from $2.60 per barrel to $3.61 per barrel April 1–May 23. The high price was offered by a Japanese concern, Sumitomo Shoji Kaisha, according to El Nacional of Caracas May 25.

Gabon. Gabonese President Omar Bongo warned foreign oil companies that "excessive exploitation" would lead to the same fate that oil companies suffered in Libya (some of which had been nationalized), it was reported Oct. 9. ELF-Gabon, an oil company controlled by the French government, had turned over 10% of its shares to Gabon, it was reported at the same time.

Greece. The first significant discovery of gas and oil in Greece was reported by Colorado Greece Oil Corp., a unit of Coastal States Gas Corp. and operator for a five-company U.S. group, the Wall Street Journal reported Jan. 11. The firm said it had successfully tested a gas well 11 miles off the Greek coast in the northern Aegean Sea. Greece imported all its fuel.

Guatemala. Two companies, the U.S. based Shenandoah Oil Corp. and Norway's Saga Petroleum A/S & Co. said they had agreed with a Luxembourg firm, Basic Resources International S.A., to explore and develop a 933,000 acre petroleum concession in Guatemala's Peten basin, to which Basic had beneficial rights, the Wall Street Journal reported March 27.

Shenandoah and Saga said a Basic subsidiary, Recursos del Norte Ltd., had discovered oil on the concession in northern Guatemala and had successfully completed a second well there. They said a third well was being drilled.

The companies said each would have a 25% interest in the venture, with the Basic unit holding the remaining 50%. Shenandoah would be the operator and the Basic unit would act as petroleum contractor in Guatemala.

Mexico. Mexico planned to develop coal mines of Oaxaca and Veracruz states and to increase production at other mines in hopes that coal could replace oil as Mexico's main fossil fuel source of electric power, it was reported Aug. 17.

North Sea. *Great Britain*—Oil production in Britain's sector of the North Sea could reach 70–100 million tons by 1980, according to a government report published May 14. The top figure would be equivalent to Britain's current imports, costing £1 billion and totaling two-thirds of the nation's estimated 150 million tons of oil consumption in 1980. Production from the five established fields was estimated at 40–50 million tons.

A consortium led by the U.S. oil firm Mobil announced July 11 that the Beryl field in the North Sea, discovered in 1972, was a commercial prospect with an eventual expected yield of more than 100,-000 barrels of oil daily. The consortium, Mobil North Sea, was composed of the

British Gas Corp., Mobil, and the U.S. firms of Amerada and Texas Eastern.

The Shell and Esso oil exploration consortium, called Shell UK Exploration and Production, July 4 announced a new oil strike northeast of the Shetland Islands and close to its Brent field. The announcement said the strike was of "possible commercial size." It was the fourth major oil discovery by Shell and Esso in the North Sea.

The Occidental Oil Group had announced March 8 that its recently discovered Piper North Sea oil field, about 100 miles east of Wick, was a "commercial find with the potential of being a major oil accumulation." The announcement followed the drilling of a second well in the field. Occidental had a 36.5% holding in the group, with British-based firms owning the remaining shares.

After the announcement of the discovery of an extension to its Brent oil field located 100 miles northeast of the Shetland Islands in the North Sea, the Shell and Esso exploration consortium boosted by 50% Oct. 10, to 1.5 billion barrels, its estimate of oil reserves in the field. Analysts said the expanded field might yield 450,000 barrels a day, or about 20% of Britain's current oil consumption. Production from the field was expected to begin in 1976.

The discovery of a big new oil field named Hutton, located 75 miles northeast of the Shetland Islands, was announced Dec. 5 by the National Coal Board, in partnership with two U.S. oil firms, Continental and Gulf Oil. The field was said to have a potential output of 300,000 barrels a day.

Norwegian developments—The Norwegian state oil company, Statoil, signed an agreement May 18 with the Phillips Petroleum Co. of the U.S. for construction of a 220-mile oil pipeline from the Ekofisk field in the Norwegian sector of the North Sea to Teeside, England, and a 268-mile natural gas pipeline from the same area to Emden, West Germany.

Total cost of the project, including the pipeline and onshore facilities, was estimated at $1.5 billion. Completion of the oil pipeline was expected in 1973 and the onshore facilities by 1975. Initially, 300,-000 barrels of oil would be piped daily, but the flow rate would rise to one million barrels a day when the pipelines and onshore facilities were completed.

Statoil would have a 50% share in the new pipeline company, named Norpipe, in exchange for 10% of the investment. A consortium headed by Phillips would have the other half. Other members of the consortium were Petrofina of Belgium; Petronord, a group of French and Norwegian companies; and AGIP, the Italian state oil company.

Ownership of the pipelines would revert to Norway after 30 years. The project had drawn heavy criticism from Norwegian leftist politicians and trade unionists, who oposed Norway's oil and gas being piped ashore in foreign countries. The government agreed to the project, however, because of technical difficulties involved in landing the oil and gas in Norway.

A natural gas find wholly in the Norwegian sector of the North Sea was reported Oct. 2. Petronord, the French-led group of French and Norwegian firms, described the find as "considerable." The discovery was located east of the massive Frigg field, the latter split between the British and Norwegian sectors. The new find was the third in the area.

Peru. Firms doing only oil prospecting and exploration in Peru were exempted from tax and tribute payments by a government decree Feb. 22.

The government newspaper La Nueva Cronica announced Feb. 26 that an exploratory well dug by the Occidental Petroleum Corp. near the Ecuadorean border had begun producing more than 2,800 barrels daily.

Standard Oil of California lost a refinery and associated service stations in Peru in a tax dispute with the government, the Wall Street Journal reported March 26.

The unit, operated by the Standard subsidiary Conchan Chevron, was auctioned for alleged back taxes and fines of $5.5 million dating to 1967. The state oil concern Petroperu was the high bidder with $2.6 million, well below the $3.9 million value given the unit by a government-appointed evaluation team.

A spokesman for Standard in San Francisco said the company disagreed with Peru's tax claims and felt its properties

had been "unjustly" taken from it. Standard claimed it had been exempted from the disputed taxes by contracts signed with former governments in 1962 and 1963.

Earlier, the International Petroleum Corp., a wholly owned subsidiary of Standard Oil of New Jersey which had ceased to operate in Peru in 1968, had been declared bankrupt at the request of the International Bank of Peru, it was reported Feb. 14.

A number of U.S. companies, including Oceanic Exploration Corp., El Paso Natural Gas Co. and Signal Oil & Gas Co., had signed petroleum operations contracts with the state oil concern Petroperu, it was reported June 5. The government had predicted April 18 that crude production would reach 200,000 barrels daily in 1980, an increase of 135,000 barrels a day over current production.

Soviet Union. For the first time since World War II, the Soviet Union showed an annual decline in net petroleum exports in 1972, the New York Times reported June 7. Increased purchases from the Middle East grew faster than exports to European countries.

The imports were attributed both to political considerations and to a growth of energy consumption in the U.S.S.R. exceeding the growth in production, despite the presence of large untapped reserves.

Trade figures also showed that the Soviet Union remained a net importer of natural gas in 1972, as imports from Iran and Afghanistan outweighed exports to Europe.

U.S. & Japanese gas deals—The Soviet Union and two U.S. companies signed a preliminary agreement in Moscow June 8 on a $10 billion, 25-year project to bring Siberian natural gas to the U.S. West Coast.

The agreement, which would require $4 billion in Western financing, was signed by the Occidental Petroleum Corp. and the El Paso Natural Gas Co. Under the plan, the Soviet Union would first conduct exploratory operations to determine whether 25 trillion cubic feet of gas reserves existed in the Yakutsk area in eastern Siberia. Howard T. Boyd, chairman of El Paso, who joined Occidental chairman Armand Hammer in a Moscow news conference, said about half that total had already been proven.

The Soviet Union, possibly with Japanese help, would borrow $2 billion to build a 2,000 mile pipeline to a new liquefaction plant in the Vladivostok area. The U.S. firms would borrow a further $2 billion for a fleet of 20 tankers, to begin deliveries in 1980 at the earliest.

The entire plan depended on favorable action by the U.S. administration, which was currently considering a decision on reliance on foreign energy sources, and by Congress, which was considering tariff and credit concessions for the Soviet Union. In addition, gas prices and volume remained to be negotiated.

Though the deal was still highly tentative, it was believed that Soviet Communist Party General Secretary Leonid Brezhnev wanted a signed accord before his trip to the U.S. to encourage additional support from American businessmen for Soviet development projects.

Three weeks later another accord was signed in Moscow June 29 with the U.S. firms Tenneco, Inc., Texas Eastern Transmission Corp. and Brown & Root, Inc., all of Houston. Under the pact, preliminary in nature, the U.S. Export-Import Bank and private Western banks would provide loans totaling up to $3 billion for construction of a pipeline and liquefaction plant. The project would ship 2 billion cubic feet of gas daily to U.S. Atlantic coast ports from West Siberia.

Another agreement, signed in memorandum form by Japanese and Soviet officials in Tokyo July 17, provided for development of natural gas facilities in the Yakutsk area by the Soviet Union, Japan and the U.S. American and Japanese banks were to lend $150 million for the project, which would supply the U.S. and Japan with 1.5 million tons of liquefied gas over 20 years.

Soviet gas line opens—The Soviet Union opened a natural gas transmission line to West Germany Oct. 1. The link was established at the Bavarian town of Waidhaus, on the Czechoslovak border, at the western end of a trans-Czechoslovakia pipeline that transported gas from producing fields in the Soviet Ukraine and Central Asia. This pipeline system would transport Soviet gas pending completion of a pipeline from the vast West Siberian fields through Central Russia.

The gas deliveries were agreed in 1970 and 1972 under an arrangement in which West German steel pipe for the transmission lines would be repaid with gas.

East Germany had announced June 4 that it had authorized construction of an oil pipeline to serve West Berlin. It was believed the East Germans would build a branch line from a new pipeline bringing Soviet crude oil to East Germany across Poland.

Venezuela. The government announced that it would increase oil prices by as much as 10% to compensate for the U.S. dollar devaluation, it was reported Feb. 21.

The price increase, the second in four months, followed a Mines Ministry report Jan. 6 which showed 1972 oil production had dropped by 9.27%, or 330,000 barrels daily, compared with 1971 production. U.S. companies had reportedly decreased production to protest certain provisions of the 1971 oil reversion law. But the president of Creole Petroleum Corp. had announced his company would increase production in 1973, the London newsletter Latin America reported Jan. 5.

Mines Minister Hugo Perez La Salvia asserted Jan. 15 that Venezuela would not sign a treaty with the U.S. for development of the Orinoco tar belt without assurances of preferential treatment for all Venezuelan oil on the U.S. market. The Orinoco belt contained oil reserves estimated at 700 billion barrels.

As a result of President Rafael Caldera's recent visit to Buenos Aires, Venezuela and Argentina would coordinate their policies on hydrocarbons, with Venezuela providing oil and gas products not produced in Argentina and purchasing Argentine equipment and machinery for its oil industry, it was reported Feb. 16.

A Venezuelan negotiating team returned from Rio de Janeiro Feb. 1 after completing preliminary talks with Brazilian officials on the sale of Venezuelan crude oil to Brazil in exchange for refined derivatives.

Other Areas

European policy split. A meeting of the energy ministers of the nine European Economic Community (EEC) members broke up May 23 in Brussels without reaching agreement on a common EEC energy policy.

France had blocked approval of a resolution, drafted May 22, proposing an action program to prevent a critical energy shortage and "mutual overbidding and confrontation" by oil consuming nations. Instead, the ministers directed the EEC Commission to formulate new energy proposals by the end of 1973.

The resolution, based on an EEC Commission paper released April 19, had called for better relations with both oil consumer and producer countries, an organized EEC energy market, greater and more effective use of nuclear energy and natural gas and definition of the future of coal.

Most members had given priority to coordinating a policy with the other oil consuming nations, particularly the U.S. and Japan. However, France contended that a unified oil policy within the community should be agreed first.

The failure to reach a common position weakened the negotiating authority of EEC Energy Commissioner Henri Simonet during a five-day visit to the U.S. that ended June 5. Following talks with government officials and businessmen in Washington and New York, Simonet told a press conference June 4 that government officials had been "receptive" to suggestions for cooperation between the U.S., the EEC and Japan, although some were "skeptical" such cooperation could be achieved.

Other developments:

The European Parliament's energy, research and technology committee presented a report in Strasbourg May 8, stressing the urgency of developing a comprehensive EEC energy policy to prevent critical shortages by 1980 that would "quite literally" cause "the lights [to] go out all over Europe." It predicted a growing role for coal by the year 2,000 to compensate for depleted oil supplies.

(The British National Coal Board announced May 31 that coal producers in Belgium, France, Germany, Netherlands, Spain and Britain had adopted a plan that called for expansion of coal production in new coalfields, shipment of the largest possible supplies of indigenous coal to power stations and more scientific and technical research on coal production and utilization.)

The European Atomic Energy Community (Euratom) was reported June 2 to have announced its first purchase of natural uranium for enrichment by the U.S. Atomic Energy Commission. Previous purchases had been made separately by European companies, Euratom had until now leased enriched uranium from the U.S. The purchase, made from the Urangesellschaft of Frankfurt, West Germany, was unofficially reported to be worth over $16 million.

Belgium. Belgium's Socialist-headed government won a vote of confidence, 136–40, in Parliament May 9 at the end of a debate over a project to build a joint Belgian-Iranian oil refinery at Lannaye, near Liege. Opposition deputies had called for the resignation of Finance Minister Willy Claes in the affair.

The dispute stemmed from the fact that only three out of a total 6,000 shares in the project were owned by Belgians all prominent Socialists, who were consequently named to the board doing the feasibility study of the project.

The Socialist coalition partners—the Social Christians and Liberals—had balked at the financing as well as at the composition of the board. To retain their support, Premier Leburton had announced earlier in May an increase of the state's holding in the scheme to 50% and a promise to change the composition of the board doing the feasibility study to insure a better political balance.

The EEC Commission barred Belgium from giving over $10 million in subsidies and property tax exemptions to subsidiaries of the Exxon Corp. and the Shell Oil Co. over a three-year period, according to unidentified sources Sept. 18. The aid would have been designed to help Exxon Europe and Shell International extend refining capacity in Antwerp. The subsidies were deemed contrary to EEC rules. French officials had argued that the subsidies would give the two oil companies an unfair advantage and that Belgium already had more than adequate refining capacity.

Japan. The Joint Japan-United States Committee on Trade & Economic Affairs, meeting in Tokyo July 16–17, discussed a five-point oil cooperation plan proposed by U.S. Secretary of State William P. Rogers, calling for cooperation between oil producing and consuming nations, joint measures to cope with fuel shortages and multinational efforts to expand oil and gas production.

A communique July 17 disclosed agreement by the two nations on a continuing exchange of views on the problem of insuring future energy supplies.

West Germany. The Cabinet in Bonn approved Aug. 29 a program designed to secure energy supplies for the nation.

In related developments, a West German consortium of electricity producers contracted for purchase of enriched uranium from the Soviet Union; the U.S.S.R. opened a natural gas transmission line to West Germany; and West Germany agreed to build an oil refinery in Iran.

Economics Minister Hans Friderichs said Aug. 29 that the government would pursue its efforts to develop a major national oil company, Veba, to increase the security of petroleum supplies, although he assured that the government would not curtail the activities of the big international oil companies, which currently supplied 75% of Germany's oil. The program called for an increase in the nation's own oil refining capacity, an increase in supplies of natural gas, nuclear energy and lignite, and a decrease in West Germany's coal production from the current 97 million tons a year to 83 million tons in 1978.

The program proposed federal grants to the coal industry for rationalization and subsidies for coal's biggest customers, steel and electricity, to compensate them for paying more for German coke and coal than for foreign fuel. It also called for government expenditures of at least six billion marks by 1976 for a major nuclear reactor program.

Atomic Power: Safety & the Environment

A-plant debate continues. At a Jan. 3 Washington news conference, Ralph Nader and spokesmen for the Union of Concerned Scientists asked the Atomic Energy Commission (AEC) for a moratorium on all construction of new atomic power plants "until all safety-related issues are resolved." Nader promised to continue his compaign to delay massive use of atomic power, on economic and safety grounds in Congress, the courts and among electric company stockholders.

Nader and the scientists' group said the danger of "catastrophic nuclear power plant accidents is a public safety problem of the utmost urgency," and called for output reductions of up to 50% at the 29 operating nuclear reactor power plants. Although "no nuclear explosion could occur," they said, a failure in the emergency core cooling system could cause release of radioactive materials with deaths possible nearly 100 miles from the plant. Nader also said that disposal of radioactive wastes presented safety problems.

Nader charged the AEC with secrecy on the issue and with failure to publicize what he said was a belief held by a majority of AEC scientists that safety problems may exist.

In reply, William R. Gould, chairman of the Atomic Industrial Forum, said in New York that "throughout the civilized world" there was a "massive" shift to nuclear energy as a power source "because of its advantages in terms of fuel supply, economics, environmental affects and public health and safety." He said he was confident that "the extraordinary safety record of nearly 100 operating power reactors worldwide" and "the extremely conservative engineering approach" of U.S. plants would be confirmed at February hearings of the Congressional Joint Atomic Energy Committee.

Outgoing AEC Chairman James R. Schlesinger charged at a Jan. 23 hearing of the joint committee that delays in license approvals for nuclear power plants had at times been prolonged by lawyers looking for "a lucrative substitute for ambulance chasing."

Schlesinger said the AEC was the only federal agency required by law to conduct adjudicatory hearings before licensing plants, and said the hearings had delayed plants for years, although they often covered issues resolved long before in earlier cases.

Reactor safety disputed. The Environmental Protection Agency (EPA) rejected as "inadequate" Feb. 16 the Atomic Energy Commission's (AEC) environmental impact statement on emergency safety procedures for nuclear power plants.

EPA said AEC should have explored the possibility of catastrophic accidents, such as the loss of coolant or pressure vessel failure, and suggested an independent study be commissioned in cooperation with both agencies. The environmental agency conceded that no serious accident had occurred in a nuclear plant, but said the number of such plants in the U.S. would rise from 30 at present to 1,-000 by the year 2000.

According to a Feb. 27 report, an AEC spokesman said the agency was not prepared to discuss catastrophic accidents until completion of an internal study which was expected to take another year.

The 4,500-member Federation of American Scientists asked the AEC Feb. 6 to order a reduction in reactor operating levels, as the agency had proposed in 1972, to avoid an accident that "could mean death to tens of thousands of people" through radiation release.

Although such an accident was unlikely, the federation said, the AEC should undertake "a crash program of stepped up reactor safety research," especially concerning "alternative reactor systems," which could include gas-cooled reactors. Nearly all current reactors, and most planned reactors, were watercooled.

A three-man AEC board held hearings Feb. 1-4 on environmental issues relating to power plants, including problems in mining, refining, transport and disposal of radioactive fuel. The hearings were designed to help speed licensing procedures for the 120 power plants already planned or begun by considering at one time the problems common to all the plants. Dr. Henry J. Kendall, a nuclear physicist and member of the Union of Concerned Scientists, said at the Feb. 1 hearing that the AEC had published estimates of a one in 1,000 chance of a major reactor pipe break, a risk he considered "totally unacceptable."

Suit vs. A-plants. Ralph Nader and Friends of the Earth, an environmentalist group, petitioned a federal district court in Washington May 31 to close 20 nuclear power units because of safety hazards. The suit charged the Atomic Energy Commission (AEC) with a "gross breach" of its public health and safety obligations by failing to assure that power plant cooling systems would not fail and cause the release of radioactive materials.

David Brower, president of Friends of the Earth, said "overwhelming scientific evidence" had shown that "the lives of millions of people" were threatened by operation of the plants because of "crude and untested" safety systems. The suit quoted several AEC officials as expressing doubt as to the reliability of back-up cooling systems.

In a statement on the suit, the AEC said it saw no basis for suspending operation of the units, which were at 16 power plants in 12 states. The AEC conceded there were differences in opinion within the commission on the safety systems but noted that a review was already under way to determine whether present core-cooling regulations were adequate.

The suit was dismissed June 28 by Federal Judge John H. Pratt.

Pratt ruled that the plaintiffs had not exhausted Atomic Energy Commission (AEC) procedures for challenging plant safety. Pratt also said the AEC was making a proper investigation.

Power plant cutbacks ordered. The AEC Aug. 24 ordered 10 nuclear power plants in seven states to cut back power levels 5%-25% pending studies of possible safety hazards.

The AEC said the precaution was taken because of the discovery of shrinkage in uranium oxide pellets in reactor fuel rods. With the shrinkage, the heat of the atomic process was not efficiently transferred to cooling water, narrowing the safety margin in case of cooling system failure. The shrinkage could also cause collapse of fuel rods.

The cutbacks would be in effect until the commission could evaluate new data from General Electric Co., manufacturer of the reactors.

In a related development, the AEC Aug. 30 rejected a renewed petition from consumer advocate Ralph Nader and Friends of the Earth, an environmentalist group, to close 20 nuclear power units.

In denying the petition, the AEC stood by its earlier position that existing interim regulations covering core-cooling systems provided "reasonable assurance" of protection of public health and safety.

Reactor impact report ordered. A three-judge federal appeals panel in Washington ruled unanimously June 12 that the Atomic Energy Commission (AEC) must prepare a formal environmental impact statement for the entire program of projected liquid metal-fast breeder reactor nuclear power plants, envisioned as the eventual answer to growing power needs.

The ruling overturned a lower court decision on a suit by the Scientists' Institute for Public Information to force the AEC to assess the long-term effects of breeder reactor use as required by the National Environmental Policy Act of 1970 (NEPA). The lower court had ruled against the institute because of a lack of tangible impact in the near future.

Writing for the appeals panel, Judge J. Skelly Wright said the program presented "unique and unprecedented environmental hazards." The institute had argued that since such reactors "breed" plutonium—an extremely toxic radiological metal—and that construction of over 1,000 such plants had been projected, the potential harm of such systems should be studied before construction started.

Although the AEC had issued an impact statement for a demonstration reactor near Oak Ridge, Tenn., Judge Wright said the commission had taken an "unnecessarily crabbed approach to NEPA in assuming that the impact statement was designed only for particular fa-

cilities rather than for analysis of broad agency programs."

Wright noted that the proposed plants were expected to generate some 600,000 cubic feet of high-level radioactive wastes by the year 2000 and said the problems "attendant upon processing, transporting and storing these wastes, and the other environmental issues raised by the widespread deployment [of such plants] warrant the most searching scrutiny under NEPA."

The decision also required that, in addition to the environmental impact report, the AEC make "a detailed statement" on alternatives to the breeder reactor program.

Columbia reactor decision upheld. A three-judge panel of the U.S. Court of Appeals July 5 declined to review the AEC's decision to grant Columbia University a license to operate a small nuclear reactor on its New York City campus. Community groups opposing the research reactor had asked the court to review a May 1972 opinion by the AEC's Atomic Safety and Licensing Appeals Board that the reactor would not be "inimical to the health and safety of the public."

A-cargo transport curbed. The AEC and the Department of Transportation (DOT) issued new rules April 2 to regulate ground and air transport of radioactive materials.

The ruling would transfer to the AEC control of handling and transport of all fissionable material and large packages of other materials. DOT would retain control of small packages, for which it would design new handling safety standards.

The AEC had announced a ban Jan. 30 on air transport of significant amounts of fissionable uranium and plutonium. The move was designed to forestall the possibility of theft by hijackers and illicit manufacture of atomic bombs.

AEC to review waste policies. Atomic Energy Commission (AEC) Chairman Dixy Lee Ray ordered a re-evaluation of nuclear waste management programs after a controversy developed over leaks from waste storage tanks at the AEC facility at Hanford, Wash., it was reported Aug. 28. Dr. Ray conceded that radioactive waste management had at times been "sloppy" and "negligent," but she said public health had not been endangered.

The 115,000-gallon Hanford leak had been confirmed June 8. The AEC reported July 31 that an investigation had found questionable waste practices being followed by a private contractor, the Atlantic Richfield Hanford Co. The AEC said the leak had gone undiscovered for six weeks because a company supervisor had not read reports showing waste levels dropping steadily in the holding tank.

The AEC reported, however, that the liquid waste had stabilized at a point underground where there would be no hazard to ground water.

Four environmentalist groups filed suit in federal district court in Spokane Aug. 1 seeking to prevent resumption of the Hanford operation after repair of the leaks. The suit accused the AEC of failing to file the required environmental impact statement for the Hanford operation and of violating radioactive wastes sections of the Atomic Energy Act of 1954.

The plaintiffs alleged that, in addition to accidental leaks of high-level wastes, low and intermediate-level wastes had been deliberately disposed of in the soil, endangering water supplies. The AEC said the high-level wastes had been contained and the other wastes posed no dangers. The suit was dismissed Aug. 17 after the AEC agreed to file an environmental impact statement.

California plant halted. The Pacific Gas & Electric Co. (PG&E) said it was withdrawing license applications submitted to the AEC and the California Public Utilities Commission for construction of an $800 million, 2 million kilowatt nuclear power plant in coastal Mendocino County, because of "unresolved geological and seismological questions" and "uncertainties" caused by approval of a state coastline protection measure in November 1972, it was reported Jan. 22.

PG&E said the U.S. Geological Survey had informed the company Jan. 8 that current knowledge of offshore geophysical characteristics was not adequate to resolve questions of site suitability.

N.Y. utilities drop power plans. Consolidated Edison Co. of New York City said July 2 it had abandoned its plan to build a nuclear power plant on an island in Long Island Sound a half mile from suburban New Rochelle.

The company had bought the island from New Rochelle in 1967 but had not applied for an AEC license. A company spokesman said AEC opposition to reactors near urban areas had been a crucial factor in the decision to drop the plan.

The New York State Electric & Gas Corp. said July 14 it had abandoned plans to build a nuclear plant on the shore of Cayuga Lake near Ithaca. Environmental groups had charged that heated water discharged from the plant would damage the ecology of the lake. Company President William A. Lyons said the expected legal delaying tactics by opponents of the plant had forced the decision. The company was preparing plans to build a conventional, coal-fired plant on the same site.

Dutch halt A-planning. The Economic Commission of the Riksdag suspended for an indefinite period the planning of 13 nuclear power stations in order to study the risks involved, the French newspaper Le Monde reported May 15. A final decision on atomic stations would await the conclusions offered by 1973 conferences of atomic experts in the U.S. and Sweden.

Other Nuclear Developments

European action. The European Parliament March 16 adopted a motion calling on the European Economic Community to create its own uranium facilities and to adopt a common nuclear policy by the end of 1974.

Latin American plants. Plans were made for further expansion of atomic-power plants in Latin America.

The Canadian government and an Italian engineering firm were awarded a $220 million contract to build Argentina's first nuclear power plant, it was reported March 21. A Canadian government firm, Atomic Energy of Canada Ltd., would receive about $100 million to design and supply the nuclear reactor, while Italy's Societa Italimpianti per Azioni would get about $120 million to design and build the plant. The two firms outbid two U. S. concerns and a West German firm.

The Wells Fargo Bank of San Francisco and the U.S. Export-Import Bank had signed a $54.2 million loan to pay for a $60.2 million sale of U.S. equipment to Mexico for construction of Mexico's first nuclear power plant at Laguna Verde, on the Gulf of Mexico, it was reported Sept. 21.

New Spanish plant. General Electric Co. of the U.S. said it had been given a contract by Hidroelectrica Espanola S.A. to built a 957,000 kilowatt nuclear power plant in Spain, it was reported Feb. 1.

The plant, to cost over $100 million, would go into operation in 1977. GE had completed a 440,000 kilowatt nuclear plant in Spain in 1971, and Spanish companies had ordered three more major nuclear plants from American manufacturers in 1972.

Colorado A-blast tests for natural gas. Three 30-kiloton (equal to 90,000 tons of TNT) nuclear devices stacked in a steel well casing were detonated simultaneously May 17 more than a mile below the Piceance Creek basin in Rio Blanco County, in northwest Colorado.

The blast, part of the U.S. Atomic Energy Commission's (AEC) Plowshare Program for developing peaceful uses of atomic energy, was aimed at breaking up sandstone formations in an effort to release natural gas trapped deep underground. It was the third such experiment in Plowshare's gas stimulation program.

AEC officials said initial checks after the blast indicated there had been no above-ground radiation leakage. The blast measured 5.3 on the Richter scale.

The experiment, called Project Rio Blanco, was jointly sponsored by the AEC and the CER Geonuclear Corp., a private Las Vegas concern. Cost of the test was estimated at $7.5 million. Future project plans called for opening the cavity created by the May 17 explosion in 3 6 months and "flaring" or burning off 300 million 800 million cubic feet of gas.

Environmentalists had attempted to halt the test with a suit charging it might contaminate water underground, fracture oil-bearing shale layers and produce gas that would be much more radioactive than normal gas. Colorado District Court Judge Henry E. Santo had rejected the arguments May 14 and authorized the AEC to proceed with the test.

Opponents of the blast were concerned that exploitation of the gas deposits in the Piceance Creek area would require an estimated total of 140 280 nuclear blasts, while 5,600 12,620 tests would be required to fully develop the gas fields under the combined area of Wyoming, Utah, Arizona, New Mexico and Colorado. Three hundred trillion cubic feet of gas were believed trapped in the five-state region.

U.S.-Soviet pact. President Nixon and Soviet Communist Party General Secretary Leonid I. Brezhnev, meeting in Camp David June 20, signed an agreement on cooperation in nuclear energy research.

The agreement on peaceful research cooperation went far beyond previous programs in providing for construction of joint research facilities "at all stages up to industrial-scale operations," and increased exchanges of information and personnel. Work would be concentrated in controlled thermonuclear fusion, fast breeder reactors and the fundamental properties of matter.

U.S. Atomic Energy Commission Chairman Dixie Lee Ray said the agreement could lead to evidence of the feasibility of fusion electricity generation "in the next two to five—at the most 10—years," although industrial use would probably not come until 2000 or 2010.

Uranium enrichment method reported. South Africa was said to have perfected a new uranium enrichment method, different from the currently practiced gaseous diffusion process and the experimental centrifugal method, the London Times reported Aug. 20. An experimental plant had obtained a few tons of enriched material and a larger factory was under construction to produce several hundred tons a year, according to the Times.

The report was based on an account of progress by the Uranium Enrichment Corp. published in a semiannual publication of the South African Atomic Energy Board. According to that account, capital investment in the plant was less than 65% of a comparable gaseous diffusion plant, although energy consumption was as large as that used in the diffusion process.

Germans buy Soviet uranium. A West German consortium of electricity producers was reported Oct. 15 to have signed in Moscow the previous week an agreement for the purchase of enriched uranium from the Soviet state-owned trading company Techsnabexport.

The deal involved a reported 700 tons of separative work units (the basic industrial measure of energy used to enrich uranium), with the first consignment to be delivered in 1974–75, the second consignment in 1976 77 and possible options on further supplies.

The financial terms were not disclosed, although the Rheinish-Westfalische Elektrizitaten-Werken, the West German firm heading the consortium, said the price was lower than that charged by the U.S., until now the sole supplier of enriched uranium to West Germany and other members of the European Economic Community (except France, which had previously placed a uranium order with the Soviets). The contract was also signed by the director of the European Atomic Energy Community, the first time the director had signed a contract with the U.S.S.R. The purchase had been approved by the EEC Commission Oct. 4.

The uranium would be used for the initial core of three nuclear power plants in West Germany that would have a combined annual output of 3,250 megawatts of electricity.

Arab Oil Embargo & Aftermath

Arab 'Oil Weapon' Unleashed

Following the outbreak of the fourth Arab-Israeli war, Arab oil-producing countries retaliated against Israel's supporters by first curbing petroleum production and then embargoing oil to the U.S. and the Netherlands. Big increases in the prices of the curtailed oil supply were announced simultaneously by most producing countries. The use of the Arab "oil weapon," however, had only a minor effect, initially at least, on the U.S., Israel's staunchest and most important supporter. The most serious effects were felt by the countries of Western Europe and Japan, which indicated quickly that they would try to appease the Arabs in an effort to restore their oil supplies.

Arabs cut oil output. Ministers of 11 nations of the Organization of Arab Petroleum Exporting Countrties agreed in Kuwait Oct. 17 on a coordinated program of oil production and export cuts in an attempt to force a change in the Middle East policy of the U.S.

The largest producer, Saudi Arabia, announced Oct. 18 it was immediately slashing oil production by 10%, and would cut off all shipments to the U.S. if it continued to supply arms to Israel and refused to "modify" its pro-Israel policy.

The reduction would continue until Nov. 30, after which further reductions would be announced.

It was revealed later that Saudi Arabia had actually imposed an embargo on oil to the U.S. Oct. 17, and other Arab oil countries did likewise.

The Persian Gulf state of Abu Dhabi announced Oct. 18 it was stopping all oil shipments to the U.S. and would do the same to any other country that supported Israel.

The Kuwait meeting had announced that each country would reduce production by 5% each month over the previous month until Israel withdrew from the territories occupied during the 1967 war and agreed to respect the rights of Palestinian refugees, which were not defined.

The ministers' statement said the cutback was not intended to "harm the Arabs' friends," although they acknowledged some hardships would result in Europe, which they hoped would bring pressures on the U.S. to change its pro-Israel policy.

Officials in France, which had given the Arabs diplomatic support, were reported Oct. 17 to have received assurances that deliveries to France would not be affected, although an Oct. 18 report said they were investigating the possibility of fuel rationing. But the Wall Street Journal reported Oct. 18 that oil company officials in New York assumed the cutbacks would

199

come across the board. Selective embargoes would probably be impossible to enforce, they said, since oil could be shipped to neutral countries and re-exported to embargoed countries.

The different interpretations of the decision reached in Kuwait were prompted by a reported division at the Kuwait conference, with Kuwait, Saudi Arabia and Egypt opting for a more moderate policy and Libya, Algeria and Iraq urging radical measures. Each country was apparently interpreting the conference recommendation in its own way. The other nations at the conference were Abu Dhabi, Bahrain, Dubai, Qatar and Syria.

The nations at the conference had been producing about 19.6 million barrels daily, about 40% of production in non-Communist countries. Only one million barrels of crude oil had been sold directly to the U.S., where it constituted about 6% of current daily consumption of 17.4 million barrels. But the U.S. also consumed nearly another million barrels of oil products processed in third countries from Middle Eastern crude, which would be affected if the production cut were across the board.

Saudi Arabia had been exporting over 500,000 barrels of crude a day to the U.S. A 10% cut, combined with the cessation of Abu Dhabi's daily U.S. shipments of 180,-000 barrels, would cause an immediate loss to the U.S. of over 230,000 barrels, about 1.3% of consumption. Coupled with predictions that the U.S. would need increasingly large Middle East shipments to meet growing needs, the cuts were expected to have an immediate impact on the tight U.S. fuel supply situation.

The cutbacks were expected to have a far more serious effect in Europe, which filled well over half its oil needs from Arab sources, and Japan, which received 45% of its oil from Arab countries. Both regions had two- or three-month stocks of crude or refined oil on hand or enroute. But the cutoff, combined with a substantial price increase by the Arab countries and Iran Oct. 17 which supplied an additional 40% of Japanese needs, would add to inflationary pressures in those countries.

The New York Times reported Oct. 17 that Japanese government officials and oil company executives expressed private bitterness at the Arab moves, in light of the long-standing compliance by most Japanese firms with the economic boycott against Israel. M.A.H. Luns, Secretary General of the North Atlantic Treaty Organization (NATO) said Oct. 14 that a halt in Arab shipments of oil would "come very close to a hostile act." Talks had been conducted within the Organization for Economic Cooperation and Development (OECD) in Paris recently to work out allocation arrangements in case of shortages, but no agreement had been reached by the outbreak of the war.

Oil prices increased. The six largest Persian Gulf oil-producing countries announced in Kuwait Oct. 17 a 17% increase in the price of their crude oil, and a 70% increase in taxes to be paid by oil companies on oil produced and sold by the companies.

Talks begun Oct. 8 between the producers and the companies in Vienna had broken down Oct. 12 after the companies refused demands for substantial re-

Major Oil Producing Middle East Nations

	Proved oil reserves (in barrels)	Current daily production (in barrels)	Annual revenues	Daily production after planned increases
Saudi Arabia	145 billion	8 million	$4.4 billion	10 million
Libya	30 billion	2.3 million	$2.1 billion	2.3 million
Iran	65 billion	5.9 million	$3.4 billion	8 million
Iraq	29 billion	2 million	$1.3 billion	3 million
Kuwait	65 billion	2.8 million	$2 billion	2.8 million
Abu Dhabi	21 billion	1.3 million	$934 million	2.7 million

Source: oil industry officials.

vision of the 1971 Teheran agreement, which had called for limited annual price increases through 1975. The exporting nations had argued that world inflation had been exceeding the price increases in the agreement, and charged that the companies had benefited from the increase in market prices.

The six nations—Iran, Saudi Arabia, Iraq, Abu Dhabi, Kuwait and Qatar—accounted for over one half of non-Communist countries' oil exports. They said the "action was unrelated to the Middle East war." Iran, a neutral in the conflict, was not a participant in the curb on exports to the U.S.

The countries set a market price on their own oil of $3.65 per barrel for light crude, up from $3.12. Their own oil constituted about 25% of production. The tax reference price, on which the companies paid taxes at a 60% rate on the oil they produced and sold themselves, was raised to 1.4 times the actual market price, resulting in a tax increase of $1.25 a barrel.

(Ecuador announced Oct. 17 it would increase its tax reference price from $3.60 to $5.33 a barrel.)

Arabs impose total ban against U.S. Arab oil-producing states carried out their threat to embargo all petroleum exports to the U.S. for its support of Israel.

Libya Oct. 19 ordered a complete halt in shipments of crude oil and petroleum products to the U.S. At the same time Tripoli raised the price of its oil for other importers from $4.90 to $8.25 a barrel. Libyan exports to the U.S. had totaled about 142,000 barrels a day of crude oil and indirectly 100,000 barrels of petroleum products, or about 1.4% of total U.S. consumption.

Col. Muammar el-Qaddafi, Libyan leader, threatened Oct. 22 to stop oil exports to Western Europe because their U.N. representatives had approved the Security Council's cease-fire resolution. In an interview with the Paris newspaper Le Monde, Qaddafi asserted that Libya would "ruin" Western European trade with the Arab world and would stop buying its arms. The Libyan leader denounced the cease-fire "imposed" by the U.S. and Soviet Union and expressed opposition to the stated war objectives of Egypt and Syria. He said those two states should not limit their aims to recovering the Arab territories lost to Israel in the 1967 war, but should "liberate all the Palestinians from the Zionist yoke."

Riyadh radio announced Oct. 20 that Saudi Arabia had decided to halt all oil exports to the U.S. in view of Washington's aid to Israel. The broadcast did not indicate when the embargo would go into effect or what action would be taken to prevent Saudi oil from reaching the U.S. through a third country.

Algeria had announced it would reduce by 10% its annual oil production, it was reported Oct. 20.

The Arab boycott of U.S. markets became total when Kuwait, Bahrain, Qatar and Dubai announced a cutoff of their supplies Oct. 21.

Kuwait Oil Minister Abdul Rahman Atiki Oct. 23 rejected an Iraqi request to nationalize U.S. oil interests. Atiki said the current curtailment was sufficient "to let the world feel our suffering."

(In reprisal against U.S. support of Israel, Iraq Oct. 7 had announced the nationalization of two major American oil firms—Mobil and Exxon. Baghdad said the action was taken because "aggression in the Arab world necessitates directing a blow at American interests in the Arab nation.")

Non-U.S. developments—Iraq retaliated Oct. 21 against the Netherlands' support of Israel by nationalizing the Dutch share of the Basra Oil Co. At the same time, Iraq denounced the Oct. 17 decision by the Organization of Arab Petroleum Exporting Countries to cut oil production. The Baghdad statement said the cutback would only harm Western Europe and Japan, which it said were friendly to the Arab cause.

Japanese Foreign Minister Masayoshi Ohira said Oct. 19 that his country, which imported 90% of its oil, would not be affected by the Arab cutoff. Yashuhiro Nakasone, Japan's minister of international trade and industry, had warned Oct. 12 that his country might have to impose oil consumption curbs if the Middle East war continued. Japan had a 55-day supply of oil and another 20-day supply on its way to Japan in tankers.

Japanese businessmen decided Oct. 12 to establish a Japan-Saudi Arabia Cooperation Organization. The function of the group would be to provide Japanese financial and technical aid to Saudi Arabia, which supplied Japan with 17% of its oil.

Several non-Arab oil producers, including Iran, Indonesia and Venezuela, said Oct. 23 they would not increase their own production despite the Arab cutback. Shah Mohammed Reza Pahlevi of Iran said his country was already producing a maximum capacity of about five million barrels a day.

Italy's state-owned fuel firm Ente Nazionale Idrocarburi (ENI) signed an agreement in Baghdad Oct. 27 calling for construction of a 400-mile oil pipeline as a spur from the Er-Rumahive field in Iraq to the pipeline system that pumped Iraqi crude to the Mediterranean.

Arabs set further oil cuts. Saudi Arabia announced Oct. 30 it would reduce oil production by an additional 5% Nov. 1, in accordance with the plan approved by the Arab oil producers in Kuwait Oct. 17.

The Saudi delegate to the Arab League, Taher Radwan, said the cut brought Saudi production 15% below the level before the latest Middle East war, but United Press International reported that total output had been reduced 26½%.

The New York Times reported Oct. 26 that major oil companies estimated that shipments of oil from the Arab countries had been cut by about four million barrels a day, or about 20% of the pre-war flow. The effect of the cutback was expected to become more severe because of a normal increase in demand during the winter. The Times said at least one Arab nation had demanded written proof of the destination of oil loaded onto each ship, apparently to prevent shipments to embargoed nations.

Libya announced Oct. 30 it was suspending oil deliveries to the Netherlands, the seventh Arab country to do so. The actions would result in a two-thirds reduction of Dutch crude petroleum imports.

The Netherlands imposed a ban on Sunday pleasure driving Oct. 30, and called on other European Economic Community (EEC) members to abide by rules providing for free movement of oil within the community. Foreign Minister Max van der Stoel said at a news conference that his government had followed a strictly neutral Middle East policy, and claimed to have information that the Arab nations might have planned the boycott before the outbreak of the war as a means of pressuring European countries that imported their oil through Dutch ports. The Netherlands had imported 139 million tons of crude oil in 1972, but re-exported 116 million tons, mostly as refined products, to Belgium, West Germany and the Scandinavian countries. The Arab governments had reportedly been angered by Dutch offers to aid in the transit of Soviet Jewish emigrants to Israel and by the outspoken Dutch view that Israel did not have to withdraw its forces from all lands occupied in 1967.

The Oil Committee of the 24-member Organization for Economic Cooperation and Development (OECD) decided after a meeting in Paris Oct. 25–26 that the oil situation did not "at present" warrant the activation of an oil-sharing program. However the countries agreed that a long-term program should be evolved, including provision for adapting economies to the eventual exhaustion of world oil supplies.

France had vetoed the reactivation of the oil industry international advisory board, composed of fifteen major oil companies, to avoid the appearance of pressuring Arab producers. The OECD pledged to maintain the customary trade in refined products, as a concession to the U.S., despite the action of some EEC members to restrict exports.

Sheik Ali Khalifa al-Sabah, head of the price commission of the Organization of Petroleum Exporting Countries, said in an interview reported Oct. 30 that Arab government officials feared that the U.S. might resort to military moves to assure continued deliveries of Arab oil.

Arabs widen production cuts. The Organization of Arab Petroleum Exporting Countries (OAPEC) announced Nov. 5, following a two-day meeting in Kuwait, that each of its 11 members would reduce their oil production in November to 75% of the September output, with an additional drop of 5% in December.

The 25% cutoff would include the complete embargoes already imposed on shipments to the U.S. and the Netherlands for their pro-Israel policy. The conference statement said the cutbacks would be arranged so that countries friendly to the Arab cause would not be deprived of their normal flow of oil. The drop in production represented a sharp advance on the sanctions decided by the OAPEC ministers at their Oct. 17 meeting in Kuwait.

The conference decided to send Saudi Petroleum Minister Sheik Ahmed Zaki al-Yamani and Algerian Industry and Power Minister Belaid Abdesselam on a tour of Western Europe to explain the organization's views and the decisions taken at the conference.

Yamani had said Nov. 4 that Saudi Arabia was "tracking down every last barrel of oil that reached the United States." He said his country had computer records on the destination of all Saudi oil shipments, including refineries in Trinidad, Puerto Rico and Canada, that shipped products to the U.S.

The 11 OAPEC nations were Abu Dhabi, Algeria, Bahrain, Dubai, Egypt, Iraq, Kuwait, Libya, Qatar, Saudi Arabia and Syria.

The Middle East Economic Survey, a Beirut publication, said Nov. 3 that Saudi Arabia had divided countries into three categories for the purpose of oil supplies: embargoed, exempt or most favored, and not exempt but not embargoed either. Among the most favored nations listed were France, Spain, Jordan, Lebanon, Malaysia, Pakistan, Tunisia and Egypt.

The Middle East Economic Survey reported Nov. 6 that since the start of the Arab-Israeli war Oct. 6 Arab oil producers had cut production by 28.5%, or 5.83 million barrels a day.

In related developments, Nigeria announced Nov. 3 that it raised the posted price of its crude oil from $4.29 to $8.31 a barrel, effective Oct. 20. Nigeria produced two million barrels of oil a day.

Saudi Arabia announced Nov. 6 it had halted work on further expansion of the Arabian Americ Oil Co. (Aramco) in the country. The government said no authorization for expansion had been issued since Oct. 17, when Saudi Arabia imposed an oil embargo on the U.S. Indonesia announced Oct. 30 a 20% increase in the export price of oil. Nearly 80% of the nation's oil exports went to Japan. The National Iranian Oil Co. said it would increase its crude oil price to Japanese importers by 26% retroactive to Oct. 16, it was reported Oct. 27.

Saudi seeks Aramco control. Saudi Arabian Petroleum Minister Sheik Ahmed Zaki Yamani confirmed Nov. 17 that he was seeking "substantial revision" of the participation agreement with Arabian American Oil Co. (Aramco) that would give his government 51% control of the U.S.-owned firm. The current Saudi share of Aramco was 25%.

The Middle East Economic Survey, a Beirut publication, reported Nov. 19 that Saudi Arabia had set Aramco's November output level at 6,198,000 barrels a day. This was a 25% reduction from September and 32% lower than the original November target of 9.1 million barrels a day.

EEC urges troop pullback. In an attempt to avert an Arab oil boycott threatened against nations adopting a pro-Israeli policy, the foreign ministers of the European Economic Community (EEC) adopted a joint statement Nov. 6 calling on Israel and Egypt to return to the cease-fire lines of Oct. 22—before Israeli troops completed their encirclement of the Egyptian III Corps.

The statement, approved at the end of a two-day meeting in Brussels, called on Israel to "end the territorial occupation which it has maintained since the conflict of 1967" and declared that peace in the Middle East was incompatible with "the acquisition of territory by force."

The ministers did not act on a Netherlands request for a pooling of oil resources by EEC members if the Netherlands began to run out of oil because of the Arab oil boycott against it. Instead, the ministers issued a separate statement that said the Council of Ministers was "conscious of the interdependence of the economies of the member states" and asked the EEC's permanent bodies "to follow attentively the situation resulting from the scarcity of crude oil."

France and Britain had been the most vigorous opponents of a joint EEC oil pool because they feared such a policy might jeopardize their privileged position with the Arabs on assured petroleum supplies. French Premier Pierre Messmer had said Nov. 3 that France would insist on a common European energy policy as a condition for oil sharing. He had pointed to the lack of European solidarity over construction of a plant to make enriched uranium, with Britain, West Germany and the Netherlands following one approach and France another.

Individual oil actions taken—While the EEC refused to adopt a joint policy to counter the Arab oil threat, individual member nations took actions to conserve their petroleum supplies.

Belgium adopted a ban on Sunday pleasure driving Nov. 5 and "requisitioned" all oil stocks, installations and personnel Nov. 7. It also ordered suppliers to meet all normal demands of oil buyers and canceled the annual Tour de Belgique auto race.

Luxembourg Nov. 7 ordered gasoline stations to close Saturdays and Sundays.

Belgium and Luxembourg imported much of their Arab oil through the Netherlands, the main European target of an Arab petroleum boycott. The Netherlands' ban on Sunday pleasure driving had gone into effect Nov. 4.

Denmark Nov. 6 issued decrees lowering speed limits, closing schools on Saturdays and barring sales of high-octane gasoline.

The Irish government was reported Nov. 4 to have told public offices to reduce heating and ordered the national airline, Aer Lingus, to reduce cruising speed on transatlantic flights. It also appealed to homeowners to conserve heating supplies and urged motorists to reduce traveling speeds.

France suggested Nov. 7 that drivers voluntarily limit speeds and lower home and office thermostats by two degrees.

The British government Nov. 5 reactivated the Oil Supplies Advisory Committee, suspended since the 1956 Suez crisis, to keep the government informed on supplies. The committee consisted of representatives of all major oil companies operating in Britain. The government said gas rationing was not necessary at the present time. Peter Walker, secretary for trade and industry, referred to "assurances from Arab states" on oil supplies. The British government had asked the public Oct. 24 to cut down on home heating.

OPEC membership expanded. The Organization of Petroleum Exporting Countries (OPEC) accepted Ecuador as its fifth non-Arab member and Gabon as an associate member at a meeting in Vienna Nov. 19. Ecuador had been serving as an associate member for the past five months.

The OPEC announced it had established a committee that would link oil prices more closely to the soaring costs of other international commodities such as wheat, soybeans and oil.

Arabs ease Europe oil curb. The Organization of Arab Petroleum Exporting Countries (OAPEC) announced at a meeting in Vienna Nov. 18 that its member states would cancel the scheduled December 5% reduction in exports for most countries of the European Economic Community (EEC). The total embargo against the Netherlands and the U.S., however, would continue.

An OAPEC communique made public after the meeting said that "in appreciation of the political stand taken by the Common Market countries in their [Nov 6] communique regarding the Middle East crisis," the organization had decided "not to implement the 5% reduction for the month of December" for the EEC member nations.

Japanese economy endangered. Japan faced a serious threat to its economy because of the Arab oil cutbacks, a government official warned Nov. 12.

Eimei Yamashita, vice minister of the Ministry of International Trade and Industry, made the statement after submitting an emergency action plan to Prime Minister Kakuei Tanaka to cope with the energy crisis. He said the fuel shortage might reduce Japan's economic growth rate for the current fiscal year ending March 1974 from 10% to 5%.

Japan's crude oil imports from all sources would be 20% below the planned levels in December and the drop could be greater in the first three months of 1974, Yamashita predicted. He said, however, that he expected the "matter of oil cuts to be solved in four to six months." About 83% of Japan's oil imports came from the Middle East—45% from the Arab countries and 38% from Iran.

The emergency action program outlined by Yamashita called for a nation-wide appeal for voluntary savings of fuel and energy, including restrictions on non-essential automobile use, letters of "administrative guidance" to 10 major industries requesting that they reduce consumption of oil and electricity by specific amounts, and enactment of legislation formally allocating fuel supplies and possibly imposing price controls.

The Japanese Foreign Ministry had said Nov. 8 that Tokyo had rejected a request by Arab nations to sever diplomatic relations with Israel, impose an official embargo on trade with Israel and provide military assistance to the Arab states. Ministry spokesman Mizuo Kuroda said the Arab states, which he did not identify, had informed Japan that compliance with their request would place Japan in the "friendly" nation category, assuring it of uninterrupted oil supplies.

Emergency fuel plan adopted—Japan announced an emergency rationing plan Nov. 16 to conserve oil and electricity in industry and urged the public to save fuel voluntarily at home, on the road and at work.

The program, to go into effect the following week, would require industry to reduce oil and electricity consumption by 10%. The guideline to the public for saving energy included limiting the amount of advertising lights, keeping room temperatures below 68 degrees, and lowering highway speed limits to under 50 miles per hour.

According to government estimates, the conservation measures would save 60%-80% of total national consumption.

An accompanying government announcement on the rationing program said an unspecified "strong policy will be carried out to prevent prices from rising through taking advantage of the emergency."

Kissinger visits Japan. U.S. Secretary of State Henry A. Kissinger conferred with Premier Kakuei Tanaka and other Japanese officials in Tokyo Nov. 15–16. The discussions dealt almost exclusively with the adverse effects of the Arab oil cutoff on Japan's economy.

One of the Cabinet members who met with Kissinger, International Trade and Industry Minister Yasuhiro Nakasone, said Nov. 16 that Kissinger had offered no assurances when the oil crisis would be resolved or when the Arab states would renew full exports.

Singapore cuts U.S. military fuel. Singapore stopped all fuel supplies to U.S. military forces in the Pacific in response to an Arab threat to cut off oil exports to Singapore, industrial sources in the country reported Nov. 14. The U.S. embassy in Singapore confirmed the same day that "Certain companies in Singapore are no longer able to supply petroleum to the United States Department of Defense under their current contracts."

Prime Minister Lee Kuan Yew was reported to have ordered refineries in Singapore Nov. 12 to stop supplying the American market. His action followed a statement by a Saudi Arabian diplomat to Singapore newsmen that "Our policy is clear—no supplies of our oil are to go to any United States military buyers."

U.S. warns Arabs on oil embargo. Secretary of State Henry. A. Kissinger warned Nov. 21 that the U.S. might have to consider retaliatory action if the Arab oil embargo continued "unreasonably and indefinitely."

Speaking at a Washington news conference, Kissinger said the U.S. "has full understanding for actions . . . taken [by Arab oil producers] when the war was going on, by which the parties and their friends attempted to demonstrate how seriously they took the situation." But he said "those countries who are engaging in economic pressure against the United States should consider" whether their continued embargo was now appropriate while peace efforts were in progress. Asserting that the U.S. commitment to a peaceful Middle East settlement would not be influenced by the oil cutoff, the secretary stated, "We will not be pushed beyond this point by any pressure." Any U.S. reprisals would be taken "with enormous reluctance and we are still hopeful that matters will not reach this point," Kissinger declared.

Saudis warn U.S. against reprisals— Saudi Arabian Oil Minister Ahmed Zaki Yamani warned Nov. 22 that his country would cut its oil production by 80% if the U.S., Western Europe or Japan took any action to counter the Arab oil embargo.

Responding directly to Kissinger's remarks about possible unspecified American reprisals, Zamani said if the U.S. attempted to use military means, Saudi Arabia would blow up its oil fields. He cautioned Western Europe and Japan against joining the U.S. in any move, saying "your whole economy will definitely collapse all of a sudden."

Zamani made his statement in a television interview in Copenhagen.

U.S. rejects retaliation on cutbacks— The U.S. had no plans for the present to retaliate against Arab nations for their oil embargo against the U.S., Administration officials said Nov. 19.

State Department spokesman George S. Vest was asked whether the U.S. was still willing to sell Saudi Arabia military equipment in view of that country's suspension of oil exports to the U.S. He replied that there was "no basic change in policy" but that the matter would remain under review.

Secretary of Agriculture Earl L. Butz told a news conference that proposals had been received from "many quarters" for a halt to U.S. grain shipments to Arab nations in reprisal for the oil ban. Butz said that such action "would simply irritate the situation, make negotiations more difficult and would not put any pressure on the Arab countries."

Butz said American grain exports to the Arab states were "not high enough to be significant, and in view of the fact that the Russian nation has a much easier grain situation than a year ago they could very easily make up the deficit of anything we cut off."

Shah Mohammed Riza Pahlevi of Iran called on the Arabs to end their oil boycott in a Beirut weekly newspaper interview published Nov. 22. "Since you have accepted the cease-fire and moves toward a peaceful settlement, why are you continuing to shut off oil supplies and reducing production?" the shah asked.

New European oil action. The Arab oil squeeze had forced European nations to take more action to insure energy supplies:

The Swiss government Nov. 21 announced a ban on the private use of automobiles, planes and boats for the next three Sundays. It also announced plans to cut deliveries of gasoline and other motor fuel by 20% and heating oils by 25% from levels of a year ago.

Italy adopted the most stringent oil conservation measures of any European nation Nov. 23. The government raised the price of gasoline by 8% to the equivalent of $1.33 a gallon, a record level in Western Europe; banned driving on Sundays and holidays; asked stores and offices to close by 7 p.m.; ordered all gasoline stations to close Saturday afternoons and Sundays and lowered speed limits on highways; reduced deliveries of heating oil by 20% of last winter's level and increased its price by 55%; asked cities to cut street lighting by 40%; and ordered movie theaters and television stations to close earlier.

A ban against pleasure Sunday driving went into effect in Denmark Nov. 25. The government had previously ordered a 25% cut in sales of fuel oil and curbed use of electricity.

Dutch Premier Joop den Uyl announced Nov. 30 that gas rationing would begin in the Netherlands Jan. 7, 1974. The government later announced that the allotment would be restricted to two and a half gallons a week for each motorist.

Norway announced Dec. 5 it would introduce gas rationing Jan. 7, 1974, the same day Sweden began rationing. Norway also banned Sunday driving Dec. 5, the eighth European nation to do so.

Britain Dec. 5 ordered a compulsory speed limit of 50 m.p.h. on all highways. The government said it would cut street lighting by about 50% and impose new restrictions on heating and lighting of commercial premises. The government continued to insist that rationing was not yet required despite a public run on gasoline that began Dec. 3. The shortage in oil supplies had been intensified by a government decision Dec. 4 to divert more fuel to oil-fired power stations for the duration of the coal miners' slowdown.

France imposed a 75 m.p.h. speed limit on previously unlimited superhighways and reduced limits on all other roads to 56 m.p.h. Nov. 30. It also ordered a cut in airline flights and lighting for public buildings and advertising, and imposed an 11 p.m. cutoff time for week night television programs. The fuel restrictions were regarded as largely a symbolic gesture of solidarity with neighboring countries more affected by the Arab oil cutbacks. Premier Pierre Messmer was reported to have proposed Dec. 2 and 3 that Western European oil-importing countries cooperate directly with the oil-producing countries, cutting out the major international oil companies. He also called for government organizations to control the internal markets in consumer countries so those markets would not be "under the sway of the most powerful companies." (France was also accelerating the schedule for installing nuclear power plants, with one government minister predicting that 40% of the country's needs would be met from this source by 1980, it was reported Dec. 11.)

The West German government announced a major effort Nov. 22 to substitute coal for oil in industry. Chancellor Willy Brandt warned Nov. 29 of a possible recession as a result of the oil shortages and called for greater efforts to develop alternative sources of energy. The government proposed Dec. 5 to subsidize oil purchases by low-income families.

West German Economy Minister Hans Friderichs Nov. 19 had banned all private driving for the next four Sundays and reduced the speed limit on highways to about 62 miles an hour and on other roads to 50 miles an hour.

Friderichs said the measures should "save up to 13% of a month's gasoline consumption and thus conserve energy for our industry and places of work." He said Germany expected its Arab oil imports to be cut by 15%.

U.S. urges Arabs end oil ban. Secretary of State Henry A. Kissinger said Dec. 6 that the U.S. again had called on the Arabs to end their oil embargo.

Speaking at a Washington news conference, Kissinger said since the U.S. was actively engaged in efforts to achieve peace in the Middle East, "We continue to believe that discriminatory measures against the United States and pressures are no longer appropriate." The secretary refused to divulge what retaliatory action, if any, the U.S. might take if the Arab oil ban were not lifted.

"What the United States might do if other countries treat us unreasonably . . . we will leave until that situation arises," he said. Kissinger stressed the need for cooperation rather than confrontation between the Arabs and the U.S. as "we are approaching the negotiations" at the Geneva peace conference, scheduled to begin Dec. 18.

Kissinger meets Saudi on oil ban— Saudi Arabian Petroleum Minister Sheik Ahmed Zaki al-Yamani conferred in Washington Dec. 5 with Secretary of State Kissinger and other American officials on the Arab oil embargo against the U.S. Yamani told newsmen after his meetings that his country was prepared to lift its total ban as soon as the Israelis "start withdrawing" their forces from occupied Arab lands. "I think that when the Israelis decide to withdraw, and there is a timetable for that, there will be a timetable to increase production step by step in a manner which corresponds with the timetable of the withdrawal," Yamani said.

On the question of Arab recognition of Israel, Yamani said, "So many Arab leaders have announced they are prepared to sign a peace treaty with Israel that the question of the existence of Israel is not an issue. . . . It is not tied to the oil issue in any case."

Yamani had arrived in the U.S. Dec. 3 for a week-long stay as part of a tour of Western nations to explain the Arab use of their "oil weapon." Yamani, accompanied by Algerian Minister of Industry and Energy Belaid Abdessalem, had toured West European countries Nov. 26–Dec. 1. They had met with French officials in Paris Nov. 26, with Belgian officials in Brussels Nov. 30 and with Dutch leaders and representatives of the European Economic Community in Brussels Dec. 1.

Dutch Economic Affairs Minister Ruud Lubbers said Dec. 1 after his meeting with the two Arab oil representatives that their demands for his government's condemnation of Israel and a call for a complete Israeli withdrawal in exchange for a resumption of oil exports to the Netherlands was "out of the question." Lubbers added: "We are not going to buy oil on a Saturday morning after making a declaration that could be misunderstood by others."

A statement issued by the Dutch government Dec. 4 for the first time referred to Israel's occupation of Arab lands as "illegal." A Foreign Office spokesman clarifying his government's Middle East policy, said the EEC statement of Nov. 6, endorsed at the time by the Netherlands, "meant in principle that Israel should move out of the Arab territories she is holding, occupied since the 1967 six-day war. Fully adhering to that position the Dutch government considers that the Israeli presence in occupied territory is illegal."

Prior to endorsement of the EEC statement, the Dutch had interpreted the 1967 U.N. Security Countil resolution on Israeli withdrawal not to necessarily mean that Israel was to pull back to the pre-1967 borders.

Arabs announce new cutback. Oil ministers of nine Arab nations agreed at a meeting in Kuwait Dec. 9 to reduce petroleum production by about 750,000 barrels a day starting Jan. 1, 1974. The new cutback, 5% of the current level of output, would largely affect Western Europe and Japan. It would bring the total cutback since September to about 28.75%.

A statement issued by the conferees said if Israel signed an agreement to withdraw its troops from occupied Arab lands and if the U.S. guaranteed that pullback, "the ban on oil exportation to the U.S. shall be lifted as soon as the implementation of the withdrawal time schedule begins."

A spokesman for the Kuwait Oil Ministry was quoted as saying that West Europe and Japan must "offer something

more concrete and put more pressure on the United States and Israel" if they wanted to avoid more oil cutbacks.

The Kuwait meeting was attended by oil ministers of Abu Dhabi, Algeria, Bahrain, Egypt, Kuwait, Libya, Qatar, Saudi Arabia and Syria.

The Soviet Union denied speculation in the West European press that Moscow was involved in the Arab oil embargo. The newspaper Sovestskaya Rossiya Dec. 4 singled out the London Times and the Frankfurter Allgemine of West Germany for "starting an anti-Soviet ballyhoo resorting to concoctions and lies" and placing the blame "at the wrong door." The newspaper added: "Everyone knows who is to blame for the present energy crisis in Western Europe. They are the imperialist and Zionist circles stubbornly resisting a just Middle East settlement."

Saudi minister restates aims—Saudi Arabian Oil Minister Sheik Ahmed Zaki al-Yamani restated Arab demands Dec. 9.

In a television interview in Washington, Yamani again said the Arabs would lift the oil embargo against the U.S. "when Israel accepts the withdrawal from the occupied Arab territory, and that acceptance is guaranteed" by the U.S. He said "there is nothing between the United States and Saudi Arabia as a bilateral dispute." He asserted that if "the political obstacle" to peace was removed, "What is left is an economic question and I think it is possible that Saudi Arabia will meet most of the world demand for oil."

Italy seeks closer Arab ties. A meeting of Italy's top political leaders Dec. 12 called for closer relations with Arab oil-producing countries. The statement was drawn up after a 12-hour national strategy meeting that included Premier Mariano Rumor and the leaders of the four parties that comprised his government.

The conference discussed what action Italy should take in the coming months to cope with the energy crisis. The statement urged "a joint European initiative and strategy" to deal with the fuel shortage and the Middle East dispute. The leaders also stressed Italy's commitment to European unity "within the framework of our alliances."

Raffele Girotti, president of the state-owned fuel combine Ente Nazionale Idrocarburi (ENI), had called on the government to adopt "a new foreign policy" that would satisfy the Arab oil-producing states. (The Communist party, too, had submitted a motion in Parliament Dec. 2 urging that the ENI be given "the necessary means" to increase its search for crude oil sources and that Italy develop a "foreign policy of active collaboration with the countries of the Middle East and the Mediterranean.")

U.S. firms to build Iran refinery. Five U.S. oil companies and the government-owned National Iranian Oil Co. signed a memorandum of understanding for the construction of a petroleum refinery on the Persian Gulf in Iran, it was reported Nov. 8.

The U.S. companies were Apco Oil Co., Cities Service Co., Clark Oil & Refining Co., Commonwealth Oil Refining Co. and Crown Central Petroleum Corp. The projected refinery would have a capacity of up to 500,000 barrels a day for export. Operations would start in 1977.

A New York Times report from Teheran Nov. 11 said that despite the Arab oil embargo, Iran was rapidly increasing its output and was expected to produce about six million barrels a day by the end of 1973. This would be 270,000 barrels a day more than produced in 1972. Premier Amir Abbas Hoveida was quoted as saying that Iran "shall not use oil as a political weapon."

Iran oil auctioned at higher prices. The Iranian government announced Dec. 11 that more than 80 million barrels of crude oil were sold to unidentified U.S., European and Japanese firms at $14–$17.40 a barrel. This was $11–$12 a barrel higher than December's posted prices for Iranian oil. The oil, auctioned by the Iranian National Oil Co., was to be delivered over the first six months of 1974.

Kissinger urges joint efforts. U.S. Secretary of State Henry A. Kissinger, in a major address in London Dec. 12, called for a massive joint effort by the U.S., Europe, Canada and Japan to develop long-term solutions to the world energy crisis exacerbated by the Arab oil squeeze.

Speaking to the Pilgrim Society, an Anglo-American friendship group, Kissinger proposed creation of an "energy action group" of "senior and prestigious individuals" to develop within three months "an initial action program for collaboration in all areas of the energy problem." He said the producing nations should be invited to join the group in any aspects directly affecting them.

The goal of the proposed energy group should be "the assurance of required energy supplies at reasonable cost," Kissinger said. He listed four possible areas of cooperation: conservation of energy through more rational use of existing supplies, discovery and development of new sources of energy, new incentives to producers to increase supply, and a coordinated international program of research on technology to use energy more efficiently and to discover alternatives to petroleum.

Kissinger said the U.S. was "prepared to make a very major financial and intellectual contribution to the objective of solving the energy problem on a common basis." He said the current energy crisis "can become the economic equivalent of the Sputnik challenge of 1957," a reference to the Soviet launching of a satellite that spurred the U.S. space program. Kissinger added that "this time, the giant step for mankind will be one that America and its closest partners take together."

The energy proposals were made as part of a general speech on issues facing the Atlantic allies.

(Kissinger's energy proposals seemed to undercut a French proposal, reported Dec. 12, for a summit meeting between the European Economic Community and Arab nations to discuss financial, technological and energy relationships.)

Kissinger tours Mideast capitals— Kissinger visited the Middle East capitals of Algiers, Cairo, Riyadh, Damascus, Amman, Beirut and Jerusalem Dec. 13–17 to discuss last-minute procedural and substantive problems involved in the Geneva peace conference and other matters related to the region.

Kissinger's first stop was in Algiers, where he conferred for two hours Dec. 13

with Algerian President Houari Boumedienne.

Kissinger flew to Cairo and met Dec.
13–14 with Egyptian President Anwar
Sadat.

At his meeting in Riyadh Dec. 14 with
King Faisal, Kissinger and the Saudi
Arabian monarch discussed the diplomatic details preparatory to the Geneva
parley and the Arab oil embargo against
the U.S. Aides with Kissinger said they did
not expect the ban to be lifted immediately. They hinted, however, that
they had received indications that the embargo would be removed when the Geneva
talks came to grips with substantive issues
after Jan. 1, 1974.

Kissinger's six-hour meeting in Damascus Dec. 15 with President Hafez al-Assad
was the first high-level discussions between
Syrian and U.S. officials since 1967 and
the first visit to Damascus by a U.S. secretary of state since 1953.

EEC agrees on unity in oil crisis. The
leaders of the nine member nations of the
European Economic Community (EEC)
agreed at their summit conference in
Copenhagen Dec. 14–15 to develop a common energy policy to deal with the oil
crisis.

The conference began Dec. 14 with the
unexpected arrival in Copenhagen of the
foreign ministers of Algeria, Tunisia, the
Sudan and the Union of Arab Emirates,
who later met with EEC foreign ministers.
The Arabs reportedly stressed that the
EEC nations would have to take a
stronger pro-Arab position if they wanted
an end to the oil squeeze. The Arabs were
not officially invited to appear in
Copenhagen.

Danish Foreign Minister K. B. An-
dersen, speaking for the rest of the EEC,
warned the Arabs in a second meeting
Dec. 15 that the oil squeeze could cause
severe damage even to friendly European
nations and that it was alienating public
opinion, which could adversely affect
Arab interests.

The final summit communique reaffirmed the EEC's Nov. 6 Middle East
declaration calling for Israel to withdraw
from Arab territories occupied since the
1967 war. It also called for the conclusion
of Middle East peace agreements, including international guarantees and es-

tablishment of demilitarized zones. The
EEC position would be conveyed to
United Nations Secretary General Kurt
Waldheim, an apparent concession to the
Arab demand that the U.N. be responsible
for the Middle East situation.

In an accompanying energy statement,
the leaders said the oil crisis was "a threat
to the world economy as a whole" and
warned it would hurt developing as well as
developed nations. Guidelines for a common EEC energy policy included a call for
statistics from the EEC Commission on
the energy situation by Jan. 15, 1974, staff
proposals by the end of January and decisions by foreign ministers on the orderly
functioning of the common market for
energy before Feb. 28.

The leaders also called on the 23-nation
Organization for Economic Cooperation
and Development, which included the
U.S. and Japan, to study possible solutions for the energy problems of consuming nations.

Among other recommendations were
calls for energy restrictions throughout
the EEC so that the people of all nine nations would share deprivations; negotiation of agreements on wide-scale cooperation with oil-producing countries under
which Europe would help with economic
and industrial development in exchange
for stable energy supplies; and harmo-
nious development of alternative energy
sources, including a European capacity to
produce enriched uranium. Strongest ad-
vocates of a joint energy policy were West
German Chancellor Willy Brandt, who
submitted a 12-point plan, and Dutch
Premier Joop den Uyl.

Iraq increasing oil output. Iraq announced Dec. 18 that it planned to increase its present oil output of 2.1 million
barrels a day to 3.5 million barrels by 1975
despite the production cutoff of nine other
Arab oil-producing states.

Oil and Minerals Minister Saadun
Hammadi said "An oil production reduction means that we punish all countries of
the world rather than just Israel and its
backers." He said Iraq's national oil company was trying to help "friendly countries" that "have stood by us" in the Arab
confrontation with Israel. Hammadi listed
those friendly nations whose representatives had recently visited Iraq to discuss

their fuel problems as Austria, Brazil, Bulgaria, India, Pakistan, Poland, Spain and Tanzania.

Arabs double price, ease embargo. Oil producers of six Persian Gulf states announced after a two-day meeting in Teheran Dec. 23 that they would double the price of a barrel of oil effective Jan. 1, 1974. The decision was followed by an announcement by Arab petroleum producers in Kuwait Dec. 25 that starting Jan. 1 they would ease their embargo, but would continue the total ban against the U.S. and the Netherlands.

The gulf states of Iraq, Iran, Saudi Arabia, Kuwait, Abdu Dhabi and Qatar raised their posted price for crude oil to $11.65 a barrel from the current price of $5.11. The oil ministers' communique said the government income under the new price system would be $7 a barrel, compared with the present revenue of $3.09.

The six states were meeting under the aegis of the Organization of Petroleum Exporting Countries with five other OPEC members—Algeria, Indonesia, Libya, Nigeria and Venezuela—attending as observers.

Shah Mohammed Riza Pahlevi of Iran said after the meeting, "The industrial world will have to realize that the era of their terrific progress and even more terrific income and wealth based on cheap fuel is finished. They will have to find new sources of energy and tighten their belts."

Following a two-day meeting in Kuwait, the nine members of the 10-member Organization of Arab Petroleum Exporting Countries (OAPEC) announced Dec. 25 that starting Jan. 1 they would cancel a scheduled further 5% cut in production and would increase their output by 10% instead. Production, however, would remain 15% below what it had been in September when the oil countries began their cutbacks. Western Europe and Japan, hardest hit by the cutoff, would be the principal beneficiaries of the Arab decision.

The move was announced by Saudi Arabian Oil Minister Sheik Ahmed Zaki al-Yamani. He said the OAPEC had acted to ease their oil ban because Arab pressure had brought about favorable shifts in some countries' positions in the Arab-Is-raeli conflict. Yamani cited the change in the neutral policies adopted by Japan and Belgium, both of whom had called on Israel to withdraw from occupied Arab territory.

The countries that took action in Kuwait were Saudi Arabia, Kuwait, Abu Dhabi, Algeria, Libya, Bahrain, Qatar, Syria and Egypt. The 10th OAPEC member, Iraq, had expressed opposition to cutting back oil production. Iraqi Oil Minister Hammadi said in Kuwait Dec. 25 that the oil embargo was not affecting the U.S.

The shah of Iran had called on the Arabs Dec. 20 to end their oil embargo. He warned, "Playing with the oil weapon is extremely dangerous. What's the use of all that money in the bank if the whole system crumbles?"

Five more nations raise prices—Five more oil-producing nations adjusted to the pattern of the Persian Gulf states and sharply increased their prices.

Venezuela announced Dec. 28 that it was raising the posted price of a barrel of oil from the December level of $7.74 to $14.08. The new price became effective Dec. 31.

Libya, Nigeria, Bolivia and Indonesia announced Dec. 31 a 60%-80% increase in the price of their crude oil. New prices (former prices in parentheses): Indonesia—$10.80 a barrel ($6); Nigeria—$14.60 ($8.31); Libya—$18.77 ($13.40); and Bolivia—$16 ($7.44).

Japanese official tours oil states. Japanese Deputy Premier Takeo Miki returned to Tokyo Dec. 28 after a 19-day tour of the Middle East oil-producing states of Saudi Arabia, Egypt, Iran, Syria, Iraq, Kuwait and Abu Dhabi. Miki's mission was aimed at persuading those countries to reclassify Japan as a friendly nation to assure its "full oil needs."

After reporting to Premier Kakuei Tanaka, Miki said the premier "agreed with my view that our promises must, by all means, be carried out." The only commitment Miki was known to have made on his trip was disclosed after a meeting in Cairo with Egyptian President Anwar Sadat Dec. 18. Miki had offered Egypt a 25-year loan of $140 million to assist in widening and deepening the Suez Canal.

The Japanese government Dec. 13 had again called on Israel to withdraw to the Oct. 22 cease-fire lines as a first step toward a total pullback from occupied Arab territories.

Egyptian pipeline set. Egypt and four oil-producing countries signed an agreement in Cairo Dec. 29 to finance the construction by Bechtel, Inc. of the U.S. of a $400 million petroleum pipeline between the Red Sea and the Mediterranean.

The contract established a company of joint shareholders to provide funds for the project. Saudi Arabia, Kuwait and Abu Dhabi would each contribute $60 million, Qatar $20 million and Egypt $200 million.

U.S. Policy

American leaders responded to the Arab oil embargo by pledging intensified efforts to make the nation self-sufficient in energy fuels. The plans being considered included more intensive exploitation of coal resources, development of such secondary petroleum resources as shale and tar sands and research on other alternative energy sources. As expected, pressure began to build up in favor of relaxing environmental restrictions on the various branches of the energy industry. And allocations and rationing were again proposed.

U.S. energy requirements, in oil equivalents, rose from 34.4 million barrels a day in 1972 to a record 35.9 million in 1973.

The total capability of the electric utility industry went up from 395.5 million kilowatts at the end of 1972 to a record 437.6 million at the end of 1973. The number of kilowatt-hours of electricity generated increased from 1.747 trillion in 1972 to a record 1.87 trillion in 1973, although the rate of increase in electricity consumption began to slow down at the end of 1973.

Impact of oil cutback assessed. In the aftermath of the Arab oil cutback, representatives of 21 major U.S. oil companies with foreign oil interests urged the Administration Oct. 30 to activate an emergency industry committee authorized to allocate foreign oil supplies.

The Emergency Petroleum Supply Committee had been established during the Arab-Israeli war in 1967 and had existed on a standby basis since then.

The group also offered new assessments of the Middle East situation which prompted government officials to revise earlier estimates of the cutback's impact on the nation's fuel supply. Experts had set the loss at 1.6 million barrels a day in testimony Oct. 24 before the Senate Interior Committee, but new figures put the deficit at 2 million barrels a day or 12% of the total 17 million barrels a day used by the U.S.

The estimate reflected expected losses in crude oil and refinery products coming from plants in Europe and the Caribbean. Administration officials cautioned, however, that the shortfall could reach 3 million barrels a day by midwinter.

Fuel used by the Pentagon was in serious short supply, according to the Washington Post Oct. 26. An estimated 10% of the fuel needed to power ships and planes already was lacking. Nearly 50% of the Pentagon's supplies came from overseas sources, and 20% of its oil was bought in the Middle East. Most of the jet fuel used by planes in Southeast Asia was refined from Saudi Arabian crude oil, but the Defense Department said flights had not yet been curtailed in Southeast Asia despite the shortage.

Consumers in the Northeast and Middle Atlantic states, particularly industrial users burning low sulfur oil from the Middle East, were expected to feel the greatest effects of the fuel shortfall.

Oil company profits rise sharply. The nation's major oil companies reported dramatic annual increases in third quarter earnings for 1973.

Reports of sharply rising profits were compiled by the New York Times Oct. 25. Figures show net earnings for the July-September period and the percentage gain over the third quarter of 1972: Exxon—$638 million, 80%; Texaco—$307 million, 48%; Mobil—$231 million, 64%; Gulf—$210 million, 91%; Standard Oil (Indiana)—$147 million, 37%; Shell—$82 million, 23%; Atlantic Richfield—$60 million, 16%; Phillips—$54 million, 43%; Continental—$54 million, 38%; Getty—$32 million, 71%; Marathon—$31 million, 35%; Cities Service Co.—$27 million, 61%; Ashland—$24 million, 17%; Standard Oil (Ohio)—$18 million, 14%.

Oil price spiral continues. Prices for gasoline and distillate fuels were at record levels after the Cost of Living Council lifted its freeze on the price of petroleum products Nov. 1. Thirteen major oil companies raised their prices by Nov. 7 following the dramatic recent rise in imported oil prices.

Phillips Petroleum Co.'s 3¢ a gallon hike of gasoline and distillate prices announced Nov. 1 was the largest single increase for either product ever recorded, according to industry spokesmen, but Standard Oil Co. (Ohio) announced a 4¢ a gallon increase in the price of gasoline Nov. 2. Gulf Oil Co. announced its fourth increase in two months Nov. 7 when prices were raised 1¢ a gallon.

Exxon Corp. said Oct. 31 its U.S. division would resume gasoline allocations to East and Gulf coast customers. The allocation system had been abandoned in July. Standard Oil of Ohio confirmed Nov. 1 that it was reducing fuel deliveries to bus and trucking companies, airlines and railroads by 25%.

The Labor Department's Wholesale Price Index (WPI) declined a seasonally adjusted .3%, or 3.6% at an annual rate, in October, but higher fuel costs were reflected in industrial prices, which were 9.1% above October 1972 levels. Total fuel costs were 24.9% higher than in 1972, and refined petroleum products, which rose 7.2% (unadjusted) in one month, rose 40.4% in 12 months. Natural gas and electricity prices gained .9% at wholesale levels during October.

The WPI then rose a seasonally adjusted 1.8% during November, or 21.6% at an adjusted annual rate.

A record increase in the price of petroleum products—19.3% in a single month and 232% on an adjusted annual basis—accounted for 75% of the index's advance. (Prices for refined petroleum products, such as gasoline and home heating oil, increased 34.7% during November, or 89.1% above the November 1972 level.)

The closely watched industrial commodities index posted its biggest gain in 27 years, rising 3.2%, or 38.4% at an adjusted annual rate, as a reflection of surging petroleum and other prices.

Energy conservation steps. Commerce Secretary Frederick B. Dent launched the Administration's "savEnergy" campaign Oct. 25 with an urgent appeal to the business community, which used 70% of the nation's energy, to conserve fuel. (The industrial sector consumed 43% of total energy resources.)

Dent suggested several steps that businessmen could implement to reduce the estimated 20%–40% wastage in the nation's fuel supply. Conservation measures included energy audits to review fuel consumption and costs, energy conservation goals based on the President's plan to reduce consumption 5%, and education campaigns aimed at "convincing employes, suppliers, customers and the community of the need for conservation." Businessmen also received a list of 33 ways to reduce fuel consumption and save money in the operation of plants and offices.

Other conservation steps were being implemented by states. Oregon banned the use of electricity for decorative and commercial lighting displays Sept. 23. Washington, where weather conditions had reduced the water level in its major dams, ordered all state agencies to reduce electrical consumption 10% and asked the voluntary cooperation of other state residents, according to the Wall Street Journal Sept. 10. Christmas displays also were being cut back throughout the country.

The General Services Administration, which served as the federal government's principal purchasing agent, was planning to buy its first block of compact cars, the Journal reported Oct. 19.

FPC backs emergency gas deregulation. The Federal Power Commission (FPC) Nov. 2 reaffirmed its earlier decision that interstate pipeliners of natural gas had to pay whatever prices the market could bear during a 180-day period beginning in mid-September.

The FPC allowed the emergency purchases to proceed after reviewing public comment.

The U.S. Court of Appeals in Washington had ordered a stay Oct. 3 in the first ruling which had deregulated the price of natural gas at the wellhead. The court blocked the decision until the FPC acted on a request by four consumer organizations that the case be reconsidered. The group regarded the FPC's action as an attempt to circumvent price regulations without holding public hearings.

The same appeals court Aug. 24 had reversed another FPC ruling permitting natural gas producers to set higher rates for the Texas coastal area in 1971. The FPC had failed to prove that higher rates would have the effect of increasing supplies, the court ruled.

Nixon's energy-saving plans. In a nationally televised address Nov. 7, President Nixon presented the Administration's wide-ranging energy conservation proposals for meeting the crisis precipitated by the Arab oil embargo. Nixon asked the nation to "face up to a very stark fact." "We are heading toward the most acute shortages of energy since World War II," shortages of at least 10% of anticipated demand and possibly 17%, he predicted.

Calling on citizens to make sacrifices by using "less heat, less electricity, less gasoline," Nixon announced a three-stage program to deal with the crisis.

Under provisions of immediate executive orders:

■ Industries and utilities using coal ("our most abundant resource") would be prevented from converting from coal to oil, a cleaner fuel but in short supply. Some reconversion from oil to coal also was planned.

■ Jet fuel allocations would be reduced, forcing an estimated 10% cutback in air travel and rescheduling of some flights.

■ Allocations of heating oil for homes, industries and schools would be reduced 15% from the usage level in 1972. (Previous allocation plans had called for distribution on the basis of 100% usage in 1972.) Nixon asked "everyone to lower the thermostat . . . by at least six degrees

so that we can achieve a national daytime average of 68 degrees." Factories and offices were asked to reduce heating levels by 10 degrees by "either lowering the thermostat or by curtailing working hours." "Incidently," he said, "my doctor tells me that in a temperature of 66–68 degrees you're really more healthy than when it's 75–78, if that's any comfort."

■ Temperatures in federal offices would be set at 65–68 degrees. Federally owned cars would travel no faster than 50 miles per hour except in emergencies, the President said.

Nixon asked the Atomic Energy Commission to reduce the time required to make new nuclear power plants operative from 10 years to six years.

He urged state and local governments to take similar steps reinforcing these federal level measures, as well as make other efforts to curb necessary electric lighting, slightly alter the school year, stagger working hours, encourage greater use of mass transit and car pools, and reduce highway limits to 50 miles per hour. If adopted nationwide, this action alone could save over 200,000 barrels of oil a day, Nixon said.

Other Administration recommendations required Congressional action in the form of an Emergency Energy Act. The proposed legislation, which Nixon said must be passed before the Congress adjourned in December, would include:

■ An immediate return to daylight saving time on a year round basis.

■ "The necessary authority to relax environmental regulations on a temporary case by case basis, thus permitting an appropriate balancing of our environmental interests, which all of us share, with our energy requirements, which of course, are indispensable."

■ "Authority to impose special energy conservation measures, such as restrictions on the working hours for shopping centers and other commercial establishments."

■ Approval and funding for increased exploration, development and production from naval petroleum reserves, such as Elk Hills, Calif., where production could reach 160,000 barrels a day within two months, according to Nixon.

■ Authority to reduce speed limits throughout the country and adjust schedules of planes, ships and carriers.

If the temporary shortage of crude oil products persisted, Nixon warned, rationing or taxation measures could become necessary.

The energy crisis, which was exacerbated by the Arab states' cutoff of oil exports, did not originate with the Middle East war, Nixon said, but with the "peace and abundance" known since the end of World War II. "We are running out of energy today because our economy has grown enormously and because in prosperity what were once considered luxuries are now considered necessities."

To maintain the nation's level of growth and wealth, Nixon asked for a bipartisan response from Congress to a matter of bipartisan concern.

That bipartisan solution had not been forthcoming, according to Nixon, because Congress had failed to act on seven energy proposals sent to it in April.

The Administration had already increased its plan for a five-year, $10 billion program of energy research and development, Nixon said, but a greater long range effort was needed. The nation had responded to the great challenges of "rapidly producing an atomic capability" in World War II" (the Manhattan Project) and "focusing our scientific and technologic genius on the frontiers of space," he observed, and said another priority project was required.

He called on the nation to dedicate itself "in this bicentennial era" to Project Independence" with the goal of achieving self sufficiency in energy supplies by 1980.

"Let us set as our national goal, in the spirit of Apollo and with the determination of the Manhattan Project, that by the end of this decade, we will have developed the potential to meet our own energy needs without depending on any foreign enemy—foreign energy sources," Nixon said.

"We have an energy crisis, but there is no crisis of the American spirit," he continued.

"Let us find in this time of national necessity a renewed awareness of our capacities as a people, a deeper sense of our responsibilities as a nation, and an increased understanding that the measure

and the meaning of America has always been determined by the devotion which each of us brings to our duty as citizens of America."

Funding in excess of the Manhattan Project outlays already had been requested, Nixon said, adding that the U.S. possessed half the world's known coal reserves, "huge untapped sources of natural gas," oil in the continental shelf and oil shale in western deposits, as well as "the most advanced nuclear technology known to man" and "some of the finest technical and scientific minds in the world.

"In short we have all the resources we need to meet the great challenges before us. Now we must demonstrate the will to meet that challenge," Nixon said.

Love holds news briefing—In a briefing for reporters, John A. Love, director of the White House Energy Office, said he would work closely with Congress for passage of the emergency legislation.

If an increase in the gasoline tax were required to reduce consumption, Love said, rates could go as high as 30¢ a gallon from their present level of 4¢ a gallon.

Love added that the U.S. had no plans to export fuel to Japan and European nations also affected by the cutoff of Arab oil supplies, although Interior Secretary Rogers C. B. Morton had indicated a willingness to help Nov. 1.

States act to save fuel. New Jersey Nov. 8 became the first state to act on President Nixon's urging, reducing speed limits to 50 miles an hour. Several hours before Nixon addressed the nation on the energy crisis Nov. 7, Gov. William J. Cahill had announced that New Jersey would ease clean air standards to permit burning of high sulfur content fuel. Restrictions would remain stricter, however, in the five highly industrialized, densely populated northern counties.

By Nov. 9, the New York Times reported that 29 states had ordered various limited reductions in the driving speed of state-owned cars or reduced temperatures in state-owned buildings, but 27 of those polled said legislation would be required to mandate state-wide speed limits for all drivers. California, which Nov. 6 reduced the limit from 70 to 65 miles per hour, effective Dec. 1, refused to consider a further cutback, contending that major traffic jams would result.

In most cases, a reduction in speed limits was combined with other energy saving measures, including reduced electrical lighting, lowered temperatures in state-operated buildings, and increased use of car pools and rescheduled work hours for state employes.

In one of the most dramatic responses to the current energy crisis, Maryland Gov. Marvin Mandel declared a state of emergency Nov. 14, one day after signing a sweeping emergency powers bill authorizing an immediate inventory of the state's energy sources, $2,500 fines for those who refused to cooperate and establishment of a list of priorities for fuel allocation.

Massachusetts Gov. Francis W. Sargent Nov. 13 also announced a drastic 11-point program of fuel conservation and eased anti-pollution laws.

A spokesman for President Nixon said Nov. 8 that heat and lighting at the White House had been cut and that similar restrictions had been ordered for the presidential monuments in Washington.

The fuel shortage was affecting the nation in other ways. Christmas mail deadlines were advanced Nov. 11 because cutbacks in airline schedules would result in added delivery delays. The lighting displays for Christmas trees in New York City and Washington also would be affected as officials announced planned electrical reductions of 20%–25% Nov. 9.

The Civil Aeronautics Board Nov. 9 permitted the airline industry to negotiate schedules to offer joint service plans and reduce redundant and competitive flights. However, "pooling" agreements, in which airlines flying the same routes controlled capacity, divided profits and eliminated competitive marking practices, were rejected.

Fears were raised Nov. 14 that a fuel cutback for buses and subways could impede federal and state efforts for increased use of mass transit facilities. Authorities hoped to curb excessive use of single passenger cars, particularly by commuters. Another unexpected effect of the oil crisis would be higher food prices, the Wall Street Journal reported Nov. 11. The agriculture complex, made up of fertilizer makers, farmers, food processors and others, accounted for nearly 30% of annual fuel consumption in the U.S., according to the Journal, and any cutback in production caused by reduced fuel supplies would increase food prices.

The petroleum crisis also threatened other areas of the economy which were indirectly dependent on petroleum by-products. Areas facing raw material shortages were the cosmetics industry, hosiery makers and manufacturers of synthetic fabrics, the plastics industry and producers of antifreeze and asphalt.

Gas rationing debated. A controversy appeared to be developing within the Administration over the advisability of adopting gasoline rationing plans.

Interior Secretary Rogers C. B. Morton said Nov. 11 that the odds were "better than 50–50" that gas rationing would be instituted within the "next two–three months" and that the program would remain in effect for a "year or two" under a coupon system that was last used during World War II.

Charles DiBona, deputy director of the White House Energy Policy Office, appearing with Morton on ABC's "Issues and Answers," echoed his warnings. "I think if the present [Arab oil] cutoff continues," he said, "the probabilities of having gas rationing before the winter is over are very high."

Energy Office Director John A. Love told a meeting of the American Petroleum Institute Nov. 12 that a gas rationing program could be implemented by Jan. 1, 1974, when prices could be at record levels of $1.20 a gallon because of soaring prices for imported oil. (The API released a report during its annual meeting showing that in the first 10 months of 1973, imports were 31.4% above 1972 levels. Imports accounted for 35.5% of the nation's daily petroleum consumption, according to the report.)

The API's advisory group, the National Petroleum Council, Nov. 13 joined the call for an immediate gas rationing system to prevent serious harm befalling the nation's growth and employment goals.

Those opposed to such a drastic measure included Treasury Secretary George P. Shultz, who said Nov. 13 that talk of a forthcoming rationing program was "panicking the people" into "overreacting," with consequent ill effects on the stock market and other barometers of confidence in the nation's economy.

"It's a wild thing to make work," Shultz cautioned. "I think we should explore every other avenue of solution before diving into the tank."

White House Domestic Affairs Adviser Melvin Laird Nov. 14 supported Shultz and other economists. He favored a tax of 20%-30% on gasoline sales.

Herbert Stein, chairman of the President's Council of Economic Advisers, a "free market" economist like Shultz, said in a speech in Bonn, West Germany Nov. 14 that a "sharp, large rise" in the price of gasoline could prove a more effective way of discouraging gas consumption than a rationing program.

"The direct effect of doubling the price of crude oil would be equivalent to an increase of about 3% in the price of all goods and services," Stein said. If this increase were spread over 10 years, until 1980 when it was hoped the U.S. could achieve self sufficiency in fuel supplies, the annual increase would be as little as .4%, he declared.

Stein also urged the Administration to "insulate" industry from the worst effects of the petroleum shortage by mandating 15% fuel reductions in home heating, private auto driving and commercial use.

Nixon opposes gasoline rationing. President Nixon told questioners at his meeting Nov. 17 with members of the Associated Press Managing Editors Association that the public would "resent" gas rationing, and that he hoped such drastic measures could be avoided. Although a contingency plan had been drawn up to prepare for the possible need of gas rationing measures, Nixon said, "Our goal is to make it not necessary." But he warned, "I'm not going to pledge to this audience and I'm not going to pledge to the television audience that rationing may never come. If you have another war in the Mideast, if you have a complete cut-off and not a resumption of the flow of oil from the Mideast, or some other disaster occurs, rationing may come."

Nixon mentioned his own effort to save fuel. "I came down here in a plane here today—Air Force One—I asked them if I couldn't take the Jetstar. They said, no, it doesn't have communications. So I've had to take the big plane. But we did one thing that saved half the cost. We didn't have the backup plane. The Secret Service didn't like it, communications didn't like it, but I don't need a backup plane. If this one goes down, it goes down—and then they don't have to impeach."

When asked why the Administration "didn't anticipate the energy crisis several years ago," Nixon noted that he had prepared the first energy message ever sent to Congress and that he had followed up his legislative requests in April with renewed calls for Congressional action.

"Now, I'm not saying here the Congress is to blame, the President should have done something. What I do say is that the President warned about it and the Congress did not act, even though we warned two years ago. The President warned in April. The Congress did not act. And now it's time for the Congress to get away from some of these other diversions, if they have time, and get on to this energy crisis."

The debate over gasoline rationing continued. Secretary of Health, Education and Welfare Caspar Weinberger Nov. 17 rejected both a rationing plan and its alternative, imposition of large sales taxes on gasoline.

Rationing would entail a "whole new extension and addition to government power and authority, besides being a terribly annoying and intrusive sort of activity," Weinberger said. His opposition to tax penalties was based on a belief that high taxes on consumers would not cause gasoline supplies to increase.

Sen. William Proxmire (D, Wis.) and Sen. Mike Mansfield (D, Mont.) Nov. 18 joined David Rockefeller, chairman of Chase Manhattan Bank, in calling for immediate establishment of a gasoline rationing program.

The senators agreed that rationing was an undesirable alternative, but emphasized that it appeared to be the only solution to a worsening energy crisis. Rockefeller agreed with Administration opponents of rationing that it should be considered only as a "last resort," but he added, "I'm not sure but that we're not already there."

Rockefeller's support for rationing was based on a fear that the nation's economy would show a zero growth rate in 1974 if the oil shortage reached 3 million barrels a day.

The National Petroleum Council Nov. 15 also warned that a recession could result from the fuel shortage.

Sunday gasoline sale ban urged. The Cabinet-level Special Energy Action Group, chaired by President Nixon's energy adviser, John A. Love, Nov. 19 urged the President to ban Sunday gasoline sales and limit gasoline purchases during Saturday evenings. The group also recommended that Nixon order a 10% across the board cutback in heating oil use by industrial consumers. Deliveries to residential customers would be cut back by an amount equal to reducing thermostat settings six degrees below the 1972 average, according to the plan.

Other proposals recommended to the President for adoption as voluntary measures (but subject to mandatory enforcement if needed) included a reduction in outdoor electrical advertising, 50 mile per hour nationwide speed limits and an end to the use of natural gas for outdoor lamps.

Nixon's fuel-saving orders. In a nationally televised address Nov. 25, President Nixon ordered "strong, effective counter-measures" to offset the winter's expected 17% fuel shortage.

Among the steps taken by the Administration to meet the energy crisis:

■ A 15% nationwide reduction in delivery of gasoline by refiners to wholesalers and retailers. (Proposed regulations would be issued in December regarding a gasoline allocation program at the wholesale and retail levels.) He was ordering the cutback in gasoline production, Nixon said, so that refiners could concentrate their production efforts and petroleum supplies on the goal of increasing heating oil supplies. The estimated savings were put at 900,000 barrels a day.

■ To conserve the reduced supply of gasoline, Nixon said he was "asking" gasoline filling stations not to sell gasoline between 9 p.m. Saturday and midnight

Sunday every weekend beginning Dec. 1. Until Congress gave final approval to the emergency energy bill passed by the Senate and legislated a ban on Sunday gasoline sales, Nixon said the station closings would be conducted on a voluntary basis. The halt in gasoline sales applied to cars, trucks, pleasure boats, private aircraft and recreational vehicles, with an estimated saving of 50,000 barrels a day. Potential fuel savings could reach 25%–30% if public consumption and industry production were reduced, Nixon said.

■ In other gasoline conservation steps, Nixon said he would impose a maximum speed limit of 50 miles an hour for automobile travel, pending final action on the energy bill, but "intercity buses and heavy duty trucks which operate more efficiently at higher speed and therefore do not use more gasoline will be permitted to observe a 55 mile per hour speed limit." (Savings were put at 200,000 barrels a day.)

■ Jet fuel supplies, which already had been cut back 10%, were subject to an additional phased cut of 15%, requiring a further limitation of flight schedules and an increase in passenger loads. Beginning Dec. 1, domestic airlines would be allotted 5% less than their 1972 levels and international airlines would be reduced to 1972 levels.

The full 25% cut (compared with January 1972 levels) would take effect by Jan. 7, 1974. Fuel for high priority aviation such as air taxi service and industrial usage would be curtailed 20%, business flying, including corporate jets, cut 40%, and fuel for personal pleasure and instructional flying restricted to half of previous levels.

■ Another fuel conservation measure, also dependent on the energy legislation before Congress, involved a reduction in outdoor lighting, particularly Christmas decorations. Nixon said he would order a curtailment in ornamental outdoor lighting for homes and elimination of all commercial lighting except that which identified places of business when Congress authorized him to act.

"As just one example of the impact which such an initiative can have, the energy consumed by ornamental gas lights alone, in this country, is equivalent to 35,-

000 barrels per day of oil, and that is enough fuel to heat 175,000 homes," Nixon said.

■ Nixon said "final plans" had been developed for allocation of scarce heating oil. There would be 10% reductions in supplies for industrial use, 15% for home use and 25% for commercial use, effective Dec. 27. Priority would go to fuel production activities, public passenger transportation, food production and processing, and essential community services. (The regulations were published Nov. 27.)

According to the President, "the reduction for homeowners alone will result in a saving of some 315,000 barrels of heating oil a day which is enough to heat over one and a half million homes every day."

Nixon concluded his speech by urging Congress, state and local governments and individual citizens to cooperate in the "spirit of discipline, self-restraint and unity" in conserving fuel, minimizing the hardships of the energy crisis and reducing disruptions to the economy.

The nation's "overall objective" in meeting the energy crisis, Nixon said, "can be summed up in one word that best characterizes this nation and its essential nature. That word is independence. . . . In the last third of this century, our independence will depend on maintaining and achieving self-sufficiency in energy. What I have called Project Independence 1980 is a series of plans and goals set to insure that by the end of this decade Americans will not have to rely on any source of energy beyond our own."

In other action announced by the White House following the President's address, electric power would be diverted from uranium enrichment facilities if the energy crisis required it. The Emergency Petroleum and Gas Administration Executive Reserve, a standby group of industry personnel established under the Defense Production Act of 1950, would be partially activated to plan and administer emergency programs.

The Office of Petroleum Allocation would administer and enforce the fuel conservation program. An estimated 5,-500 persons, including 500 oil industry executives, would be employed by the new office, according to the White House.

Final regulations preventing industries and utilities from switching from coal to

oil would take effect Dec. 7, pending public comment on the proposals.

In other Administration action announced Nov. 16, three priority users of diesel fuel—farmers and ranchers, mass transit systems and oil dealers—were designated to obtain all the fuel needed for the next 60-day period.

Congressional, industry comment—House Speaker Carl Albert (D, Okla.) issued a 22-page statement Nov. 24 blaming President Nixon for trying to "rewrite history" and shifting the responsibility for fuel shortages to Congress. He said the energy crisis was "deeply rooted" in the Administration's policies of impoundments of research funds, support for oil import quotas and delays in implementing stockpile and allocation programs.

"The American people are now being asked to pay with their patriotism the price exacted by the misguided policies and negligence of the Nixon Administration," Albert declared.

According to the Washington Post Nov. 25, Sen. Henry Jackson (D, Wash.), sponsor of several energy-related bills, also reacted sharply to Nixon's accusations that Congress had failed to heed Administration warnings of a forthcoming energy crisis and had delayed action on necessary legislation.

Because Nixon "has never been willing to face up to the realities of the situation, Congress at every turn has had to force the Administration to act, either by proposing, enacting or threatening to enact appropriate legislation," Jackson said.

Senate Majority Leader Mike Mansfield's (D, Mont.) remarks Nov. 18 also reflected the partisan nature of the debate over solutions to the energy crisis. Nixon "is great on messages, he is great on rhetoric, but when it comes to legislation, he is wanting. . . . The Administration has been very backward in doing anything about the crisis which confronts the country today and the Administration knows it," Mansfield said.

Donald Cook, chief executive of the American Electric Power Co., the nation's largest investor-owned electric utility system, Nov. 26 called on the

President and Congress to "stop playing politics" with the energy crisis.

According to the New York Times Nov. 27, most of the nation's economists, businessmen and oil industry experts blamed both the Administration and legislators with failing to face the realities of the current fuel shortage.

M. A. Adelman, a leading oil economist at the Massachusetts Institute of Technology, criticized the President's latest conservation measures as inadequate, an opinion which reflected the views of many experts, according to the Times. "President Nixon is dragging his feet because he knows the necessary measures will be unpopular. He is so unpopular now that he feels he cannot afford to do anything that will alienate more people," Adelman said.

The reaction to the ban on Sunday gasoline sales varied widely among gasoline station owners, some of whom had already planned to close on weekends because of short gasoline supplies. But there was widespread criticism of the action from businesses dependent on automobile travel, especially the ski industry.

Stocks in motels, recreation areas and other leisure-based industries closed down sharply Nov. 26 on the New York Stock Exchange.

Proxmire replies to Nixon speech. Sen. William Proxmire (D, Wis.) Dec. 2 delivered a nationally televised response to President Nixon's energy address. The spokesman for Congressional Democrats criticized Nixon's proposals as inadequate to avoid severe winter fuel shortages. Proxmire called for fuel rationing to reduce hardships.

"We need it now," Proxmire said. "It would provide some gas at present prices to everyone with a car and enough heating oil to keep from freezing in every home."

Proxmire also accused Nixon of underestimating the severity of the energy crisis.

Military faces severe fuel shortage. Pentagon officials said Nov. 15 that because the military had purchased 50% of its oil supplies from Arab states whose U.S. sales were now under embargo, the Pentagon would be forced to divert an esti-

mated 300,000 barrels a day from the reduced fuel supplies available for domestic consumption. (Under the Defense Production Act of 1950, the armed forces had first priority on the nation's supplies.)

The impact on consumers' use would not be "serious," officials said, because the replacement needs amounted only to 1.8% of the total U.S. requirements of 17.9 million barrels a day.

There also was no crisis in military preparedness, according to the Pentagon, because untapped fuel reserves remained available to them. Despite the assurances, spokesmen said military flying hours had been ordered cut by 18% and steaming days reduced by 20%. The 6th and 7th Fleets in Mediterranean and Pacific waters and ground troops in Europe would reserve priority supplies, according to the Pentagon.

Pentagon commandeers oil supplies. Faced with the reluctance of oil companies to sell to the military, the Pentagon said Nov. 27 it had requested the Department of Interior to order 22 oil producers to deliver 19.6 million barrels of petroleum products to military users in November and December.

The action, taken under authority granted the Defense Department in 1950 by the Defense Production Act, came after domestic oil suppliers refused to bid on contracts offered by the Pentagon. The 1950 law empowered the President to requisition materials critical to national defense.

According to Pentagon spokesmen, the oil companies formerly sold their excess production to the military on a long-term contractual basis. The oil companies contended they currently lacked sufficient supplies to bid on military contracts, the Pentagon said.

New York eases anti-pollution laws. New York City and state officials, reacting to a critical shortage of low sulphur content fuel oil, authorized the city's Consolidated Edison Co. to burn oil containing more than the .3% sulphur limit allowed under anti-pollution laws.

The city's Environmental Protection Administration acted Nov. 19 to temporarily ease the sulphur restrictions, but officials warned that burning of the dirtier fuel oil would result in an increase of at least 10% in the amount of sulphur dioxide in the air. An estimated 50% of the city would have unhealthy levels of the gas, in contrast to the current level of 20%, spokesmen said.

Consolidated Edison's request that the burning of coal be allowed was denied on grounds that an intolerable level of filth would be added to the city's air; however, state officials ruled Nov. 27 that the power company could burn coal in two generating stations within the city. (Sulphur restrictions were relaxed at the same time.)

Fuel shortage evaluated. Rep. Les Aspin (D, Wis.) Nov. 17 released a Library of Congress study of the energy shortage which predicted that the fuel oil shortage could reach 35% of demand, or a 6.1 million barrel daily shortfall this winter.

The figures were based on an "exceptionally cold winter" and a continued embargo by Arab oil sources, but it was "unlikely," according to the report, that mild weather or an end to the Arab cutoff could reduce the shortage to 20% or less.

Another report by the Senate Permanent Subcommittee on Investigations released Nov. 7 blamed the policies of the Administration and the oil industry for the 1972 winter heating oil shortage and the 1973 spring gasoline shortage.

Federal import policies were held responsible for blocking the importation of adequate supplies of foreign crude oil, according to subcommittee chairman Sen. Henry M. Jackson (D, Wash.), a sponsor of several major energy bills.

The oil industry was blamed for refusing to recognize as early as 1969 that an energy crisis was developing. The oil shortage was traced back four years when Texas and Louisiana oil fields, which produced two-thirds of the nation's domestic supplies, suffered the largest decline in crude oil reserves in history. The drop was related to the industry's serious overestimation of the fields' capacities. This lack of candor, which was directly related to oil producers' opposition to an end to federal quotas and increased importation of foreign petroleum supplies, continued to characterize the industry, the report declared.

As evidence, the study said the 10 largest oil companies in the U.S. showed a crude oil reserve decline in 1972, but nevertheless reported that their oil stocks were sufficient to meet demand.

"At the same time, inventory levels of heating oil and gasoline were being drawn down because of increased demand and because refineries operated at reduced levels of capacity," the report stated.

The industry also was criticized for its failure since 1969 to construct new refineries to produce gasoline and heating oil. The August 1971 price freeze ordered by President Nixon also contributed to the shortage, according to the report, because gasoline prices were frozen at seasonal highs and heating oil prices at off-season lows.

"It was not surprising to find the industry more anxious to convert its crude oil into gasoline than into fuel oil," the report continued. "The profits were in gasoline and the price controls, regardless of their many possible merits, were in fact a factor in the creation of the alarmingly low fuel oil stocks the industry held late in the summer of 1972."

Circumstances had worsened because refineries operated well below capacity during the first four months of 1972, the report charged. The refineries failed to meet production schedules required to replenish heating oil stocks and to meet the seasonal demand for gasoline.

"Although refinery utilization jumped to 94.9% in June of 1973 and has continued for several months at approximately 94%, the increase in capacity utilization was too late to avert the gasoline shortage which occurred during the peak summer driving season," the report concluded.

New York sues 7 oil firms. New York Attorney General Louis Lefkowitz announced Dec. 2 that an antitrust action had been initiated against seven major oil companies—Exxon Corp.'s USA division, Shell Oil Co., Gulf Oil Corp., Sun Oil Co., Texaco, Inc., Mobil Oil Corp. and Amoco Oil Co., a subsidiary of Standard Oil Co. of Indiana.

The defendants were charged with "engaging in a common course of conduct" and making an "illegal understanding and arrangement" to fix gasoline prices for the purpose of driving rival independent gasoline dealers out of business.

Unlike antitrust suits brought against a number of major companies by Florida and Connecticut which sought to break up the firms' allegedly monopolistic structure, the New York suit emphasized price behavior exhibited by the companies, particularly in regard to rebates (for high "official" wholesale prices) granted on an allegedly discriminatory basis to selected gasoline dealers.

In another aspect of the suit, Lefkowitz noted, "The major companies do not compete with each other on the basis of price, but instead make advertising claims stressing the alleged unique qualities of their products. Nevertheless, gasoline of any given octane rating is an essentially standard product, as evidenced by the fact that the companies exchange products with each other."

New England faces voltage cut. Connecticut Gov. Thomas J. Meskill (R), chairman of the New England Governors Conference, announced Nov. 21 that there would be a 5% voltage reduction in the six-state region from 4 p.m. to 8 p.m. daily, beginning Nov. 26. It was the first fuel saving measure of its kind in the country.

(Nixon Administration officials warned Nov. 21 that brownouts, or reductions in voltage, and "rolling blackouts," which would deny power altogether to communities on a rotating basis, could occur during the winter, particularly along the eastern seaboard.)

In a fuel-saving step applied only within Connecticut, Meskill ordered state agencies to reduce gasoline consumption by 10%. Beginning Dec. 1, state workers would require gasoline stamps to purchase fuel for official use. (Gasoline stamps had not been used in the U.S. since World War II.)

Airlines, auto industry affected. Domestic and international airline fares were allowed to rise as a result of higher fuel costs. The Civil Aeronautics Board (CAB) Nov. 21 approved a 5% increase in domestic fares, effective Dec. 1, that was based on surging airlines costs.

The CAB also said it would monitor the carriers' financial records "to insure the continued provision of vital services at fair prices and to measure the effect of the [new] fare increase on traffic." Critics had charged that airlines, which had overextended their flight capacity during 1973 and had reported a subsequent decline in revenues, would profit from the forced reduction in scheduled flights by filling more seats on the planes and laying off personnel.

The CAB-approved fare increase was the fifth in a series since 1969.

Airlines flying international routes announced Nov. 23 that an average 6% fare increase was needed "to meet the recent drastic increase in fuel costs." CAB approval of the proposed increase would be required before the fare hike could take effect Jan. 1, 1974. The proposed increase was intended to cover higher costs only through March 1974 and industry spokesmen warned that further increases would be needed.

A spokesman for the Air Transport Association Nov. 27 assailed the Administration's planned cutbacks in aviation fuel as "arbitrary and inequitable slashes" that would endanger a "vital national asset affecting the country's total economy." Since mid-October, spokesmen said, the industry had canceled an estimated 1,000 of its total 15,000 scheduled daily flights. Further forced reductions required by the Administration's allocation plan would have a "discriminatory" effect on the industry, it was charged.

The nation's largest domestic carrier, United Air Lines, announced Nov. 27 that 950 employes, including 300 pilots, would be laid off by Jan. 7, 1974, when the airline expected to eliminate 100 flights a day. The furloughs had been planned before the President announced new cutbacks in jet fuel deliveries and airline spokesmen warned that further personnel layoffs could occur.

The Cessna Aircraft Co. said Nov. 28 that layoff notices had been sent to 2,400 of its 11,000 employes because of an anticipated drop in the sale of private planes and executive jets.

Trans World Airlines had planned to furlough 100 pilots and 303 flight attendants Dec. 1, the Washington Post reported Nov. 28, but a strike had grounded the airline and affected all union personnel since Nov. 5.

American Airlines announced Nov. 28 that 214 of the carrier's 3,700 pilots would be laid off Jan. 2, 1974 because of flight reductions. Ninety-six daily flights already had been canceled because of the jet fuel shortage and an additional 16 daily departures would be eliminated as a result of the latest Administration-ordered cutback in fuel allocation, American officials said.

Other major airlines curtailing departures were Delta Air Lines, which disclosed Nov. 28 that an estimated 10% of its 1,500 daily flights had been eliminated. Nearly 8% of Eastern Airlines' flights would be canceled during the first two weeks in December, spokesmen said Nov. 29, in hopes that heavy cuts during light travel periods would allow a resumption of service during the peak holiday travel time.

Auto industry employes also were directly affected by the energy crisis. General Motors Corp. (GM), whose big car sales had dropped an estimated 25% recently because of the gasoline shortage, announced Nov. 23 that 16 of its 24 assembly plants in the U.S. and Canada would be totally or partially closed for one week beginning Dec. 17.

The shutdown was designed to cut production of large and intermediate-sized cars by 79,000 units, which was 3% of the total 1974 models scheduled for manufacture by GM in the calendar year. Total output during December would be reduced to 32% below December 1972 levels.

An undetermined number of workers would be laid off, according to GM, but observers believed the shutdown could involve tens of thousands of employes.

GM had not curtailed production since early in 1970 when 16 plants were idled because of a major slump in car sales. Current sales figures for all cars had declined to a level 10% below the rate of a year earlier.

Chrysler Corp. became the second of the "Big 3" auto makers to announce plant closings. Officials said Nov. 29 that six of its seven assembly plants in the U.S. and Canada would close for three days during the first week of 1974. The seventh plant, which would be converted from big car to small car production, was scheduled to shut down for two weeks after Jan. 1, 1974. An estimated 38,000 workers would be idled, according to the company.

The closings were necessitated by sharp drops in the sale of big cars and inability to obtain parts for other vehicles, spokesmen said.

Ford Motor Co. and American Motors Corp. said Nov. 28 that no immediate interruptions in production were anticipated but Ford officials confirmed plans to convert some big car assembly plants to the production of smaller autos.

Stein fears economic consequences— Herbert Stein, chairman of the President's Council of Economic Advisers, Nov. 29 issued the Administration's first public warning that the energy crisis could result in a 6% unemployment rate, sluggish "real" economic growth of about 1%, a 1.5% reduction in consumer spending and increased inflationary pressures in 1974, threatening a national recession.

The forecast was based on a premise that the Arab oil embargo, which had "introduced major new uncertainties into the economic picture," would continue throughout 1974, Stein said.

"One cannot rule out the possibility of larger negative effects" if the Administration's plans to concentrate its allocation cutbacks in the consumer sector proved unworkable, he warned.

In the first sign that the Arab oil embargo was affecting the operation of U.S. oil companies, Mobil Oil Corp. announced Nov. 27 that it was "mothballing" a refinery in East Chicago, Ind. The plant, which had a 47,000 barrel a day capacity, would cease operation Dec. 31.

According to the American Petroleum Institute Nov. 28, the processing of foreign crude oil by U.S. refineries had declined for the fifth consecutive week. Foreign petroleum supplies averaged 3,-262,000 barrels a day in the week ending Nov. 27, down from 3,795,000 in the week ending Oct. 19, before the Arab embargo took effect.

Other areas of the economy expected to be affected by the petroleum shortage included the record industry, which was dependent on vinyl supplies; the plastics industry, which produced toys, automobile accessories, furniture, packaging material and numerous other consumer items; sporting goods and boat manufacturers, which used fiberglass material; and the pharmaceutical industry. Drug makers warned Nov. 29 that the U.S. faced a critical shortage of penicillin, cortisone and other vital drugs whose manufacture required the use of petroleum-based solvents unless the government diverted more crude oil to the petrochemical industry.

Global trade in all major commodities—grain, raw materials and consumer finished goods—also was threatened by the energy crisis, the Wall Street Journal reported Nov. 30. Helen Bentley, chairman of the Federal Maritime Commision, warned that a "critical shortage" existed in bunker fuel used by ships. She called for a meeting in Washington with representatives of the shipping industry to "rationalize" sea shipments, which, she said, could drop by "20%, at least, perhaps even more."

Shippers would be asked to establish a priority system for the movement of goods, reduce the number of ship departures, slow speeds and increase cargo capacity, the Journal reported.

Oil shale project approved. Interior Secretary Rogers C. B. Morton Nov. 28 approved a commercial leasing plan for experimental development of oil shale deposits on six tracts of federal land in Colorado, Utah and Wyoming. The first lease sale would be Jan. 8, 1974, to be followed by sales in each of the next five months.

Morton estimated there were reserves of at least 600 billion barrels of crude on federal shale land, but emphasized that development would have no immediate effect on current petroleum shortages. Large-scale production, Morton said, could result if the prototype project proved successful, but not until the late 1980s.

Conceding that development would "incur some environmental costs and risks," Morton contended that "rigorous and comprehensive" controls had been set and that potential benefits outweighed the risks.

Morton said four of the plants in the initial project would process the shale on the surface, but two in Utah would attempt to heat the shale underground and pipe oil to the surface.

In a related development, Occidental Petroleum Corp. Chairman Armand Hammer said Nov. 23 that his company had developed a process for extracting oil directly from underground shale formations. A chamber would be blasted in the formation, then natural gas injected and fired. Oil would be separated from the rock by the heat and pumped to the surface.

FPC action. The Federal Power Commission (FPC) took two steps designed to ease the current shortage of natural gas. In an announcement Dec. 7, the FPC allowed two distributors of natural gas to purchase their supplies from affiliated producers. The action, termed "backward integration" because the usual pattern involved producers seeking transmission and distribution opportunities, would not violate antitrust laws, FPC officials declared.

The FPC gave conditional approval Dec. 31 to a New Jersey firm's plan to import liquefied natural gas from Algeria equivalent to 4.76 trillion cubic feet over the next 22 years. It would be the largest importation project of natural gas in the nation.

The American Gas Association reported Dec. 30 that sales of natural gas to industrial users during 1973 dropped 1.9% from the previous year. Although it was the first decline since 1938, the AGA also reported that dollar revenues from industrial customers rose 7% during the year because of sharply higher prices.

Love quits in power struggle. John A. Love, President Nixon's principal adviser on energy matters and director of the White House Energy Policy Office, announced his resignation Dec. 3. Love's decision to leave the Administration was precipitated by Nixon's intention to replace him as "energy czar" with Deputy Treasury Secretary William E. Simon, a decision which resulted from a power struggle within the executive branch over the controversial issues of gasoline rationing, mandatory fuel allocation and future fuel conservation policies.

News of the planned shakeup was published Dec. 1, when widespread dissatisfaction was reported within the Administration and in Congress with Love's performance because of his inability to swiftly carry out strong, affirmative actions throughout the period of worsening fuel shortages.

Love reportedly was angered by the charges, which were based on "informed sources" within the Administration. Instead of remaining "on the team" in a reduced policy-making role in which his office and functions eventually would be absorbed by Simon's agency, Love informed the President of his decision to resign as a presidential assistant and director of the White House Energy Policy Office. (Love's chief aide, Charles DiBona, resigned at the same time.)

Nixon accepted Love's resignation with "deep regret," a White House spokesman said Dec. 3. "The President thinks that Governor Love has done an outstanding job under difficult circumstances in implementing the positive actions taken thus far by the Administration to meet the energy crisis," the spokesman said.

In an interview with the Denver Post Dec. 3, Love (a former Colorado governor) responded to charges that he had provided ineffectual leadership during the energy crisis. "To be honest, it's been difficult to try to do anything meaningful and even to get the attention of the President," Love said. The President had failed to appreciate the seriousness of the situation, Love said, citing Nixon's appearance Nov. 26 before the Seafarers' Union when he "said it's [the fuel shortage] all going to be over in a year." Love broadened his criticism Dec. 4 to include "most" other Administration officials who "didn't realize the depth of this problem and were not reacting swiftly enough."

Love said he would not remain in Washington "twiddling my thumbs" in a post rendered "superfluous" by Simon's pending apointment as executive director of a Cabinet-level Emergency Energy Action Group and as director of a new Federal Energy Office within the White House.

Love also confirmed reports that his ability to act decisively had been hampered by a power struggle in which he had failed to win support for gasoline rationing against the opposition of Treasury Secretary George P. Shultz and other economists, who favored higher gasoline prices and a heavy gasoline sale tax as methods to inhibit consumer demand.

Upon his return to Denver Dec. 4, Love admitted he was "battered and bruised" by the political in-fighting in Washington and confirmed reports that he had been importuned to remain within the Administration during Simon's tenure in order to present a united policy front to the public and to minimize any ill-feeling caused by the appointment and reorganization.

When he rejected the appeal, Love said, presidential chief of staff Alexander M. Haig Jr. offered him the Canadian ambassadorship, a post Love also refused.

Simon's appointment was announced at a press conference Dec. 4 by President Nixon, who also announced the creation (by executive order) of the new Federal Energy Office. A Federal Energy Administration, which Simon would head, would be established later by statute, Nixon said.

Simon, a former partner with the New York investment banking house, Salomon Brothers, had headed the Administration's Oil Policy Committee until October when Love was given authority over the group. Simon would remain as deputy Treasury secretary in addition to assuming the new energy policy making role as a presidential counselor.

Simon said the Administration would reach a decision "by the end of December" on gasoline rationing. He added that a greatly increased federal excise tax on gasoline, or a combination of rationing and a tax increase, were also options he was considering.

The aim of any government-sponsored conservation program, Simon said, would be a reduction in private gasoline consumption by 30% during the first three months of 1974. Simon indicated that the Administration would not allow gasoline prices to rise by more than 40¢ a gallon, the amount needed to bring about a 30% reduction in demand.

In other energy-related matters, Simon said he favored construction of a second Alaskan oil pipeline. He also warned that the government's austerity program would "require some price increases or tax on natural gas and electricity in addition to the allocation program on home heating oil."

Simon said he did not favor "punitive" tax actions, such as a reduction in the 22% oil depletion allowance; however, Treasury Secretary George P. Shultz had said Dec. 2 that he supported a "plowback" clause which would prohibit use of the depletion allowance unless the oil or natural gas production company put an equivalent amount of funds into research and development efforts. The effects of such a measure, Shultz said, would serve to increase the supply of petroleum products and limit "windfall" profits realized by U.S. companies during the Arab oil embargo.

Sens. Henry M. Jackson (D, Wash.), Jacob Javits (R, N.Y.) and Edmund S. Muskie (D, Me.), members of the Senate Government Operations Committee, Dec. 6 urged Simon to implement a rationing system quickly, but he continued to indicate his opposition to the "last resort" measure because of the administrative problems and tendencies toward black marketeering attendant in any rationing program.

If the decision were made at the end of December to initiate such a program, Simon told the committee, "We need a minimum of 60 days to set up a rationing system. It couldn't start before March 1, [1974] at the earliest."

Reich quits—Eli T. Reich resigned Dec. 11 as head of the Office of Petroleum Allocation in a dispute with President Nixon's new director of the Federal Energy Office, William E. Simon.

Reich, a former deputy assistant secretary of defense, had retired as vice admiral Oct. 31 to direct the Administration's allocation program. He was reportedly displeased with the White House energy reorganization plan and Simon's decision to reduce Reich's authority by assigning him responsibility for only the heating oil sector of the overall allocation program.

Simon's office had announced Reich's resignation Dec. 10 but Reich denied the report. He appeared for work Dec. 11 as his replacement, Frank G. Zarb, an assistant director in the Office of Management and Budget, took office as acting administrator of the allocation program.

Reich later offered his resignation to Interior Secretary Rogers C. B. Morton, the official who had hired him.

Political repercussion from Simon's earlier clash with John Love, the Administration's first "energy czar," continued. In an interview with the New York Times Dec. 8, Love, a three term governor of Colorado with no experience in the energy field, said, "To this day, I don't know why they offered me the job [as director of the now defunct White House Energy Policy Office]."

Love recounted his difficulties in winning a consensus from the President and other key Administration officials on how to deal with the worsening energy crisis. He added that his proposals were often considerably tougher than those adopted by the White House. In early August, Love said, he had argued for implementation of mandatory allocation for certain fuels, but Treasury Secretary George P. Shultz opposed it and the proposal was never presented to Nixon.

In September he sent a personal letter to Nixon asking for support in taking firm action, but received no reply. His few meetings with the President were generally unproductive, Love added.

By October when the Arab oil embargo began to take effect, Love said his efforts to greatly expand his staff were undercut, and they were finally sabotaged Nov. 30, he asserted when he first learned of the planned reorganization in which his office and powers would be absorbed into the Federal Energy Administration headed by Simon.

OEP blamed for ignoring oil shortage. The Senate Permanent Investigations Subcommittee released affidavits Dec. 13 from six oil policy officials in the Office of Emergency Preparedness (OEP) and the Interior Department whose testimony indicated that the bureaucracy within the OEP, a White House agency, was largely responsible for the Administration's failure to anticipate and plan for the oil shortage other experts knew was coming.

Among the subcommittee's charges:
■ Key officials at the OEP, including Director George A. Lincoln and his assistants "never saw" a staff report prepared March 30, 1972 warning that

domestic oil production had "peaked" while consumer demand continued to rise.

Despite that warning and other data to support fears of a worsening supply/demand curve, the OEP set oil import quotas at limits far below that required to satisfy national demand.

■ Peter Flanigan, a key White House aide, intervened at least once to oppose a large increase in foreign imports, a position also favored by oil companies which were heavy contributors to President Nixon's re-election campaign. In another example of the close relationship between politics and oil policy, the head of the Interior Department's Office of Oil and Gas warned that a pre-election "petroleum shortage identified with oil import controls could be made into a major political issue" in 1972.

Lincoln also testified that his agency had prepared a "complete" standby gasoline rationing program in 1972, but it had apparently been lost in the bureaucratic shuffle when the Administration's first "energy czar," John A. Love, took office during the summer. Lincoln testified that Love never saw the plan, which was sent to the General Services Administration, the government's housekeeping and purchasing agency.

Oil companies raise prices.

Seven major oil companies and subsidiaries announced another in a series of fuel price increases. Exxon Corp.'s USA division, Standard Oil Co. (Ohio) and BP Oil Corp., a Sohio unit, announced the price boosts Dec. 4.

Exxon raised retail and wholesale gasoline prices 1.1¢ a gallon; wholesale heating, kerosene and diesel fuel were boosted .8¢–2.5¢ a gallon (but home heating oil sold directly to the consumer was cut .5¢).

Sohio raised retail and wholesale gasoline prices 1.5¢ and 1.1¢ a gallon; home heating oil 4.3¢ a gallon. New prices for home heating oil were 28.6¢ a gallon. Per gallon pump prices for gasoline at company-owned stations were 45.4¢ for regular and 49.4¢ for premium.

BP upped gasoline prices 3¢ a gallon, but reduced wholesale home heating oil prices .5¢ a gallon. (BP prices differed from those of the parent company because each had different sources of crude oil, Sohio spokesmen said.)

Shell Oil Co., Atlantic Richfield Corp. (Arco), Sun Oil Co. and Union Oil Co. announced immediate increases in gasoline prices Nov. 30 as motorists were preparing for the first Sunday in which gasoline sales were severely restricted.

Shell raised gasoline prices 3.2¢ a gallon; Arco increased prices 2¢ a gallon and Sun instituted a 2.9¢ a gallon increase for the entire Eastern Seaboard and a 2¢ a

gallon hike for the rest of the country. Union raised prices 2¢ a gallon.

The increases caused middle range gasoline prices to top 50¢ a gallon in the Washington area.

Arco spokesmen added that because domestic supplies of home heating oil were so short, imports at sharply higher prices had risen to meet demand. According to the company, imported home heating oil had cost 14.5¢ a gallon in May, but in November was costing 41.2¢ a gallon.

Gasoline shortage.

Traffic was off across the nation Dec. 2 as more than 90% of the nation's 220,000 gasoline stations observed a voluntary ban on Sunday gasoline sales.

Traffic returned to normal levels Dec. 3 as motorists lined up for gasoline refills, lessening the fuel saving impact of one gasoline-less day during the week.

(The energy crisis had brought about a discernible change in the life styles of Washington's top officials. Among the 11 Cabinet officers, only Secretary of State Henry Kissinger continued to use a chauffeur driven limousine, the Washington Post reported Dec. 5.)

Ramifications of the energy crisis continued to be felt in the auto industry, which reported Dec. 4 that new car sales in November were 12% below November 1972 levels; however, sales of luxury cars and luxury motor yachts remained at near record levels, according to reports Dec. 5.

Ford Motor Co. announced Nov. 30 that 2,546 workers at 13 plants would be idled immediately because of a shortage of parts and a slump in big car sales. General Motors Corp. officials, who had already announced layoffs, said Dec. 3 that 137,-000 workers in the U.S. and Canada would be out of work during the production stoppage at 16 big car plants.

Airline cuts.

The nation's big airlines began to make spot cancellations in service for the first time in an effort to conserve dwindling fuel supplies, the Wall Street Journal reported Dec. 4.

Across-the-board cuts in flight schedules also continued to be announced. Eastern Airlines officials said Nov. 30

that three of its five shuttle flights—
Newark–Washington,   Newark–Boston
and New York–Montreal—were being
eliminated, effective Jan. 7, 1974. The
eight daily roundtrip flights, which had required no reservations and guaranteed
seats, would be replaced with less frequent
service, Eastern spokesmen said. The elimination of service to five other cities also
was being planned, pending approval
from the Civil Aeronautics Board.

In another cost-saving step, Eastern
said 360 of its 4,400 pilots and an undetermined number of flight personnel
would be laid off Jan. 31, 1974.

United Air Lines Dec. 5 announced it
would eliminate its entire flight schedule—
16 daily flights—to Midway Airport,
Chicago's second airport, by Jan. 3, 1974.

Truckers protest fuel crisis. Truckers
who owned and operated their own rigs
spearheaded a four-day protest Dec. 4–7
blocking key U.S. highways to protest
rising fuel costs and government-ordered
reductions in highway speed limits.

By Dec. 5, the sporadic demonstrations,
initiated by a small group of truckers
broadcasting their messages to other
owner-operators over citizen band radios,
had escalated to paralyzing proportions.
Teamster officials and representatives of
the National Association of Truck Stop
Operators disavowed the wildcat actions
Dec. 5, but requested meetings with President Nixon to discuss the truckers'
grievances.

Transportation Secretary Claude S.
Brinegar issued the government's reply
later that day, appealing for an end to the
disruptions and promising an immediate
investigation of alleged fuel price-gouging
and a review of diesel fuel allocations that
were being made available at truck stops.

The Cost of Living Council also announced Dec. 5 that the Internal Revenue
Service had been directed to patrol 11
truck stops on the nation's major interstate highways to investigate charges of
non-compliance with Phase 4 price controls on fuel.

Deputy White House Press Secretary
Gerald S. Warren Dec. 5 refused to criti-
cize the truckers although reporters noted
that when antiwar demonstrators had
blocked traffic on I-95 in New Jersey in

1971, President Nixon had assailed the
protesters.

In remarks Dec. 6, Warren continued to
disavow any White House role in clearing
the roadblocks. Asked why the federal
government had been able to act so firmly
against civil rights and antiwar demonstrators, particularly in May 1971 when
nearly 10,000 were arrested and detained
without warrants, Warren replied that the
blockading of interstate highways was a
"matter for the states" to deal with.

Truckers halted traffic with massive
"stall-ins" throughout the four-day period
on I-84 in Connecticut, I-80 in Pennsylvania, the Indiana Toll Road, I-40 which
connected Little Rock, Ark. and Mem-
phis, Tenn., as well as major roads in West
Virginia, Iowa, Illinois, Michigan, Nevada, Nebraska and South Dakota.

The principal tie-ups occurred in Delaware, western Pennsylvania and Ohio.

A seven hour stall-in involved an estimated 1,800 trucks Dec. 5 at the entrance
to the Delaware River Bridge on I-95,
which halted traffic on the main north-
south highway between New York and
Washington. No arrests were made because the bridge was cleared after
truckers held a televised news conference
to publicize their grievances. The New
Jersey National Guard was alerted Dec. 6
as sporadic tie-ups persisted.

In another major traffic jam on I-80 in
western Pennsylvania early Dec. 5, trucks
traveled 10 m.p.h. in a two-by-two
caravan 20 miles long. Truckers called off
the demonstration and others throughout
the state after meeting with Gov. Milton
Shapp, who had threatened to mobilize
the National Guard.

An estimated 1,000 drivers parked their
rigs along the Ohio Turnpike and two
other main roads leading to Columbus.
An aide to Gov. John J. Gilligan met with
the protesters Dec. 5 but traffic jams
continued throughout the night after six
arrests had been made and 300 vehicles
towed away.

Riot equipped troops were needed to
patrol the roads and, by Dec. 6, National
Guardsmen were mobilized to clear the
112-mile turnpike. Fifteen separate
blockades were counted during the previous evening and up to 4,000 truckers
were involved in the four days of protest in
Ohio, it was reported Dec. 7.

Drivers employed by trucking companies protested the lower speed limits because their salaries were linked to the number of miles covered. Independent drivers said their contracts would be limited by the time consumed in slower trips. Owner-operators also cited the cost of diesel fuel, currently above 50¢ a gallon in the Northeast, but which had been near 30¢ a gallon level during the summer.

Truckers also complained that their rigs ran more efficiently at higher speeds, but a study conducted by General Motors Corp. indicated a fuel saving of 15% if large trucks were driven at 50–55 m.p.h. rather than at 70 m.p.h.

The study, published Dec. 5, also noted that reduced speed limits would impose new costs on the trucking industry through the addition of trucks, drivers and relay stops. Those increases could total more than four times the money saved by cutting fuel consumption, the report stated.

Truckers launch strike. Independent truckers launched a two-day strike Dec. 13 that was marred by scattered reports of violence, vandalism and intimidation. Many dissident drivers also kept their rigs off the roads most of the weekend.

Much of the violence, including brick throwing and at least 35 gunfire incidents, occurred in the Ohio-Pennsylvania region, which had been the scene of the most extensive and militant job actions in the previous week.

Only one injury from gunfire was reported (in Colorado) but an unoccupied tractor-trailer cab was blown up in Arkansas and vehicles elsewhere suffered damage.

Truckers said the protest had been planned as a peaceful demonstration in which drivers would not block traffic but instead park their rigs at home and refuse to make deliveries. However, police officials in Ohio said Dec. 14 that striking truckers were "forcing" non-striking drivers "off the roadway into the truck stops and once they have them in they won't let them leave." Truck stop blockades were reported in at least 10 states, but some truck stop operators participated voluntarily in the protest.

The strike's effectiveness was difficult to gauge. Government reports showed few effects but other observers in the East and Midwest reported extensive delays in truck deliveries and reduced highway traffic. Businesses dependent on truck transportation, such as produce and livestock shippers and assembly plants, said operations were below normal because of the delivery slowdown.

The truckers' shutdown had been initiated by Michael Parkhurst, a former driver and editor of Overdrive Magazine, an independent, anti-Teamsters publication published in Los Angeles. The strike spread, as in previous protests, by an informal communications network of citizen band radios and clear channel stations.

Teamster President Frank Fitzsimmons, who claimed to represent the grievances of all truckers while opposing the non-union drivers' strike, met Dec. 7 and 8 with William Simon and other Administration energy officials to seek increased fuel allotments for the industry. He also met briefly Dec. 8 with President Nixon but no progress was reported in heading off the planned strike.

Parkhurst and other independent drivers also met Dec. 7 with Transportation Secretary Claude Brinegar and George M. Stafford, chairman of the Interstate Commerce Commission. Stafford announced that ICC procedures would be streamlined to allow truckers and railroads to raise their rates to reflect higher fuel costs 10 days after they filed an application with the commission.

Parkhurst said he was "dissatisfied" with the concession because "desperate" conditions required stronger actions.

Distillate fuel price hike OKd. The Cost of Living Council (CLC) said Dec. 5 that refiners of middle distillate fuel would be allowed a 2¢ a gallon increase as an incentive to increase production of home heating oil.

It was estimated that by January 1974, consumers would be paying 7% more for fuel to heat their homes as a result of the CLC decision. The 15.4% price increase also affected users of diesel and jet fuel and kerosene.

Further price increases designed with the double purpose of encouraging greater output of certain fuels and discouraging fuel consumption in other areas could be expected, CLC officials warned.

In a parallel move, the CLC announced a 1¢ a gallon rollback in gasoline prices paid to refiners. (Motorists were not expected to benefit from the wholesale price adjustment because increases in crude oil prices passed along to consumers would offset any drop in gasoline costs.)

John T. Dunlop, director of the CLC, said the dual aim of the price reduction was to prevent refiners from earning "very substantial windfall profits" and to remove their incentives for producing large supplies of gasoline. It was hoped that crude oil supplies and manpower used to refine gasoline would be diverted to increase production of home heating oil, Dunlop said, because the nation's 250 refineries currently were operating "substantially below the physical limits of their capacity to produce distillate fuels.

According to the CLC, the adjustments in its price control mechanism would bring about a 14% reduction in gasoline output and expand distillate fuel supplies 9% to nearly 1.5 million barrels a day. Current gasoline production represented 45% of the total refinery output. Production of distillate fuels represented 25% of total refinery operations.

Allocation plans proposed. Energy Office Director William E. Simon outlined broad fuel allocation plans Dec. 12. He emphasized that the aim of the proposals, which were scheduled to take effect Dec. 27, was not to ration fuel but "to insure equitable distribution at the wholesale level."

Those with priority rights to gasoline supplies were the Defense Department; essential community services, such as police, firefighters and health care; farms, manufacturing, passenger and freight transportation; mail service; and producers of energy, especially electric utilities. They would be able to obtain at the wholesale level all the fuel required.

"Other business" buyers in bulk (at wholesale) would be entitled to 100% of the gasoline purchased in the corresponding month for the 1972 base period.

All other gasoline customers, including filling stations, would receive the leftover supplies, or "not more than 90% of their 1972 use." Motorists would experience a 15%–25% shortage of gasoline at the pumps (compared with "normal" periods), officials said.

Simon had originally said gasoline producers would be asked to reduce their first quarter output to 75% of production in the same three-month period of 1972. The figure was revised later that day with the corrected cutback set at 5% rather than 25%. (The 5% reduction against 1972 output equaled a 15% reduction compared with the anticipated level of gasoline output in the first quarter of 1974.)

Spokesmen for Simon said the error represented both a "misprint" and "miscalculation" by energy planners.

All refiners, regardless of size, would be required to share available crude oil supplies. The petrochemical industry held priority rights to 120% of their 1972 consumption, although doubts were expressed that the full amount of petroleum byproduct supplies would be available.

Heating oil and diesel fuel consumers were divided into two categories. Public passenger transportation and energy producers were awarded 100% of their current requirements. Health care services and other vital community users were entitled to 110% of 1972 use. Comprising the second category were industry, manufacturing, farming, food processing and cargo and mail hauling, which had rights to 110% of 1972 consumption. All others received awards of 100% of 1972 consumption (Retail heating oil allocations remained unchanged.)

The aviation industry's allocations also were subdivided. Long haul domestic carriers could get 95% (of their 1972 supplies) until Jan. 7, 1974 and 85% thereafter. Small regional carriers would get 95% until Jan. 7 and then 90%. International carriers were awarded 100% until Jan. 7 and 85% thereafter. Commercial and industrial users in the general aviation category were entitled to 90%, business and executive users 80%, personal pleasure and instructional users 70%.

Spokesmen for Simon revealed Dec. 12 that he believed that a special investigative force doing field work was required for his office. Investigators would have enforcement authority for the allocation program as well as powers to make compliance checks on federal price ceilings, a function currently performed by the Internal Revenue Service.

The Federal Energy Office was given authority to set petroleum price controls by President Nixon, effective Dec. 17. The executive order, issued Dec. 12, superseded the Cost of Living Council's price setting authority for petroleum products.

Fuel saving reported—President Nixon appeared briefly at Simon's news conference Dec. 13 to express gratitude for the national cooperation which had resulted in a significant 7.8% reduction in petroleum demand during November. Nixon also announced a 15% drop in gasoline consumption in the previous week.

The appearance was noted as part of a White House effort to depict the President as actively involved in energy policy deliberations in order to counter the public image of Administration disarray that emerged from the Love-Simon dispute.

The cutback in gasoline demand, as well as the 1.1 million barrel a day drop in petroleum usage in November (compared with anticipated demand) resulted from the Administration's energy conservation efforts and mild weather, officials said. (An 11.3% reduction in petroleum demand was reported for the last week in November.)

Simon also announced further energy saving steps at his news conference:

■ A mandatory cutback in electric lighting in commercial and industrial facilities. Initially, the action would be voluntary, but it would be given statutory authority with enactment of the emergency energy bill.

■ Simon tentatively proposed elimination of lights on major highways and freeways, with the exception of interchanges and ramps.

■ He ordered government agencies to allot priority parking spaces to car pools and directed federal agencies to cut automobile travel 20%. Official limousine use was also severely restricted.

■ Simon announced that a new licensing system to "monitor" (rather than halt) petroleum exports had been developed.

■ Portable electric space heaters were banned in government offices. Simon appealed to the public to restrict their use.

■ He asked individual industries to establish energy audit committees to develop specific conservation techniques.

Another controversial step aimed at conserving fuel was announced Dec. 11 by Energy Office Director Simon. He banned outdoor ornamental lighting for private residences and commercial establishments during the holiday period. Indoor lighting was not restricted. (The ban would be voluntary until Congress passed the emergency energy bill requested by President Nixon, Simon said.)

Economic dislocations feared. Herbert Stein, chairman of the President's Council of Economic Advisers, told the Joint Economic Committee of Congress Dec. 11 that the energy shortage would raise the 1974 inflation rate 3% above the expected increase of 4.5%–5% in the cost of living. Stein's figures were based on an estimated 50% boost in fuel prices during 1974.

Despite his warning, Stein continued to support the Administration position that price increases for petroleum products were needed to stimulate exploration that could result in a long-term solution to the nation's declining supplies of domestic fuel.

The energy crisis already had caused major economic dislocations in three industries. Affected were truckers and employes of the nation's airlines and auto makers.

Paul R. Ignatius, president of the Air Transport Association (ATA), said Dec. 10 that because of the Administration's cutback in jet fuel allocation scheduled to take effect Jan. 7, 1974, the industry would be forced to make 1,600 additional cuts in daily flights, ground 275 aircraft and lay off 25,000 employes.

Since Nov. 1, 1,000 daily flights had been cut and more than 2,500 workers laid off or given furlough notices, the ATA said.

Curtailment plans reported by individual airlines included Pan American's announcement Dec. 10 that it would seek Civil Aeronautics Board (CAB) approval to suspend all passenger and most cargo service on international flights to two cities in the U.S. and 10 cities in Europe and Latin America.

Other routes would receive less frequent service, according to Pan Am's plan to reduce its 1974 fuel requirements from the 1.28 billion gallons planned for use to 974 million gallons.

A Trans World Airlines spokesman said Dec. 9 that 350 more pilots would be furloughed in the first half of 1974. The layoffs were in addition to the 100 pilots given furlough notices in October.

"Selective" cancellations in about 250 flights were announced for United Air Lines' early December schedules to conserve fuel during peak holiday periods, spokesmen said Dec. 6.

Chrysler Corp. announced Dec. 12 that it was extending the plant shutdown planned for Jan. 7–11, 1974 to two more facilities. More than 2,000 employes would remain on extended layoff after the plants reopened because daily production rates would be reduced.

The fuel shortage also was causing schools and colleges to reschedule holidays so heat could be conserved over a long Christmas recess, the Washington Post reported Dec. 12. Shutdown of three days to six weeks were being planned in schools from Maine to Iowa, according to the Post.

The New England area was especially hard hit by the fuel shortage. The New England Power Pool announced Dec. 11 that utility companies in the region would receive 9 million barrels (or 24%) less oil than required for the winter.

Auto, aviation industries affected. There were continued economic repercussions from the energy crisis in the nation's automaking and aviation industries.

In the largest single layoff yet announced by the industry, General Motors Corp. (GM) officials said Dec. 28 that 86,-000 workers would be laid off at 10 plants across the country because of a sales drop in medium sized cars.

The cutbacks, which would start in January and February, affected 48,000 hourly and salaried workers scheduled for temporary furloughs lasting up to 10 days, and 38,000 workers (6% of GM's total work force) laid off indefinitely because of job elimination.

In announcing the cutbacks, GM Chairman Richard Gerstenberg said 1974 car and truck sales were expected to run 8%–12% below levels set in 1973. According to the Wall Street Journal Dec. 31, GM's new and extended plans to slash production would result in a 15% cutback

in manufacture of large cars during the first quarter of 1974 compared with production goals set earlier, or a 20% reduction when compared with 1973 output.

Airlines also planned extensive layoffs and further curtailments in flight schedules. Among the announcements:

Eastern Airlines said Dec. 13 that 3,760 employes including 440 pilots would be laid off in January, bringing the total personnel cutbacks to more than 4,000 (and a total of 800 pilots); Allegheny Airlines announced Dec. 13 a 30% reduction in daily flights that officials said Dec. 17 would be carried out by laying off 1,288 workers, or 15% of the carrier's work force, effective Jan. 7, 1974; Pan Am officials Dec. 13 confirmed reports that 1,000 flight attendants, nearly 20% of the total employed, would be furloughed after Jan. 10, 1974 to bring about a 25% scheduling reduction during 1974; the Pan Am layoffs were supplemented Dec. 20 by reports that 175 management and nonunion personnel from the New York headquarters also would be furloughed in January; Ozark Airlines announced Dec. 20 that about 300 employes, 12% of the work force, would be furloughed as January departures were cut 23% (following a 7% cutback in daily departures during December).

International airlines using bonded fuel unaffected by the U.S. allocation program were especially hard hit by the jet fuel shortage; British Overseas Airways Corp. (BOAC) was forced to cancel nearly every flight from U.S. cities, with the exception of New York, during the last week of December, because supplies of jet fuel in London were nearly exhausted, officials said Dec. 13. Other airlines flying the London route also reported severe shortages.

The Federal Energy Office became embroiled in a dispute with the Pentagon over the Administration's announcement Dec. 19 that 1.5 million barrels of jet fuel would be diverted from military use to civilian airlines.

FEO Director William Simon reaffirmed the decision Dec. 23 despite a joint Pentagon and FEO press release the previous day stating that as a compromise, the amount diverted to civilian use would be reduced to 900,000 barrels.

The Pentagon contended that national security could be endangered if its petroleum supplies were tampered with. In a letter to Simon released Dec. 22, Defense Secretary James R. Schlesinger warned that war reserves of fuel already were 20% below desired levels and that by January, the deficit would total 30%. FEO officials claimed that the diversion of military fuel would result only in a 5% fuel loss by the Pentagon.

Pentagon concern was generated in part by the destruction of South Vietnam's major petroleum depot in a Viet Cong rocket attack Dec. 3 and the Defense Department's announcement Dec. 12 that U.S. war reserves would be used to supply South Vietnamese and Cambodian troops with their daily "minimum military requirements."

The 23,500 barrels drawn on daily would indirectly deplete civilian fuel supplies because it was planned that war reserves would be replaced by oil commandeered from domestic supplies.

The Senate Dec. 13 reacted to the Defense Department's announcement by voting to cut off funds for fuel shipments by the Pentagon to Indochina. The amendment was retained in the bill's final version, passed Dec. 20 by both houses of Congress, which set Pentagon appropriations for fiscal 1974 at $73.7 billion.

The House Dec. 15 also included a ban on Pentagon petroleum shipments destined for military use in South Vietnam, Cambodia and Laos, voting 201–172 in favor of the amendment attached to the emergency energy bill. A similar ban on oil shipments to Israel was defeated at the same session, but no action was taken on the legislation in final form.

In other developments, Trans World Airlines officials acknowledged Dec. 19 that the company was training Saudi Arabian commercial pilots in domestic jets that burn 1,800 gallons of fuel an hour, despite the Arab oil embargo which was spearheaded by Saudi Arabia. The training program had been in effect since 1946, officials said, adding that there were no plans to discontinue it.

The Civil Aeronautics Board (CAB) Dec. 26 approved a revised fare increase package for North Atlantic air routes. The new proposals by the International Air Transport Association, increased the popular excursion fare by up to 12% and dropped the youth fare, effective Jan. 1, 1974. The CAB had acted Dec. 23 in passing a revised fare package for north and central Pacific routes, raising passenger rates an estimated 3%.

The increases would take effect in addition to a 4%–6% increase in international passenger and cargo fares approved by the CAB Dec. 20 because of higher fuel costs incurred by the airlines.

Exports of oil products prompt debate. The nation's continued export of petroleum products during a period of worsening fuel shortages and the Arab oil embargo prompted debate and public scrutiny of charges that oil companies were profiteering with their shipments abroad as well as sending refined petroleum products to the Middle East countries responsible for the embargo.

According to the New York Times Dec. 8, the U.S. had exported 236,000 barrels a day, or 1.4% of total U.S. consumption and 2.5% of domestic production, in the 10 months ending Oct. 31. During October, however, the rate had increased to 242,000 barrels a day and the export of fuels in shortest supply at home—gasoline, distillates and residual, were increasing. (Newspaper reports of Dec. 6–8 were based on Commerce Department data.)

According to the Commerce Department, gasoline exports for the first 10 months of 1973 were 68% higher than in the same period of 1972; distillate fuel shipments soared 180%; crude climbed 7% and residual exports were up slightly. Seven billion cubic feet of natural gas were being exported monthly, most of it under long term contract to Japan.

Rep. Les Aspin (D, Wis.), whose discovery of the rise in petroleum product exports had prompted the Commerce Department to investigate the charges, was critical of the shipments. Even if only 1.4% of domestic supplies were siphoned off for profit abroad, the percentage was significant when compared with the 17% fuel deficit expected during the winter, Aspin declared.

Aspin also raised the profiteering charges. According to Aspin, Brazil bought 246,000 barrels of gasoline in October at 28¢ a gallon when the domestic price was about 20¢ a gallon. Other large sales were reported in Great Britain and the Netherlands, the European countries hardest hit by the Arab action.

Most of the petroleum products being sold abroad—4 million barrels of a total 7.5 million barrels—were lubricating oils and other greases. These products were not in short domestic supply, but critics pointed out that they were derived from the same crude oil currently under embargo.

According to the Wall Street Journal Dec. 6, 7,000 barrels a month in automotive, diesel and marine engine lubricating oil were delivered to Saudi Arabia, the leading Arab oil producer, during the first 10 months of 1973. Kuwait received 1,400 barrels a month, and other Arab countries also received supplies, the newspaper reported.

Alaska pipeline approved. Final Congressional approval of a bill authorizing construction of the trans-Alaska oil pipeline came on a House vote of 361–14 Nov. 12 and a Senate vote of 80–5 Nov. 13. President Nixon signed the measure Nov. 16.

The bill which emerged from conference committee retained a section barring court review of the environmental impact of the project. Any challenge to the constitutionality of the provision would have to be brought within 60 days of enactment and be tried ahead of other pending matters in federal district court. A district court decision would be directly appealable to the Supreme Court.

Nixon's approval of the final bill had been in doubt because of the inclusion of three provisions not related to the pipeline.

One allowed regulatory agencies to obtain financial data from corporations without approval from the Office of Management and Budget (OMB).

The second allowed the Federal Trade Commission (FTC) to seek injunctions against unfair trade practices and represent itself in court if the Justice Department failed to act on an FTC request within 10 days.

The third required Senate approval of the heads of the Energy Policy Office and the Interior Department's Mining Enforcement and Safety Administration. OMB Director Roy L. Ash had said he

would recommend a veto of the bill if the extraneous provisions remained.

Congress acts on legislation. The Senate approved a bill Nov. 14 requiring the Nixon Administration to include crude oil and gasoline under the existing provisions of the government's mandatory fuel allocation system. The vote was 83–3. The House had passed a final version of the measure Nov. 13 by 348–46 vote. The bill, the Emergency Petroleum Allocation Act of 1973, was signed by President Nixon Nov. 27.

The bill also extended the government's power to regulate the price of petroleum products until March 1975, nearly a year after expiration of the Economic Stabilization Act.

The Administration had not supported the bill, which forced the President to utilize an existing bill giving him authority to allocate crude oil and refined products. However, Nixon's signature of the bill was expected.

Daylight savings approved. Both houses of Congress approved by voice votes Dec. 14 a conference committee version of a bill to put most of the nation on year-round daylight savings time for two years. President Nixon signed the bill Dec. 15, making it effective Jan. 6, 1974.

Alaska and Hawaii would be exempt from the requirement. The final bill included Senate provisions allowing states split by time zone boundaries to have uniform time throughout the state and allowing the President to exempt a state if the governor certified—before the effective date—that the requirement would cause undue hardship.

The measure, which would put the nation on year-round daylight savings for the first time since World War II, could save an estimated 95,000 barrels of fuel a day, or 3% of the U.S.' estimated fuel shortage, according to supporters of the bill.

'Windfall' profits tax proposed. President Nixon made a brief appearance before reporters Dec. 19 to announce that the White House would propose legislation to levy an "emergency windfall profits tax" on the oil industry. The measure was actually a misnomer because profits—revenues minus costs—would not be taxed. Instead, an excise-type levy would be placed on crude oil prices that rose above a certain base.

The proposal was designed to offset excess profits accruing to producers as a result of the scarcity of crude oil and the recent enormous increases in the price of petroleum products without diminishing the industry's incentive to spend money on exploration for new oil reserves.

The proposal also had a "fadeaway" provision in which the tax would decrease gradually and then die, probably within five years—a reflection of the Administration's belief that rising oil prices would have reached equilibrium by that time and profits would no longer be considered excessive.

In a parallel announcement Dec. 19, the Cost of Living Council (CLC) authorized an immediate $1 a barrel increase in the price of crude, a 23% boost over the current price of $4.25 a barrel. Retail prices were expected to rise by 2.3¢ a gallon for both gasoline, whose average national price was 44.6¢ a gallon, and home heating oil, which cost 30.7¢ a gallon on the national average, CLC spokesmen said.

The full effect of the increase would not be felt until February 1974. Under current CLC rules, producers could institute one-third on the increase Jan. 1, 1974 and the remainder Feb. 1.

Treasury Secretary George P. Shultz conducted the White House news briefing after President Nixon left without answering any questions. According to Shultz, the windfall profits tax would take effect when oil prices topped $4.75 a barrel. The tax would rise progressively, reaching a maximum 85%..

Schultz estimated that producers eventually would have to charge $7 a barrel to balance supply and demand, but he warned that while new supplies of petroleum were being sought in the interim, prices could reach $9 a barrel before conditions stabilized.

The CLC action would apply only to "old" oil—equal to the volume the producer pumped in 1972. (Old oil represented about 75% of current production.) The price of "new" oil, or the additional output, was not regulated by government price control. New oil was selling for $6.17 a barrel ($7.17 with the CLC-approved increase). The price of imported crude was about $6 at the end of November.

The CLC's $1 increase in crude prices would yield a 9.5¢ tax if the proposal were law. Shultz said that no decision had been reached on how to allocate the tax revenues, which could total $5 billion during the first year of operation.

Congress delays legislation. Congress adjourned Dec. 22 without resolving major differences in the emergency energy bill that was reported out of conference committee early Dec. 21. The principal stumbling block to its passage was a windfall profits provision opposed by the Administration and oil state senators.

The measure, which would have required price rollbacks if it could be proven that a seller in any sector of the petroleum industry had made windfall profits because of the energy crisis, was deleted by the Senate Dec. 21 after several senators led a filibuster against the provision.

Stripped of that section, the Senate passed the modified bill but the House rejected the compromise version 219–34. House members were angered by the Senate's capitulation to the few senators who staged the filibuster with the support of emissaries from the FEO who had established a "command post" in Vice President Gerald Ford's Congressional office to direct the opposition to the windfall profits provision. (The Administration supported a tax on windfall profits only in the area of crude oil.)

Congress' failure to act on the legislation left President Nixon without statutory authority to relax clean standards, order the conversion of public utilities from oil to coal, ban Sunday gasoline sales, limit gasoline purchases to 10 gallons per motorist or take other important emergency steps during the fuel shortage.

Despite the impasse, President Nixon issued a conciliatory statement Dec. 22 regretting the Congressional delay but professing to understand the difficulty in reaching agreement on the complex legislation during the rush to adjourn.

Action was pending on these energy-related issues:

Auto emission standards—The Senate Dec. 17 passed a bill (85–0) delaying compliance with auto emission standards for hydrocarbons and carbon monoxide until the 1977 model year. House-Senate conferees had included an amendment with similar provisions in the emergency energy bill.

Energy research and development—By a vote of 82–0 Dec. 7 the Senate had approved a bill to begin a 10-year, $20 billion research and development program for non-nuclear energy sources. The Administration had proposed a five-year, $10 billion program, more than half devoted to nuclear energy.

The bill called for immediate government-industry research into production of low-sulfur industrial fuels and synthetic fuel from such sources as agricultural products and waste materials, development of geothermal and oil shale resources, improvement of environmental control systems and investigation of fuel cells for power generation.

To administer the programs, the bill proposed an independent chairman appointed by the President and representatives from various federal agencies. The Administration, according to a spokesman, objected to this setup and preferred flexible allocations of funds rather than the bill's specific proposals for each area.

The bill went to the House, where another measure awaiting action was reportedly more in line with Administration proposals.

Energy conservation—The Senate approved a bill Dec. 10 but no hearings had been held in the House on several related measures.

The broad Senate measure included among its major provisions a requirement that auto manufacturers increase gasoline mileage to 20 miles per gallon by 1984. The bill also gave the Transportation Department authority to begin enforcing unspecified minimum mileage standards by 1978. (The current average was reported to be 13.5 miles per gallon.)

Among other provisions were requirements that the Transportation Department develop an efficient, low-polluting auto within four years; that all government agencies adopt regulations cutting energy use in their operations; that the government study means of achieving a 50% improvement in fuel economy for all types of transportation; and that the Federal Power Commission conduct annual studies of utility rate structures and possible conservation tax incentives.

Speed limit lowered—Congress passed legislation Dec. 22 that would lower highway speed limits to a uniform 55 m.p.h. by authorizing the government to withhold federal highway funds from any state refusing to enact the lower limit 60 days after signature by the President.

President Nixon's request for a dual speed limit for trucks and cars was rejected.

Rationing program unveiled. The Administration's Federal Energy Office (FEO) issued a barrage of rulings in the last days of December, including a standby rationing program announced Dec. 27 by FEO Director William E. Simon.

Under the plan, licensed drivers over 18 would be entitled to average quotas of 32–35 gallons of gasoline a month. Those in densely populated areas well-served by mass transit facilities would receive "marginally" smaller allotments than those living in areas without adequate alternative mass transit systems.

The Bureau of Printing and Engraving was printing rationing coupons that could be obtained at limited local distribution points by motorists who received their authorization cards in the mail. (If needed, the plan could be implemented no earlier than March 1, 1974, Simon said.)

Coupons could be sold legally at uncontrolled prices but the coupon's gallon value would change as the federal government re-evaluated the gasoline shortfall in terms of the rationing, allocation and voluntary conservation measures in effect. Coupons would be valid for 60 days.

Gasoline stations could sell fuel only if valid coupons were presented, but those persons or groups which had made bulk purchases of gasoline in the past would receive supplies under a mandatory allocation system.

The cost of the program was estimated at $1.4 billion and would be financed by users fees. It was expected that each driver would pay about $1 a month for coupons.

Leakage, conservation efforts—Simon continued to insist that rationing was not necessary if voluntary cooperation measures were heeded. The ban on Sunday gasoline sales and cutbacks ordered under the fuel allocation system already had resulted in a 1% drop in the consumption of residual fuel oil and an 8% decline in gasoline demand, Simon said Dec. 27.

Leakage from most Arab countries engaged in the oil embargo also was a significant factor in reducing the petroleum shortfall. Simon refused to identify those nations allowing a diversion of petroleum supplies to the U.S., but according to the Washington Post Dec. 27, most Arab countries, with the exception of Saudi Arabia and Kuwait, were permitting leakage because of their need for revenue to finance the Soviet sale of arms to Egypt.

According to Simon, leakage totaled 400,000 barrels a day in the week ending Dec. 14. Consequently, overall gasoline supplies would be 13.6% below normal demand instead of 16.3% short as previously estimated. Although the U.S. required 6.8 million barrels a day, daily imports were ranging from 5.9 million barrels to 6.2 million barrels in contrast to the expected level of 4.6 million barrels.

Simon Dec. 20 asked motorists to extent their conservation efforts by imposing voluntary limits on gasoline of 10 gallons a week.

Motorists feel gasoline pinch—Simon said Dec. 27 that the Administration might be forced to implement its present standby rationing program if motorists "begin queuing up at gas stations for three and four hours at a time. I don't think we can tolerate that."

Conditions approaching intolerable limits were reached over the Christmas and New Year holidays when drivers were faced with a situation in which most stations were closed either because of the Sunday sales ban or because they had exhausted their monthly gasoline deliveries.

Most stations that were open enforced the 10 gallon sales limit while others were guilty of price gouging, which in some cases, totaled 99.9¢ a gallon.

The Justice Department announced Dec. 28 that restraining orders would be sought against stations overcharging motorists and Simon's office said Dec. 30 that 1,000 Internal Revenue Service agents were being mobilized to enforce the Administration's energy regulations.

Another price increase permitted—Officials in the Federal Energy Office Dec. 31 predicted 10¢ a gallon increases in the prices of gasoline, heating oil and diesel fuel in the first two months of 1974 as the result of a directive issued that day from Simon authorizing distributors of those fuels to cover higher operating costs by raising prices.

Wholesalers could lift prices .5¢ a gallon and retailers 1¢ a gallon but consumers would feel the full 1.5¢ a gallon boost.

The CLC Dec. 26 had delegated authority to administer certain price provisions and certain changes required in them to the FEO, retroactive to Dec. 21.

The Administration's stated purpose in allowing price increases, particularly to refiners, was to provide incentive to increase their exploratory development capacities. In line with that offer, major oil companies announced expanded capital spending programs for 1974—Exxon Corp. planned a 73% increase in its capital budget, officials said Dec. 20. Exxon's announcement that it planned to spend $6.1 billion during the next year (a few long term expenditures were included in the figure) followed similar announcements by Texaco and Gulf setting budgets at $1.8 billion and $2 billion.

Other FEO rulings—Final petroleum allocation rules were adopted Dec. 30 for five areas of the allocation program. The petrochemical industry was assured 100% of its current needs instead of the original proposal promising 120% of feedstocks used in the 1972 base period. Feedstock users also were allowed to pay producers 15% more than the ceiling price set by the Cost of Living Council.

Refiners' production of gasoline in the first quarter of 1974 was reduced from 95% of output in the first three months of 1972 to 89%. As a result, motorists could expect 20% fewer supplies of gasoline than would have existed if there were no Arab oil embargo, Simon warned.

Any "refiner, importer, wholesale purchaser or user" was prohibited from accumulating petroleum stockpiles that exceeded "customary inventories" required by " normal business practices."

Certain general aviation users of jet fuel received increased allotments. Simon had said Dec. 27 that major aviation users could also expect more generous allotments. Allocation to trunk airlines were up from 85% to 95% of fuel use in the 1972 base period; regional carriers and air taxi services were increased from 90% to 100%.

Final rulings on allocations of motor gasoline, petroleum middle distillates, diesel fuel and propane had not been issued yet, but all rules were expected to take effect Jan. 15, 1974. By law, the entire allocation program was scheduled to become effective Dec. 27. Because of the complexities involved in devising final rules, Simon had appealed to Congressional leaders Dec. 21 for authority to order implementation of the system on a staggered basis from Jan. 11–26, 1974, but his request had been turned down.

Sen. Henry M. Jackson (D, Wash.) and Rep. Harley O. Staggers (D, W. Va.), chairmen of the Congressional committees overseeing the Emergency Petroleum Allocation Act, said they had no authority to grant a delay in the law.

Canada

Oil shortages seen. Energy Minister Donald Macdonald said Oct. 18 that no significant quantities of Alberta oil could be transported to Quebec and the Atlantic provinces if the cutback in Arab oil production curtailed imports to that area.

Macdonald said if the Arab nations went through with their announced cutback and applied it to Canada, rationing would be necessary within two months. He explained that, although Canada's exports of 1 million barrels of crude oil a day exceeded imports of 800,-000 barrels a day, no pipeline existed to carry Alberta crude farther east than Ontario, and truck, rail and water transport was inadequate to supply any

serious deficit. The eastern provinces had about 45 days' supplies of crude in stock.

Macdonald said only a quarter of daily imports came from the Middle East, but Venezuelan deliveries might be curtailed if the U.S. turned to that country to replace Arab oil.

New ship, pipe routes—The federal government was planning an emergency shipping route to bring Alberta oil to the East, while a partially used pipeline was being reactivated to connect Ottawa with the Alberta supply grid, in moves against expected shortages.

A government plan was reported Nov. 14 to ship up to 55,000 barrels a day of Alberta oil to Montreal and the Maritimes via the Trans Mountain Pipe Line, the Strait of Juan de Fuca, and the Panama Canal after the Great Lakes freeze. The government had protested U.S. plans to move Alaska oil in tankers through the strait, for fear of environmental damage from oil spills.

The newly reactivated Ontario pipeline, it was reported Nov. 10, would bring Ottawa up to 60,000 barrels of oil products from refineries in Toronto.

National oil firm asked—Conservative and New Democratic members of Parliament called Nov. 5 for creation of a national petroleum company, whose first task would be negotiating with Venezuela for increased oil deliveries.

Emergency oil purchase. The government announced Nov. 9 it had purchased about one million barrels of heating oil in the Caribbean and Rumania, as "insurance" against supply disruptions or an unusually severe winter.

The government was expected to pay 60¢–65¢ a gallon for the oil, double the price of domestic Canadian oil, but the price was expected to be averaged into that of existing stocks at the retail level.

Trudeau: no U.S. oil cutoff. Prime Minister Pierre Elliot Trudeau said Nov. 13 that Canada would not cut off petroleum exports to the U.S., even if a cutoff were requested by Arab nations.

In a diplomatic note to the U.S. that day, Canada said it would continue to export oil and refined products to the U.S. if they were not required in Canada. Energy Minister Donald Macdonald had said Nov. 5 that Canada might bar some refined product exports if requested by the Arabs. Most of Canada's exports were of crude oil.

Exports to U.S. cut—The National Energy Board announced Nov. 29 it was cutting the quota of oil exports to the U.S. for December to 977,000 barrels a day, down 5% from November deliveries.

James Bay injunction reversed. A three-man Quebec Court of Appeal panel Nov. 22 suspended an injunction that would have barred further work on the James Bay hydroelectric project.

The court said the interests of the Quebec public outweighed the interests of "approximately 2,000 of its inhabitants" (the local Indian and Eskimo population). Further hearings on the merits of the case were scheduled.

Federal Indian Affairs Minister Jean Chretien said Nov. 15 the federal government had provided $505,000 in research and legal expenses to the Indians. The provincial government had backed the project, which was to provide Quebec with 8.3 million kilowatts of additional power capacity by 1980.

Fuel consumption cut asked. Energy Minister Donald Mcdonald revealed the government's fuel conservation plans Nov. 26, including requests for a voluntary consumption cut of 15% by consumers. The plan had been previewed in a Nov. 22 television address by Prime Minister Pierre Trudeau.

A mandatory fuel allocation program would go into effect at the wholesale level early in 1974, directed by a board with power to impose rationing if necessary, if Parliament approved enabling legislation. Macdonald asked homeowners and businesses to reduce temperatures to 68–70 degrees during the day and 63–65 degrees at night, which, combined with other efficiencies, could reduce heating oil consumption by 10%–15%. Motorists were asked to drive more slowly and reduce auto use, to reduce gasoline consumption by 15%–25%.

The government would cut its own fuel use by 18%, Macdonald said. If all the economies were enforced, there would be no need to slow economic growth or reduce industrial output, he said.

Trudeau told Parliament Nov. 23 that the government would use public funds to secure oil supplies in shortage areas. He said the government would finance, if necessary, the extension of the Alberta pipeline system to Montreal.

The government's bill to create an energy allocation board, introduced in Parliament Dec. 3, would give the board power to ration natural gas, coal, electricity, and petroleum products such as plastics, in addition to oil. The board could order ships or trains to transport fuel, and could mandate construction by pipeline companies of branch lines.

Oil supply contracts altered. The multinational oil companies that supplied oil imports to eastern Canada were invoking force majeure clauses in supply contracts to reduce shipments on a prorated basis, causing the government to nearly double its estimate of the prospective oil shortage to 200,000 barrels a day, it was reported Nov. 28.

Energy Minister Donald Macdonald said the development, a result of the Arab oil boycott of the U.S., showed the need for a national petroleum company in Canada. He named Creole Petroleum of Venezuela, owned by Exxon Corp., and Aramco of Saudi Arabia, owned by Exxon, Texaco, Standard Oil of California and Mobil, as the suppliers involved.

Price developments—Macdonald said Canadian oil companies would have to absorb the cost of a new Venezuela oil price increase, it was reported Dec. 3. The government had been criticized for its Nov. 26 announcement that gasoline and heating oil would cost 4¢–5¢ more east of the Ottawa Valley as of Dec. 1, because of earlier international price boosts. Venezuela's new increase, announced Dec. 1, would raise prices by an average 7.6%.

Imperial Oil Ltd. of Toronto announced Dec. 4 that it was reducing the wholesale price of gasoline and diesel fuel by 1.4¢ a gallon east of the Ottawa Valley and .8¢ a gallon to the west. The reductions were made possible by a recalculation of federal sales taxes. Similar reductions were reported Dec. 7 by Texaco Canada Ltd. and Sun Oil Co. Ltd.

Fuel policy backed in confidence vote. Prime Minister Pierre Trudeau won a vote of confidence on his energy policies in the House of Commons Dec. 10 with the full support of the minority New Democratic party.

The 135–117 victory on the no-confidence motion, brought by Conservative leader Robert Stanfield, was assured by Trudeau's concessions to New Democratic demands in a major fuel policy speech to Commons Dec. 6. Trudeau announced that the price freeze on Western crude oil would not be lifted in January 1974 as previously hinted, but retained through the end of the winter. He pledged no further increase in home heating oil prices in eastern Canada.

However, Trudeau said prices for all energy sources were bound to rise "to a price high enough to allow development of the oilsands and other Canadian resources." The government planned to spend $40 million over the next five years on developing processes to extract oil from Alberta sands. (The Alberta government said Dec. 6 that it would be an equal partner with four private companies in a $1 billion oil sands project.)

Trudeau outlined plans for a government-owned National Petroleum Company, to be officially proposed to Parliament in January 1974. The company could explore for oil and natural gas, conduct research and development and invest in direct or joint venture oil projects. Eventually, Trudeau said, the company might enter the refining and distribution fields.

The chief goal of the government's energy policy was creation of a unified, self-sufficient Canadian market by 1980, with abandonment of the Ottawa Valley dividing line between domestic and imported oil consumption, in existence for 12 years. Toward that end, the National Energy Board had been instructed "to proceed with all possible haste" in plans for a major new pipeline to Montreal from the West.

Alberta's claims to greater revenues from the oil export tax would be discussed in a January 1974 federal-provincial

conference, Trudeau said, but he insisted that "the whole country should take benefit from any windfall profits." (The National Energy Board increased the tax from $1.90 to $2.20 a barrel Dec. 4, effective Jan. 1, 1974.)

The government planned "to facilitate early construction" on a Mackenzie Valley natural gas pipeline, Trudeau said, but the plan would probably be abandoned if the U.S. decided to ship North Slope Alaska gas directly across Alaska.

Trudeau said oil exports to the U.S. would continue "for some years yet," and said the U.S. would be consulted on energy plans to protect "the beneficial ties that have developed."

Other developments. Federal Energy Minister Donald Macdonald said Dec. 13 that the entire 50% federal share of the oil export tax would be invested in oil-producing provinces, probably through the proposed National Petroleum Corp., although some of the money would be used to develop other forms of energy, including atomic power.

Macdonald announced Dec. 27 that the export tax on crude oil would increase from $2.20 to $6.50 a barrel Feb. 1, 1974 to meet increased world prices for crude.

The Alberta government introduced legislation Dec. 6 to set up a provincial petroleum marketing commission. The proposed body would set the wellhead price and provincial royalty levels for crude oil, and would distribute most of Alberta's current daily production of 1.7 million barrels a day.

The Saskatchewan government introduced a bill in the provincial legislature to take over legal title to about 90% of the oil and natural gas reserves in the province, it was reported Dec. 12. The bill, in a challenge to the federal government's export tax, would set the wellhead price of crude oil and would tax oil companies at a rate of 100% of the difference between the wellhead price and world market prices. Saskatchewan produced about 210,000 barrels of crude oil a day.

Canada and Denmark reached final agreement on division of jurisidiction over the continental shelf between Greenland and Canadian Arctic islands, it was reported Dec. 15. Some experts said about

15% of Canada's oil and natural gas reserves lay within the area allotted to Canada.

Saudi Arabia was reported Dec. 20 to have placed Canada on its oil embargo list. Only one delivery of Saudi oil had been made since the embargo on pro-Israel nations was declared in October. Macdonald said Dec. 20 that all parts of Canada except the Prairie provinces had begun to experience oil shortages.

Other Areas

Brandt, Pompidou confer. West German Chancellor Willy Brandt and French President Georges Pompidou conferred in Paris Nov. 26–27 on the oil crisis, European defense and other issues facing the European Economic Community (EEC).

Pompidou agreed to a proposal by Brandt that a forthcoming meeting of EEC foreign ministers be turned into an "energy council" that would start work on a unified European policy to deal with the oil crisis. The two leaders agreed on the need for the EEC to move more rapidly toward defining a long-term energy policy that would focus on nuclear energy development.

But the talks ended without any major agreement on the oil problem, particularly Brandt's appeal for EEC solidarity to help member nations with oil supplies. Brandt, in a toast at a dinner given in his honor by Pompidou Nov. 26, had said, "None of us has the right to leave another alone with its troubles. If we accepted the weakening of one country, we would really be weakening the community itself."

EEC ministers' energy views. Finance ministers of the European Economic Community (EEC), at the end of a two-day meeting in Brussels Dec. 4, approved a modest new voluntary anti-inflation program.

Dutch Finance Minister Willem Duisenberg agreed to the plan after expressing reservations that any anti-inflation plan that ignored the energy crisis would be useless. He warned that his

government would give priority to preserving jobs over fighting inflation as the oil shortages cut industrial activity. His warning came as the nine EEC foreign ministers, also meeting in Brussels, failed to adopt joint energy measures to meet the Arab oil squeeze.

The fuel crisis and consequent fears of recession caused the Dutch guilder to drop Dec. 4 to a low of 35.52 U.S. cents, compared with 36¢ Dec. 3. The Dutch National Bank spent the equivalent of more than $100 million in reserves to halt the guilder's slide.

The danger of the oil shortages for European economies was underlined in a secret report from Henri Simonet, EEC commissioner in charge of energy, reported Nov. 30. The study, according to informed sources, predicted that continued oil shortages would cause a serious recession in Europe. It forecast a 2%–3% reduction in the EEC's combined gross national products, compared with a 4% growth rate projected before the October Middle East war and the Arab use of oil as a weapon. The report also predicted unemployment rates of 4%–5%, compared with the current 2% level and the 2%–3% rate existing since World War II. The report said EEC nations currently had 18%–20% less oil than normal for this time of year.

EEC foreign ministers, at a Council of Ministers meeting Dec. 17–18 failed to agree on the initial steps for creating a community energy policy.

Joint European project shelved—British Aircraft Corp. announced Dec. 28 that the fuel shortage had forced discontinuation of design studies on a four-nation European project to build a 200-seat airliner, dubbed Europlane.

Latin American body formed. Ministers of energy and hydrocarbons from 24 nations of the Western hemisphere held their Third Advisory Meeting in Lima, Peru Oct. 29–Nov. 2 and gave final approval to creation of the Latin American Energy Organization (OLADE).

The organization, to be based in Quito, Ecuador, would work for integration, preservation, coordinated marketing, rational use and defense of all regional energy resources, including oil, coal, and hydroelectric and nuclear power. Some officials reportedly saw OLADE as a first step toward a continental system to control exploitation of these resources by foreign interests.

The ministers also heatedly debated a report by Venezuela on creation of a Latin American organization to finance regional energy development projects. No final decision on the organization was made.

Peru made strong appeals during the meeting for hemispheric solidarity in the defense of energy resources from unfair exploitation by developed countries. Peruvian Mines Minister Gen. Jorge Fernandez Maldonado said Oct. 29 that production and export of petroleum "should not expose developing countries to blackmail, threats or aggression from societies of uncontrollable and irrational energy macroconsumption."

Creation of OLADE seemingly gave new impetus to ARPEL, an existing organization of the state oil companies of eight Latin American nations, according to the Andean Times' Latin America Economic Report Nov. 23. At a meeting in Caracas, Venezuela after the Lima conference, ARPEL voted to form a Latin American oil and gas marketing organization. ARPEL previously had limited itself to making studies and providing a forum for its companies to discuss common problems.

Australia. The Australian federal government rejected a plan to build a $400 million petrochemical complex and petroleum refinery on the northwest coast by the Burmah Oil Co. of Britain and Kanematsu-Gosho Ltd. of Japan, it was reported Nov. 29.

Western Australia Premier John Tonkin said the government had turned down the project because it would have required a large outlay of foreign capital, a practice the government had been trying to curb.

Australia had fuel enough until the year 2000, "with the notable exception of oil," according to a preliminary report issued Dec. 19 by Minerals and Energy Minister Reginald Connor.

Connor said the country's cumulative needs for fossil fuels to the end of the century could amount to 13% of black coal, 18% of brown coal and 75% of natural gas in present known reserve. "Oil requirements greatly exceed the known reserves," he said.

A major new uranium discovery in the Northern Territory was confirmed Dec. 19. Located in the Jabiluka Field, the ore reserves were said to contain 20,000 tons of uranium oxide. An additional 3,300 tons of uranium oxide had been discovered in the same area earlier in 1973.

Ecuador. The Ecuador government Nov. 10 raised posted prices for crude oil by an average of $2.05 a barrel, to $7.30 a barrel for a key grade. It was the seventh time posted prices were raised since Ecuadorean petroleum production began.

In a speech reported Nov. 14, President Guillermo Rodriguez Lara said Ecuador was now earning $1 million daily from oil exports, and that most of the money was earmarked for development.

The government had signed, after long negotiations, a contract with the U.S. oil firm OKC for exploration over 291,150 hectares of land in eastern Ecuador, it was reported Oct. 26.

The government granted rights to construct a $92 million state petroleum refinery to the Japanese firms Sumitomo Shiji Jaisha and Chiyoda Chemical Engineering, it was reported Nov. 21. The refinery, to be completed by 1976, was expected to produce 50,000 barrels a day in its initial stages.

France. International oil companies were reported Nov. 21 to have notified France they would be forced to cut deliveries to the nation by 10%–15%. The French government replied Nov. 22 with a threat to take legal action against the oil companies that diverted French-bound oil to other countries. Jean Charbonnel, French minister for science and industrial development, had disclosed Nov. 21 that France possessed more than a three-month supply of oil stocks and that supplies were coming in normally.

Finance Minister Valery Giscard d'Estaing announced new measures to combat inflation Dec. 5.

Giscard d'Estaing warned: "We are entering difficult times. An economic slowdown and a world oil crisis are looming." In a further reference to the Arab oil squeeze from which the government had previously said France would be exempted, the finance minister said, "The French economy is not isolated, nor can it be isolated. It is linked in thousands of ways first to the economies of our European partners and then to the entire planet."

The French Cabinet then approved a decree Dec. 19 to enable the government to control and allocate all energy resources, petroleum and petrochemical products. The Cabinet also approved appointment of Jean Blancard as delegate-general for energy to handle all energy problems.

Great Britain. *Economic woes*—Beset by continuing economic difficulties, the government acted Nov. 13 to curb inflation, stem the growing balance of payments deficit and conserve fuel supplies threatened by labor disputes.

Home Minister Robert Carr announced a state of emergency Nov. 13 because of labor disputes in the coal and electrical power industries. The declaration would give the government power to regulate distribution of fuels and take over public and private transport. In its first emergency power moves, the government ordered curbs on floodlighting and display advertising, effective Nov. 14, and 10% cuts in power and fuel consumption for the heating of commercial premises.

Carr said another factor in the government's fuel moves was that this was "a time of great uncertainty about the supply of oil," a reference to Arab oil curtailments stemming from the Middle East war. The government had said that Arab oil producers had assured the nation of a continued flow of oil. The fuel cutbacks in Britain, however, were expected to offset recent price increases by the Middle East oil producers and thus prevent a further worsening of the balance of payments deficits caused by higher oil costs.

Great Britain's 270,000 coal miners had banned overtime work Nov. 12 as a pressure tactic in support of their demands for sharp pay increases. The government had made a pay offer that fell far short of the miners' demands.

Brief blackouts and voltage cuts had plagued the nation since Nov. 1 when 18,000 engineers and other skilled workers began refusing to stand by for sudden power breakdowns and imposed a ban on overtime. The engineers were protesting the declining margin between their pay and that of unskilled workers.

Oil deliveries cut by 10%—The government Nov. 19 ordered an immediate 10% reduction in deliveries of all petroleum products.

The government also appealed to the public for a voluntary reduction in Sunday driving and observance of a 50-mile an hour speed limit. It also banned sale of gasoline in portable containers to discourage gasoline hoarding.

The 10% cut in gasoline supplies to service stations would be based on the September-level deliveries, while the base period for the cut on all other oil products would be from a year before.

Announcing the new action, Peter Walker, secretary of state for trade and industry, said Britain still had "substantial stocks of both oil and coal" but needed the conservation measures to insure minimum impact from industrial action by coal miners and electricity workers and from uncertainty about Arab oil production.

The moves followed a unilateral decision Nov. 16 by Esso, the local subsidiary of Exxon, to reduce deliveries to its domestic customers. Esso's action was in line with complaints by the major oil companies that the government was not facing the facts of reduced oil supplies.

The British government announced Nov. 26 it would begin to issue gasoline coupons Nov. 29, to be used if fuel rationing were introduced. It announced Nov. 27 a delay in cuts in the lead content of gasoline, scheduled to begin in January 1974 as an antipollution measure.

Emergency rule extended—The British government extended the national state of emergency for another month Dec. 12, the first day of a slowdown by 29,000 railroad engineers and firemen expected to seriously disrupt the transport of fuels.

Home Secretary Robert Carr told Parliament the extension was required because the railroad slowdown posed "an additional serious threat to the essentials of life." The job action followed the slowdowns and production cutbacks begun by coal miners and electrical power engineers in November.

The rail workers began their work-to-rule and ban on Sunday work and overtime to press demands for wage increases that the government claimed exceeded Stage 3 guidelines of the anti-inflation program.

(Coal production, according to government estimates, had been reduced about one-third and coal stockpiles at power plants were estimated to have declined by about three million tons—to 14.5 million tons—since the start of the job action in early November, the New York Times reported Dec. 13. About 70% of Britain's electricity came from coal-fired generators.)

3-day week set in Britain—Faced with crucial energy shortages resulting from slowdowns by coal, railway and electrical workers and from the Arab oil embargo, Prime Minister Edward Heath announced Dec. 13 a three-day workweek for most industry and businesses, effective Jan. 1, 1974. The government followed this announcement with an emergency budget message Dec. 17 that sharply cut public spending.

Addressing the House of Commons and then a nationwide television audience Dec. 13, Heath said, "In terms of comfort we shall have a harder Christmas than we have known since the war." For the moment, he said, "we shall have to postpone some of the hopes and aims we have set for ourselves for expansion and for our standard of living."

Among the energy conservation measures:

Continuous-process industries, such as steel and auto manufacturing, would be exempted from the three-day workweek, but their normal electricity supply would be cut by 35%. Those industries hit by the three-day order would also be permitted to work a total of only five days between Dec. 17–31. Essential industries, such as food processors and fuel suppliers, and recreational premises would be spared power and workday cuts. Television would be banned after 10:30 p.m., except for Christmas and New Year's Eve. People were urged to voluntarily heat only one room in a house.

Heath insisted the government would not yield to "grossly inflationary wage claims" and blamed trade unions more than the Arab oil squeeze for Britain's energy crisis. (The energy shortage led to unplanned blackouts lasting up to 90 minutes in London and elsewhere Dec. 15. Power cuts caused by the power workers' job action had caused blackouts Dec. 11. Meanwhile, commuter train service and the delivery of coal were seriously disrupted as the railway workers' and coal miners' slowdowns continued.)

Measures designed to reduce domestic demand were announced by Chancellor of the Exchequer Anthony Barber in the House of Commons Dec. 17.

Barber warned that the energy measures and constraints on economic growth contained in the government's measures would lead to an unemployment increase.

Gas price increased—The government approved Dec. 14 a gasoline price increase of 3 pence (about 6¢) a gallon to cover the higher rates of Arab crude oil.

Airline cuts—British Airways announced Dec. 17 it would reduce flights for the rest of December by 30%–40% on its European, domestic and transatlantic flights.

Power workers end slowdown—Electrical power engineers agreed to end their eight-week ban on overtime and emergency work, effective Jan. 2, 1974, after accepting the government's pay offer Dec. 28. The settlement would slightly ease Britain's energy crisis by facilitating a more orderly output from power stations.

The pay offer, virtually the same as one agreed in November 1972 and frozen by the government's anti-inflation program in 1972, increased the weekly pay for stand-by duty outside normal hours from the equivalent of about $11.75 to $30, and raised payments for emergency call-ups from $5.55 to $11.50 and nearly double that amount on weekends.

Layoffs estimated at 640,000—The Department of Employment said that 640,-000 workers temporarily laid off because of energy shortages had registered for unemployment benefits Dec. 31, in addition to the 450,000 persons already on the official unemployed list. Many others had been laid off but were ineligible for unemployment benefits because they had guaranteed income agreements. More

workers would be laid off with the start of the three-day workweek.

The fuel shortage stemmed mainly from the still unsettled coal miners' work slowdown; while a ban on overtime by train engineers reduced deliveries of whatever coal was available.

India. A 15-year Indian-Soviet agreement was signed in New Delhi Nov. 29 by Soviet Communist Party General Secretary Leonid I. Brezhnev and Indian Prime Minister Indira Gandhi. The accord provided for the U.S.S.R. to build 70 to 90 projects in India, including facilities for the production and refining of oil and natural gas.

The Indian government imposed heavy excise taxes on gasoline and smaller duties on other refined products.

The action was aimed at countering the latest increase in the price of crude oil by producers in the Persian Gulf area. India received 16 million tons of crude oil a year from the region and sought to cut the exports by discouraging consumption through higher prices.

Italy. Italy eased its licensing system designed to halt exports of refined oil products, it was reported Dec. 6. The oil companies had threatened to cut off crude oil supplies if the curbs were not eased. Italy had the biggest refining system of the European Economic Community (EEC), with an annual capacity of about 180 million metric tons. The export blocks were imposed in November because of the Arab oil squeeze.

The state-owned Ente Nazionale Idrocarburi (ENI) announced agreement Dec. 28 to buy Shell's oil refining and marketing activities in Italy. The accord called for Shell to supply crude oil to ENI between 1974–78.

(The government decided to construct four new nuclear power plants, in addition to the four existing ones, it was reported Dec. 25.)

Japan. *Energy action*—Japanese Premier Kakuei Tanaka said Dec. 2 that his country would seek help from Australia and other Southeast Asian nations for oil and other energy resources.

The Cabinet adopted a series of radical measures Dec. 22 to cope with the economic crisis brought on by the fuel shortage. It declared a state of emergency, ordered a 20% reduction in use of oil and electricity by major industries and approved an "austerity" budget for 1974 calling for the smallest increase in spending in four years.

The oil cutbacks were to have gone into effect Jan. 1, 1974, but the government decided Dec. 26 to delay the action in view of a decision by the Arab petroleum producers Dec. 25 to increase their output in January, a move that would benefit Japan.

In a statement accompanying his emergency declaration, Premier Kakuei Tanaka said consumption of oil would be temporarily reduced to the 1972 levels. He expressed confidence that the crisis could be overcome if businesses "refrain from raising prices, and if consumers refrain from hoarding."

The emergency program was adopted by the Cabinet after final Parliamentary passage Dec. 21 of two laws authorizing mandatory controls over energy consumption and the price and distribution of essential goods. The legislation gave the government power to fix prices and regulate processing and supply of oil and other items, ranging from steel and cement to sugar.

The Cabinet announced a "livelihood stabilization headquarters," headed by Tanaka, which announced plans to fix prices on 20 categories of goods.

In addition to the industrial cutback measures, the Cabinet ordered a ban on sale of gasoline and use of private cars on Sundays and holidays, restricted the operating hours of entertainment and service businesses and reduced television broadcasting hours.

Netherlands. The only European nation hit by a total Arab oil embargo, the Netherlands, according to the Washington Post Dec. 23, reportedly benefited from an informal crude oil allocation program imposed by international oil companies which diverted more non-Arab oil from some customers to maintain adequate supplies to hard-pressed nations. The chairman of the board of directors of Royal Dutch Shell had said in an interview, reported Dec. 5, that oil companies were trying to allocate a fair share to all countries.·

The Second Chamber (lower house of Parliament) Dec. 21 approved a bill giving the government special powers to control all aspects of the economy to meet any acute crisis stemming from oil shortages. The government was empowered to regulate wages, prices, rents and some layoffs. A 3.5-gallon weekly gasoline rationing limit was authorized for motorists, effective Jan. 7, 1974, and prison and other penalties were set for persons defying lighting restrictions. One of the prime motivations for the emergency legislation was the failure of the annual tripartite discussions among the government, employers and unions for a 1974 wage and social accord.

The Netherlands government began sending out gasoline rationing forms Nov. 25 for use if the nation had to curtail consumption more drastically. The Shell refinery at Rotterdam, the biggest oil terminal in Western Europe, announced Nov. 17 a 20% cutback in its daily output. Exxon announced the same day that its Rotterdam refinery would cut daily production by as much as 40%.

The major oil companies in the Netherlands announced Nov. 22 an average cut of 15% in supplies of raw materials to the petrochemical and chemical industry. The companies included Shell, Exxon, British Petroleum and Gulf Oil.

New Zealand. Prime Minister Norman Kirk said Nov. 17 that Australia had agreed to continue to supply his country with refined petroleum products at the current level of 70% for the next few months. New Zealand produced the other 30%. The agreement was reached in talks Kirk held in Sydney with Prime Minister Gough Whitlam. New Zealand had received 135 million gallons of refined petroleum products from Australia in 1971-72.

Philippines. President Ferdinand E. Marcos issued an order Nov. 5 restricting the sale of petroleum products to foreign users in the wake of a worsening fuel shortage resulting from the Arab oil cutoff. The Philippines' crude oil supplies

from the Middle East were reduced 16%–20%.

The export of petroleum products was banned and sale of fuel to foreign airlines and foreign shipping was reduced. In other conservation measures, the government placed a curb on unnecessary travel and planned to institute daylight saving time.

The government also planned to call on the U.S. to reduce its purchases from Philippine sources of petroleum and electricity for use at its two military installations—Clark and Subic.

Marcos had issued a decree Nov. 4 classifying the manipulation of oil supplies as a crime against national security. Gas rationing began in the Manila area Nov. 12. Schools were ordered closed in the capital region from Nov. 12 to Jan. 2, 1974 to save fuel.

Portugal. The government Nov. 8 imposed gasoline rationing, weekend closing of gas stations and gasoline product price increases of 10%–19%.

Rumania. The government imposed gasoline rationing Nov. 18, as well as speed limits and electricity and heating conservation measures.

South Korea. The Seoul government had imposed a 17% cut on oil to almost all industries, transportation and the state electric power company because of the Arab oil cutback, it was reported Dec. 15.

South Korea imported all its oil, 85% of which came from the Arab states. The country was said to be receiving only 25% of the oil it had contracted for.

The Economic Planning Board Dec. 4 had ordered an average 30% increase in the domestic market prices of all oil products because of the Arab embargo.

Soviet Union. Soviet Premier Alexei Kosygin said in a speech in Minsk, reported Nov. 17, that the Soviet energy situation was "still tense," although "incomparably more favorable" than in the West.

Kosygin reported that production of oil and natural gas had fallen below original targets in 1973 and would continue to lag in 1974, but said coal production was ahead of target. He said economizing in fuel and electrical power was a problem "not only of today but also of tomorrow." The Communist party newspaper Pravda reported Nov. 13 that committees had been set up to study "intensifying the struggle for rational and economic use of power resources," and printed a further appeal for fuel conservation Nov. 27.

Western experts, cited in news reports Nov. 15-16, said the Soviet Union might have to increase drastically its imports of Middle East oil, which stood at about six million tons, or 2% of consumption in 1972, if export policies were maintained. Exports to Communist countries, totaling 60 million tons in 1972, would probably be maintained for political reasons. Exports to the West, over 40 million tons in 1972, provided about 30% of hard currency earnings, needed to pay for imports of Western machinery and technology. Soviet domestic consumption, currently about one-third of U.S. levels, was expected to increase by roughly two-thirds by 1980, especially if the automotive industry continued to expand.

Tens of thousands of Soviet oil experts attended an October exhibit by 189 U.S. companies in Moscow, who displayed about $13 million worth of oil exploration and production equipment. Dresser Industries signed at the exhibit a five-year technology transfer agreement on prospecting offshore oil and gas reserves. Soviet reserves were estimated at about 100 billion barrels, but much of it was inaccessible with current Soviet technology or too far from the major Soviet markets.

The Soviet government had said Sept. 6 that it had reduced from 40 million tons to 25 million tons the amount of crude oil that could be available each year to Japan from the Tyumen oil field in Siberia. The government reported Nov. 25 that a significant new oil field had been tapped somewhere in the Tartar region west of the Urals.

Bonn Foreign Minister Walter Scheel urged Soviet officials to increase their exports of oil, gas and electricity to compensate for a trade imbalance, it was reported Nov. 3.

Scheel reportedly noted that Soviet deliveries of crude oil, as compensation for steel used in pipeline construction, had been less than two-thirds of quota levels in 1972. The cutback was attributed to increased Soviet and East European domestic demand and a slowdown in development of Soviet oil fields.

Venezuela. The Venezuelan government Nov. 30 raised its posted prices for crude oil and oil products by an average 50¢ a barrel, to a record $7.74 a barrel. Residual fuel shipped to the U.S. East Coast was increased by 98¢ a barrel to $8.46 a barrel. This was the sixth price increase of 1973 and the seventh since October 1972.

The posted price, or tax export value, was an artificial figure used to calculate tax payments to be made by oil companies to the government.

The increase would boost the government's income from oil to an average $4.57 a barrel, compared with $1.62 a barrel at the beginning of 1973. According to one official calculation, Venezuela would earn an extra $1.8 billion in oil revenues in 1974, the Andean Times' Latin America Economic Report noted Nov. 30.

Mines Minister Hugo Perez La Salvia had announced Nov. 15 that foreign oil companies operating in Venezuela had signed an agreement with the government, effective Nov. 1, raising royalty payments on oil production by about one-third. The pact would give the government an additional $230 million in oil revenue over one year, providing production and prices remained at current levels.

Perez La Salvia had said Nov. 5 that Venezuelan oil production would remain at the current level of 3,360,000 barrels a day despite the world energy crisis, the Cuban news agency Prensa Latina reported. Former Mines Minister Juan Pablo Perez Alfonso continued to call for a decrease in production to protect dwindling oil reserves, according to the Washington Post Dec. 5.

Congress May 28 overwhelmingly approved a law nationalizing Venezuela's internal hydrocarbons market by 1976. Only 23% of the $200 million annual market was currently controlled by the state oil firm CVP. Most gas stations were owned by U.S. and other foreign companies.

NUCLEAR POWER UNITS IN OPERATION

Name of System or Sponsor	Name or Location	Name-Plate Rating, Kw	Type of Reactor	Placed in Commercial Operation
Duquesne Light Company	Shippingport No. 1	100,000	Pressurized Water	1957
Commonwealth Edison Company	Dresden No. 1	208,675	Boiling Water	1960
Yankee Atomic Electric Company	Yankee No. 1	185,000	Pressurized Water	1961
Consolidated Edison Company of N.Y., Inc.	Indian Point No. 1	275,000	Pressurized Water	1962
Consumers Power Company	Big Rock Point	75,000	Boiling Water	1962
Pacific Gas & Electric Company	Humboldt Bay	60,000	Boiling Water	1963
Washington Public Power Supply System	Hanford	800,000	Graphite Moderated, Water Cooled	1966
Philadelphia Electric Company	Peach Bottom No. 1	46,000	High Temperature, Helium Cooled and Graphite Moderated	1967
Connecticut Yankee Atomic Power Company	Haddam No. 1	600,300	Pressurized Water	1968
Southern California Edison Company and San Diego Gas & Electric Company	San Onofre	450,000	Pressurized Water	1968
Jersey Central Power & Light Co.	Oyster Creek No. 1	550,000	Boiling Water	1969
Niagara Mohawk Power Corp.	Nine Mile Point No. 1	641,750	Boiling Water	1969
Rochester Gas and Electric Corp.	Ginna No. 1	517,140	Pressurized Water	1969
Northeast Utilities	Millstone Point No. 1	661,500	Boiling Water	1970
Wisconsin Michigan Power Company	Point Beach No. 1	502,841	Pressurized Water	1970
Carolina Power & Light Company	Robinson No. 2	739,328	Pressurized Water	1971
Dairyland Power Cooperative	La Crosse	60,000	Boiling Water	1971
Northern States Power Company	Monticello No. 1	542,730	Boiling Water	1971
Commonwealth Edison Co.	Dresden No. 2	810,000	Boiling Water	1971
Boston Edison Co.	Pilgrim No. 1	655,397	Boiling Water	1972
Commonwealth Edison Co.	Quad Cities No. 1	810,000	Boiling Water	1972
Commonwealth Edison Co.	Quad Cities No. 2	810,000	Boiling Water	1972
Florida Power & Light Co.	Turkey Point No. 3	728,300	Pressurized Water	1972
Maine Yankee Atomic Power Co.	Wiscasset	830,000	Pressurized Water	1972
Vermont Yankee Nuclear Power Corp.	Vernon	537,261	Boiling Water	1972
Virginia Electric and Power Co.	Surry No. 1	822,600	Pressurized Water	1972
Consumers Power Co.	Palisades No. 1	810,000	Pressurized Water	1973
Duke Power Co.	Oconee No. 1	886,669	Pressurized Water	1973
Florida Power & Light Co.	Turkey Point No. 4	728,317	Pressurized Water	1973
Omaha Public Power District	Fort Calhoun No. 1	481,477	Pressurized Water	1973
Virginia Electric and Power Co.	Surry No. 2	822,600	Pressurized Water	1973
Wisconsin Electric Power Co.	Point Beach No. 2	502,841	Pressurized Water	1973

West Germany. West German Economy Minister Hans Friderichs Nov. 19 banned all private driving for the next four Sundays and reduced the speed limit on highways to about 62 miles an hour and on other roads to 50 miles an hour.

Friderichs said the measures should "save up to 13% of a month's gasoline consumption and thus conserve energy for our industry and places of work." He said Germany expected its Arab oil imports to be cut by 15%.

Parliament Nov. 9 had passed a bill giving the government full control of all energy resources in case of an oil crisis.

Atomic Power

French-led group to build A-fuel plant. The French government announced Nov. 23 it would proceed with plans to construct a $1.7 billion uranium enrichment plant, beginning in January 1974, using the gaseous diffusion enrichment technique to produce fuel for nuclear power reactors. The plant was scheduled to be completed in 1979. The decision was announced in the French National Assembly by Jean Charbonnel, minister of industrial and scientific development.

The representatives of four other European nuclear energy commissions or nuclear companies announced Nov. 27 a decision in principle to join France in the gaseous diffusion project, known as Eurodif. Italy would control 22.5% of the venture and Spain, Belgium and Sweden 10% each, while France would hold 47.5%. France proposed Nov. 28 that the plant be situated in France.

The French decision followed disclosure Nov. 22 of an invitation issued Nov. 14 by the Netherlands, West Germany and Britian to join a separate project, known as Urenco, that would use the centrifugal enrichment technique. In announcing its decision, the French government said it was willing to discuss cooperation with the other consortium.

Existence of the two rival projects prompted the European Economic Community (EEC) Commission to propose Nov. 14 the parallel use of both technologies. It called for the gaseous diffusion method, already proven but enormously expensive, to be used initially, with the cheaper and smaller centrifuge plants, still at the pilot state, to be phased in later. The Commission said it also favored EEC subsidies of surplus reactor fuel production through a stockpiling program.

(The French gaseous diffusion plant would treat up to 13,500 tons of natural uranium and produce 9.3 million separative work units of enriched uranium, substantially more than EEC's estimated needs by 1980.)

The French insisted on gaseous diffusion on the grounds the rival method would not be ready soon enough to supply European needs by the 1980s. The push for European uranium enrichment facilities was spurred earlier in 1973 when the U.S., until then virtually the sole source of enriched uranium for European civilian use, set tough new contract terms for purchase and processing of the uranium.

Power plant rules tightened. The Atomic Energy Commission (AEC) voted unanimously Dec. 28 to impose stricter safety standards on heating of nuclear fuel and operation of emergency cooling systems in nuclear power plants.

The new rules set the maximum temperature for operation of uranium fuel piles at 2,200 degrees Farenheit, a reduction of 100 degrees from the old standards.

Another rule required that fuel bundles be redesigned to insure that no more than 17% of the reactor-core shielding would oxidize if it came in contact with cooling water.

The commission gave the industry six months to comply with standards.

The Union of Concerned Scientists, an intervenor in the AEC proceeding, said Dec. 28 that the ruling was "cosmetic" and represented a "continuation of the AEC's cover-up of critical safety problems."

Test accident reported. The AEC disclosed Nov. 1 that an accident had occurred Oct. 20 at its experimental gas-centrifuge uranium production facility at Oak Ridge, Tenn. The device being tested was part of a highly-secret method of separating uranium 235 from other uranium products more efficiently than by other methods.

An AEC spokesman said parts of one centrifuge, spinning at high speed, had broken into pieces and damaged nearby centrifuges. The spokesman said it would be inaccurate to call the accident an explosion. Other less serious accidents had occurred before, he said, but had not been made public.

Delaware River plant barred. The AEC Oct. 5 denied permission for construction of a nuclear power plant on a Delaware River island 11 miles from Philadelphia and 4½ miles from Trenton, N.J.

In a letter to Public Service Electric & Gas Co. of New Jersey, the AEC said "population density" near the island was the principle factor in the decision. The AEC suggested an alternate site near Salem, N.J., where two other reactors were already under construction.

EPA radiation authority curbed. Charging that the Environmental Protection Agency (EPA) had "construed too broadly" its responsibilities to set environmental standards on radioactive materials, the White House ordered the EPA to drop its plans to set radiation rules for individual nuclear power plants, the New York Times reported Dec. 12.

According to the directive, the authority to set such standards would rest solely with the Atomic Energy Commission (AEC), which was considered by critics of the nuclear power industry to have a more lenient approach to radiation protection than the EPA.

The order was in a memorandum, dated Dec. 7, sent "on behalf of the President" by Roy L. Ash, director of the Office of Management and Budget, to EPA Administrator Russell E. Train and AEC Chairman Dixy Lee Ray. The memo said its purpose was to prevent "confusion" in the area of nuclear power regulation, "particularly since nuclear power is expected to supply a growing share of the nation's energy requirements."

Ash said the AEC should proceed with its own plans to issue rules and the EPA should discontinue similar preparations to issue standards "now or in the future." The EPA would retain responsibility "for setting standards for the total amount of radiation in the general environment from all facilities combined in the uranium fuel cycle"—flexible standards which "would have to reflect the AEC's findings as to the practicability of emission controls."

The Times quoted Charles L. Elkins of the EPA's office of hazardous materials control as feeling that AEC standards were "not adequate."

Index